Genome Mining and Marine Microbial Natural Products

Genome Mining and Marine Microbial Natural Products

Special Issue Editors

Kui Hong
Changsheng Zhang
Alan Dobson

MDPI • Basel • Beijing • Wuhan • Barcelona • Belgrade

MDPI

Special Issue Editors

Kui Hong
Wuhan University
China

Changsheng Zhang
Chinese Academy of Sciences
China

Alan Dobson
University College Cork
Ireland

Editorial Office
MDPI
St. Alban-Anlage 66
4052 Basel, Switzerland

This is a reprint of articles from the Special Issue published online in the open access journal *Marine Drugs* (ISSN 1660-3397) from 2017 to 2019 (available at: https://www.mdpi.com/journal/marinedrugs/special_issues/genome_microbe).

For citation purposes, cite each article independently as indicated on the article page online and as indicated below:

LastName, A.A.; LastName, B.B.; LastName, C.C. Article Title. *Journal Name* **Year**, *Article Number, Page Range.*

ISBN 978-3-03928-090-2 (Pbk)
ISBN 978-3-03928-091-9 (PDF)

Contents

About the Special Issue Editors

Kui Hong Ph.D., is a professor in the School of Pharmaceutical Sciences, Wuhan University, China. She has studied microbiology at Wuhan University, the South China College of Tropical Crops, Nanjing Agriculture University, and Tsinghua University, where she obtained her B.S., M.S. and Ph.D. degrees. Since 2001, her research has focused on marine microbial drug discovery, especially from special marine environments such as mangroves, deep sea, and Polar Regions. She has led or participated in the national NSFC, "863," "973," and the EU "FP7" and China-Thai collaborative projects. She is a member of the Chinese Microbiology Society, Chinese Biochemistry Society, Chinese Pharmacological Society, and Chinese Pharmaceutical Association. She possesses 19 national and 1 PCT patents and published more than 100 international papers. She is now the chief scientist of the "National Key R&D Program of China," where she oversees the section "High efficient discovery and modification of marine microbial drug candidate".

Alan Dobson Ph.D., is Professor of Environmental Microbiology at University College Cork in Ireland. He studied biochemistry at the National University of Ireland, Galway, where he obtained a B.Sc. in 1981 and a Ph.D. in 1985. He then studied eukaryotic molecular biology at the Department of Molecular and Cell Biology at Baylor College of Medicine before returning to the School of Microbiology at Cork. His group is focused on gaining a fuller understanding of how microbes survive, grow, and interact in their various ecological niches; an approach which is fundamental to their exploitation for biotechnological applications. He has to date published more than 190 peer-reviewed papers. In 1992, he was awarded a Fulbright Scholarship and in 1999, he received the Royal Irish Academy Medal in Microbiology for his work in environmental microbiology. In 2005, he was awarded a D.Sc. in microbiology and molecular biology by the National University of Ireland in recognition of his contributions to research. In 2013, he was elected to the Royal Irish Academy.

Changsheng Zhang Ph.D., is a professor at the South China Sea Institute of Oceanology, Chinese Academy of Sciences, China. He obtained a B.Sc. in biology (Shanghai Jiaotong University, China, 1994) and an M.Sc. in biotechnology (East China University of Science and Technology, 1997). In 2002, he obtained his Ph.D. in chemical microbiology (Bergische University of Wuppertal, Germany) with Prof. Dr. W. Piepersberg. He carried out postdoctoral research on natural product glycosylation studies with J. S. Thorson at University of Wisconsin, Madison (2003–2008). In 2008, he joined the South China Sea Institute of Oceanology, China Academy of Sciences. His research interest is in marine microbial natural product discovery and biosynthesis. He has published more than 90 papers in peer-reviewed journals such as *Science*, *Nature Chemical Biology*, *Nature Communications*, *the Journal of American Chemical Society*, and *Angewandte Chemie*. He is now an editorial member of *Natural Product Reports*.

Preface to "Genome Mining and Marine Microbial Natural Products"

Most of the marine microbial natural products to date have been discovered using classical bioassay-guided regimes. This process is currently undergoing significant changes primarily due to of rapid developments in sequencing technology, synthetic biology, and bioinformatics. However, as increasing numbers of whole-genome sequences become available, many genomes appear to possess "silent," or cryptic, biosynthetic gene clusters. The products of these appear to be regulated by a variety of environmental factors, and therefore remain largely undetected. Genome mining has become a very attractive tool for drug discovery from marine microorganisms.

Researchers have been using different strategies to help activate these silent gene clusters from microbes, including but not limited to bioinformatic tools for gene and gene cluster identification, gene editing using the innovative CRISPR Cas 9 technology, heterologous expression based strategies, as well as activation using environmental factors. We are glad to see that these strategies are also being employed on marine microbes, which will help in the discovery of more and more new compounds. This Special Issue provides a number of interesting examples of some of the excellent work that is currently being undertaken in this arena. It also hints at the likelihood of major advances in this field in the very near future.

Kui Hong, Changsheng Zhang, Alan Dobson
Special Issue Editors

marine drugs

MDPI

Review

Current Status and Future Prospects of Marine Natural Products (MNPs) as Antimicrobials

Alka Choudhary [1], Lynn M. Naughton [2], Itxaso Montánchez [3], Alan D. W. Dobson [2] and Dilip K. Rai [1,*]

1   Department of Food Biosciences, Teagasc Food Research Centre Ashtown, Dublin D15 KN3K, Ireland; alka.choudhary@teagasc.ie
2   School of Microbiology, University College Cork, Western Road, Cork City T12 YN60, Ireland; lynn.naughton@ucc.ie (L.M.N.); a.dobson@ucc.ie (A.D.W.D.)
3   Department of Immunology, Microbiology and Parasitology, Faculty of Science, University of the Basque Country, (UPV/EHU), 48940 Leioa, Spain; itxaso.montanchez@ehu.eus
*   Correspondence: dilip.rai@teagasc.ie; Tel.: +353-(0)-1805-9569; Fax: +353-(0)-1805-9550

Received: 20 July 2017; Accepted: 23 August 2017; Published: 28 August 2017

Abstract: The marine environment is a rich source of chemically diverse, biologically active natural products, and serves as an invaluable resource in the ongoing search for novel antimicrobial compounds. Recent advances in extraction and isolation techniques, and in state-of-the-art technologies involved in organic synthesis and chemical structure elucidation, have accelerated the numbers of antimicrobial molecules originating from the ocean moving into clinical trials. The chemical diversity associated with these marine-derived molecules is immense, varying from simple linear peptides and fatty acids to complex alkaloids, terpenes and polyketides, etc. Such an array of structurally distinct molecules performs functionally diverse biological activities against many pathogenic bacteria and fungi, making marine-derived natural products valuable commodities, particularly in the current age of antimicrobial resistance. In this review, we have highlighted several marine-derived natural products (and their synthetic derivatives), which have gained recognition as effective antimicrobial agents over the past five years (2012–2017). These natural products have been categorized based on their chemical structures and the structure-activity mediated relationships of some of these bioactive molecules have been discussed. Finally, we have provided an insight into how genome mining efforts are likely to expedite the discovery of novel antimicrobial compounds.

Keywords: antimicrobial; marine natural products (MNPs); secondary metabolites; antibacterial; antifungal; genome mining

1. Introduction

Infectious diseases caused by bacteria, fungi and viruses pose a major threat to public health despite the tremendous progress in human medicine. A dearth in the availability of effective drugs and the on-going threats posed by antimicrobial resistant organisms further worsen the situation particularly in developing countries. Antimicrobial resistance accounts for at least 50,000 deaths each year in Europe and the United States and it is anticipated that drug resistant infections will be responsible for the deaths of 10 million people worldwide by 2050 [1,2]. Continuously evolving antibiotic-resistance of microbial pathogens has raised demands for the development of new and effective antimicrobial compounds [3]. For generations, humans have turned to nature as a source of invaluable medicinal products, where terrestrial and marine organisms traditionally provide the most effective remedies [4,5]. It was only after the discovery of penicillin in 1928 that microbial sources were explored as sources of new therapeutic molecules. Developments in microbial culture techniques and diving expeditions in the 1970s have largely directed the drug discovery program towards the oceans. Combinatorial chemistry developments in the late 1980s further shifted the emphasis of drug discovery efforts from nature to the laboratory bench [6]. Although the unique structural features associated with natural products cannot be matched by any synthetic library they still continue to inspire researchers in the fields of chemistry, biology, and medicine to develop/synthesize more drug-like molecules [7]. Natural products, as the name suggests, are products of secondary metabolism in nature. Traditionally, many natural products were identified as promising candidates for drug development using bioassay-guided investigations, and chemical structure elucidation techniques [5,8]. However, too often this approach led to the re-isolation of known compounds. Advances in sequencing and '-omics' technologies are expediting the identification and development of novel molecules. Today, over 60% of drugs in the market are derived from natural sources [9,10]. Among the 1562 new chemical entities introduced from the period 1981–2014, 21% are naturally derived, 16% are biological macromolecules, 10% constitute the nature mimic entities, 9% are botanical drugs, 6% constitute vaccines and 4% are unaltered natural products [11]. Several small-molecules from natural sources have been approved as antitumor, antibacterial, and antifungal agents over the period 2011–2014 [11]. In this review, we have discussed the roles played by advances in genomic sequencing and '-omics' technologies in expediting the identification and development of novel, antimicrobial marine natural products (MNPs) from biosynthetically "talented" microorganisms of marine origin.

2. Marine Natural Products (MNPs)

The ocean covers over 71% of the earth's surface and constitutes more than 90% of the inhabitable space on the planet. An estimated 50–80% of all life on earth resides in the ocean and it is home to 32 out of 33 known animal phyla, 15 of which are exclusively marine [12]. More than 20,000 natural products have been discovered in the marine environment over the past 50 years [13]. Interest in marine natural products (MNPs) based drug discovery is evident from the increase in number of isolated MNPs (from an annual number of approximately 20 in 1984 to an annual number of more than 1000 in 2010) [14,15]. From the continuing progress in the area of MNPs seven approved drugs and 12 agents currently in clinical trials have been discovered [16]. These molecules are either natural products, tailored natural products or are molecules inspired from the structure of natural products [17,18]. Marine organisms largely obtained from shallow-water, tropical ecosystems are the major sources of MNPs. Macroorganisms such as algae, sponges, corals and other invertebrates, as well as microorganisms have also contributed significantly towards the discovery of novel MNPs [19]. Marine invertebrates in particular have proven to be major sources of MNPs in clinical trials [20]. Also, mounting evidence suggests that many of the compounds originally associated with the biomass of macroorganisms such as sponges [21], tunicates [22], molluscs [23] amongst others, are not produced by the organism itself but are synthesized by symbiotic or associated microorganisms, or derive from a diet of prokaryotic microorganisms [24]. Unlike the terrestrial environment, where plants are comparatively richer in secondary metabolites, marine invertebrates and bacteria have yielded substantially more bioactive natural products than marine plants [25].

The total number of approved drugs from the marine environment steadily increased from four in 2010 to seven in 2014 [26,27]. The first U.S. Food and Drug Administration (FDA) approved marine-derived drug cytarabine (Cytosar-U®), isolated from the Caribbean sponge *Cryptotheca crypta*, reached the market in 1969 for use as an anticancer drug. Since then, six more marine natural products have moved through clinical trials and have been approved as drugs (one of which is only registered in the European Union), including the analgesic cone snail-derived peptide ziconotide (Prialt®), and the anticancer sponge-derived macrolide, eribulin mesylate (Halaven®), as well as four other products with anticancer, antiviral and antihypertriglyceridemia activities [27]. Of the 23 most recently identified marine-derived compounds, 21 are in several different stages of the clinical pipeline for use as anticancer agents, while two of them are being assessed for treatment of chronic pain and neurological disorders like schizophrenia and Alzheimer's disease [7]. In addition, a number of other compounds boasting antibacterial, antidiabetic, antifungal, antiinflammatory and antiviral properties, as well as compounds potentially affecting the nervous system, are currently being investigated for use in clinical settings and thus form part of the preclinical pipeline [27–29].

3. Chemical Entities in the Preclinical Antimicrobial Pipeline

Over the last 5 years, preclinical pharmacology has been undertaken on 262 marine compounds that are presently at various stages of clinical investigations [26]. Herein, we discuss the structural features of some of these molecules (Figure 1) which are currently under investigation for their potential as antimicrobial agents. Where noted in the text, bold numerical values correspond to their associated structures in corresponding figures.

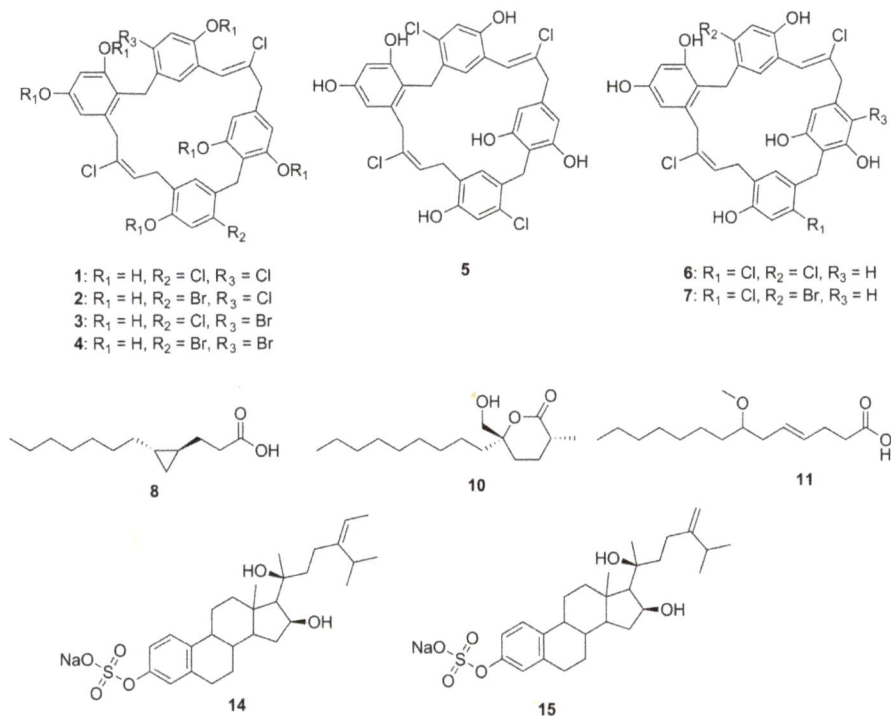

Figure 1. Marine natural products in antimicrobial preclinical studies.

Chrysophaentin A (**1**), a macrocyclic natural product (comprising two polyhalogenated, polyoxygenated ω,ω'-diarylbutene units connected by two ether bonds), was isolated from the chrysophyte alga *Chrysophaeum taylori*. This compound inhibits the growth of clinically relevant Gram-positive bacteria, including methicillin-resistant *Staphylococcus aureus* (MRSA; MIC_{50} 1.5 ± 0.7 µg/mL), multiple drug resistant *S. aureus* (MIC_{50} 1.3 ± 0.4 µg/mL), and Vancomycin-Resistant *Enterococcus faecium* (VREF; MIC_{50} 2.9 ± 0.8 µg/mL). Chrysophaentin A inhibits the GTPase activity of the bacterial cytoskeletal protein FtsZ (IC_{50} value 6.7 µg/mL), and GTP-induced formation of FtsZ protofilaments [30]. Interestingly, this compound was found to be relatively more active among its congeners, Chrysophaentins B–G (**2–7**). Analysis of bioactivity of these molecules provided insights into the pharmacophoric features of Chrysophaentins relevant to antimicrobial activity. Phenolic groups in compound **1** were determined to be crucial for activity as a hexaacetate derivative of **1** was found to be inactive at a concentration 25 µg/disk. An approximate 12-fold decrease in the MIC_{50} value of chrysophaentin D (**4**) compared to compound **1** was observed on replacement of chlorine with bromine on rings A and C. The significance of the macrocyclic structure was established following higher MIC$_{50}$ values of Chrysophaentin E (**5**) being observed towards *S. aureus* and MRSA compared to MIC$_{50}$ values observed for the chlorinated cyclic bisdiarylbutene ethers **1** and **6**. However, compound **5** was also found to be inactive toward *E. faecium* and VREF at concentrations as high as 25 µg/mL. In the case of the symmetrically linked dimers **6** and **7**, replacing a chlorine atom on ring C with bromine confers compound **6** with at least three times better activity than compound **7**. Among the tetrachlorinated macrocyles **1** and **6**, compound **1** was found to be 3–5 times more potent than compound **6**, indicating that the position of the ether linkage relative to the 2-butene unit affects activity. In fact, *ortho*-linked chrysophaentin A has been found to be more potent than the *para*-linked chrysophaentin F [31].

A small cyclopropane-containing fatty acid, lyngbyoic acid (**8**), was found to be a major metabolite of the marine cyanobacterium, *Lyngbya* cf. *majuscula*, isolated in Florida [32]. This molecule exerts antimicrobial action by disrupting quorum sensing in *Pseudomonas aeruginosa*. At a concentration of 100 µM, it inhibits the *N*-acyl homoserine lactone (HSL) receptor proteins in the organism (LasR in particular), reducing the expression of important virulence factors in the wild-type strain. The molecules inhibit the response of LasR-based QS reporter plasmids to 3-oxo-C 12-HSL. The AHL-binding site of LasR was not essential to this effect, but competition experiments indicated that compound **8** is likely to have a dual mechanism of action acting both through the AHL-binding site and independently of it. Comparison of compound **8** with related compounds (dodecanoic acid, **9**; malyngolide, **10**; and lyngbic acid, **11**; methyl ester of dodecanoic acid, **12** and butyric acid, **13**) revealed a structure-activity relationship. While compounds **9**, **10** and **11** had a similar potency in pSB1075 compared to **8**, either esterification (**12**) or shortening of the alkyl chain (**13**) reduced activity [32].

Two sulfated sterols, geodisterol-3-*O*-sulfite (**14**) and 29-demethylgeodisterol-3-*O*-sulfite (**15**), were isolated from the marine sponge *Topsentia* sp. These sulfated sterols demonstrated reverse efflux pump-mediated fluconazole resistance. They enhanced the activity of fluconazole in a *Saccharomyces cerevisiae* strain overexpressing the *Candida albicans* efflux pump MDR1, and in a fluconazole-resistant *C. albicans* clinical isolate known to overexpress MDR1. No activity for non-sulfated sterol in fluconazole-resistance reversal assay had been observed highlighting the relevance of sulfate group for MDR1 inhibition and synergy with fluconazole. Investigation of the geodisterols had provided insight into the clinical utility of combining efflux pump inhibitors with current antifungals to combat the resistance associated with opportunistic fungal infections caused by *C. albicans* [33].

In the following sections, we present a systematic overview of the marine natural products which have gained the attention of chemists and biologists over the last five years (2012–2017) as potential antimicrobial agents. The molecules are categorized according to their chemical class based on their associated structural units. Pharmacophoric features responsible for antimicrobial activity are also discussed. Table 1 lists MNPs and describes their antimicrobial potential in terms of MICs and zone of inhibition.

Table 1. Chemical classification of antimicrobial marine natural products (MNPs). MRSA: methicillin-resistant *Staphylococcus aureus*; MRSE: methicillin-resistant *Staphylococcus epidermidis* MSSA: methicillin-sensitive *Staphylococcus aureus*; MTCC: microbial type culture collection.

Compound	Source	Activity Against Pathogen
	Alkaloids	
Pyranonigrin A (**16**)	*Penicillium brocae* MA-231	*S. aureus* (MIC 0.5 µg/mL)
		V. harveyi (MIC 0.5 µg/mL)
		V. parahaemolyticus (MIC 0.5 µg/mL)
		A. brassicae (MIC 0.5 µg/mL)
		C. gloeosprioide (MIC 0.5 µg/mL)

Table 1. *Cont.*

Compound	Source	Activity Against Pathogen
Pyranonigrin F (**17**)	*P. brocae* MA-231	*S. aureus* (MIC 0.5 µg/mL) *V. harveyi* (MIC 0.5 µg/mL) *Vibrio parahaemolyticus* (MIC 0.5 µg/mL) *Alternaria brassicae* (MIC 0.5 µg/mL) *C. gloeosprioide* (MIC 0.5 µg/mL)
Rubrumazine B (**18**)	*E. cristatum* EN-220	*Magnaporthe grisea* (MIC 64 µg/mL)
Echinulin (**19**)	*E. cristatum* EN-220	*S. aureus* (MIC 256 µg/mL)
Dehydroechinulin (**20**)	*E.cristatum* EN-220	*S. aureus* (MIC 256 µg/mL)
Variecolorin H (**21**)	*E. cristatum* EN-220	*S. aureus* (MIC 256 µg/mL)
Cristatumin A (**22**)	*E.cristatum*	*E. coli* (MIC 64 µg/mL) *S. aureus* (MIC 8 µg/mL)
Cristatumin D (**23**)	*E. cristatum*	*S. aureus* (Zone of inhibition 8 mm at 100 µg/disk)
Tardioxopiperazine A (**24**)	*E. cristatum*	*E. coli* (MIC 64 µg/mL) *S. aureus* (MIC 8 µg/mL)
Hemimycalin A (**27**)	*Hemimycale arabica*	*E. coli* (Inhibition zone 18 mm at 100 µg/disk) *C. albicans* (Inhibition zone 22 mm at 100 µg/disk)
Hemimycalin B (**28**)	*H. arabica*	*E. coli* (Inhibition zone 10 mm at 100 µg/disk) *C. albicans* (Inhibition zone 14 mm at 100 µg/disk)
(Z)-5-(4-hydroxybenzylidene)imidazolidine-2,4-dione (**29**)	*H. arabica*	*E. coli* (Inhibition zone 20 mm at 100 µg/disk) *C. albicans* (Inhibition zone 20 mm at 100 µg/disk)
Peniciadametizine A (**30**)	*Penicillium adametzioides*	*A. brassicae* (MIC 4 µg/mL)
Peniciadametizine B (**31**)	*P. adametzioides*	*A. brassicae* (MIC 32 µg/mL)
Penicibrocazine B (**33**)	*P. brocae*	*S. aureus* (MIC 32 µg/mL) *G.graminis* (MIC 0.25 µg/mL)
Penicibrocazine C (**34**)	*P. brocae*	*S. aureus* (MIC 0.25 µg/mL) *M. luteus* (MIC 0.25 µg/mL)
Penicibrocazine D (**35**)	*P. brocae*	*S. aureus* (MIC 8 µg/mL) *G.graminis* (MIC 8 µg/mL)
Penicibrocazine E (**36**)	*P. brocae*	*G. graminis* (MIC 0.25 µg/mL)
Crambescidin 800 (**37**)	*Clathria cervicornis*	*A. baumannii* (MIC 2 µg/mL) *K. pneumonia* (MIC 1 µg/mL) *P. aeruginosa.* (MIC 1 µg/mL)
Xinghaiamine A (**38**)	*Streptomyces xinghaiensis*	*S. aureus* (MIC 0.69 µM) *B. subtilis* (MIC 0.35 µM) *E. coli* (MIC 0.17 µM) *A. baumanii* (MIC 2.76 µM) *P. aeruginosa* (MIC 11 µM) MRSA 5301 (MIC 5.52 µM) MRSA 5438 (MIC 2.76 µM) MRSA 5885 (MIC 5.52 µM)
Hyrtioerectine D (**39**)	*Hyrtios* sp.	*C. albicans* (Zone of inhibition 17 mm at 100 µg/disk) *S. aureus* (Zone of inhibition 20 mm at 100 µg/disk) *P. aeruginosa* (Zone of inhibition 9 mm at 100 µg/disk)
Hyrtioerectine E (**40**)	*Hyrtios* sp.	*C. albicans* (Zone of inhibition 19 mm at 100 µg/disk) *S. aureus* (Zone of inhibition 10 mm at 100 µg/disk) *P. aeruginosa* (Zone of inhibition 9 mm at 100 µg/disk)
Hyrtioerectine F (**41**)	*Hyrtios* sp.	*C. albicans* (Zone of inhibition 14 mm at 100 µg/disk) *S. aureus* (Zone of inhibition 16 mm at 100 µg/disk) *P. aeruginosa* (Zone of inhibition 9 mm at 100 µg/disk)
Ageloxime B (**42**)	*Agelas mauritiana*	*C. neoformans* (MIC 5 µg/mL) *S. aureus* (MIC 7 µg/mL)
Ageloxime D (**43**)	*A. mauritiana*	*C. neoformans* (MIC 6 µg/mL)

Table 1. *Cont.*

Compound	Source	Activity Against Pathogen
Zamamidine D (**44**)	*Amphimedon* sp.	*E. coli* (MIC 32 μg/mL) *S. aureus* (MIC 8 μg/mL) *B. subtilis* (MIC 8 μg/mL) *M. luteus* (MIC 8 μg/mL) *A. niger* (MIC 16 μg/mL) *T. mentagrophytes* (MIC 8 μg/mL) *C. albicans* (MIC 16 μg/mL) *C. neoformans* (MIC 2 μg/mL)
Adametizine A (**45**)	*P. adametzioides* AS-53	*S. aureus* (MIC 8 μg/mL) *A. hydrophilia* (MIC 8 μg/mL) *V. harveyi* (MIC 32 μg/mL) *V. parahaemolyticus* (MIC 8 μg/mL) *G. graminis* (MIC 16 μg/mL)
Adametizine B (**46**)	*P.adametzioides* AS-53	*S. aureus* (MIC 64 μg/mL)
Iso-Agelasidine B (**47**)	*A. nakamurai*	*C. albicans* (MIC 2.34 μg/mL)
(−)-Agelasidine C (**48**)	*A. nakamurai*	*C. albicans* (MIC 2.34 μg/mL)
Iso-agelasine C (**49**)	*A. nakamurai*	*S. aureus* (MIC 75 μg/mL) *E. coli* (MIC 150 μg/mL) *P. vulgaris* (MIC 19 μg/mL) *C. albicans* (MIC 5 μg/mL)
Agelasine B (**50**)	*A. nakamurai*	*P. vulgaris* (MIC 19 μg/mL) *C. albicans* (MIC 2 μg/mL)
Agelasine J (**51**)	*A. nakamurai*	*S. aureus* (MIC 75 μg/mL) *E. coli* (MIC 75 μg/mL) *P. vulgaris* (MIC 9 μg/mL) *C. albicans* (MIC 0.6 μg/mL)
Nemoechine G (**52**)	*A. nakamurai*	*S. aureus* (MIC 150 μg/mL) *E. coli* (MIC 75 μg/mL) *P. vulgaris* (MIC 9 μg/mL) *C. albicans* (MIC 0.6 μg/mL)
Brevianamide F (**53**)	*Penicillium vinaceum*	*C. albicans* (Zone of inhibition 25 mm at 100 μg/disk) *S. aureus* (Zone of inhibition 19 mm at 100 μg/disk)
N-(2-hydroxyphenyl)-2-phenazinamine (**54**)	*Nocardia dassonvillei*	*C. albicans* (MIC of 64 μg/mL)
	Terpenoids	
Puupehenol (**55**)	*Dactylospongia* sp.	*S. aureus* (Zone of inhibition 4 mm at 10 μg/disk) *B. cereus* (Zone of inhibition 4 mm at 10 μg/disk)
Puupehenone (**56**)	*Dactylospongia* sp.	*S. aureus* (Zone of inhibition 3 mm at 10 μg/disk) *B. cereus* (Zone of inhibition 3 mm at 10 μg/disk)
Penicibilaene A (**57**)	*Penicillium bilaiae*	*C. gloeosporioides* (MIC 1 μg/mL)
Penicibilaene B (**58**)	*P. bilaiae*	*C. gloeosporioides* (MIC 0.1 μg/mL)
Aspergillusene A (**59**)	*Aspergillus sydowii*	*K. pneumonia* (MIC 21 μM) *A. hydrophila* (MIC 4.3 μM)
(Z)-5-(Hydroxymenthyl)-2-(6′)-methylhept-2′-en-2′-yl)-phenol (**60**)	*A. sydowii*	*K. pneumonia* (MIC 11 μM)
Sydonic acid (**61**)	*A. sydowii*	*E. faecalis* (MIC 19 μM)
12-hydroxy isolaurene (**62**)	*Laurencia obtuse*	*B. subtilis* (MIC 46 μg/mL) *S. aureus* (MIC 52 μg/mL)
8,11-dihydro-12-hydroxy isolaurene (**63**)	*L. obtuse*	*C. albicans* (MIC 120 μg/mL) *A. fumigatus* (MIC 200 μg/mL) *B. subtilis* (MIC 39 μg/mL) *S. aureus* (MIC 31 μg/mL)
Isolauraldehyde (**64**)	*L. obtuse*	*C. albicans* (MIC 70 μg/mL) *A. fumigatus* (MIC 100 μg/mL) *B. subtilis* (MIC 35 μg/mL) *S. aureus* (MIC 27 μg/mL)

Table 1. *Cont.*

Compound	Source	Activity Against Pathogen
Napyradiomycin 1 (**65**)	*Streptomyces strain*	MRSA (MIC 16 μg/mL)
Napyradiomycin 2 (**66**)	*Streptomyces strain*	MRSA (MIC 64 μg/mL)
Napyradiomycin B2 (**67**)	*Streptomyces strain*	MRSA (MIC 32–64 μg/mL)
Napyradiomycin B3 (**68**)	*Streptomyces strain*	MRSA (MIC 2 μg/mL)
Napyradiomycin B4 (**69**)	*Streptomyces strain*	MRSA (MIC 32 μg/mL)
Dixiamycins A (**70**)	*Streptomyces* sp.	*E. coli* (MIC 8 μg/mL) *S. aureus* (MIC 8 μg/mL) *B. subtilis* (MIC 16 μg/mL) *B. thuringensis* (MIC 4 μg/mL)
Dixiamycins B (**71**)	*Streptomyces* sp.	*E. coli* (MIC 8 μg/mL) *S. aureus* (MIC 16 μg/mL) *B. subtilis* (MIC 16 μg/mL) *B. thuringensis* (MIC 8 μg/mL)
Oxiamycin (**72**)	*Streptomyces* sp.	*E. coli* (MIC 64 μg/mL) *S. aureus* (MIC 128 μg/mL)
Chloroxiamycin (**73**)	*Streptomyces* sp.	*E. coli* (MIC 64 μg/mL) *S. aureus* (MIC 64 μg/mL) *B. thuringensis* (MIC 64 μg/mL)
Xiamycin A (**74**)	*Streptomyces* sp.	*E. coli* (MIC 64 μg/mL) *S. aureus* (MIC 64 μg/mL)
Sarcotrocheliol acetate (**75**)	*Sarcophyton trocheliophorum*	*S. aureus* (MIC 1.7 μM) MRSA (MIC 1.7 μM) *Acinetobacter* pp. (MIC 4.3 μM)
Sarcotrocheliol (**76**)	*S. trocheliophorum*	*S. aureus* (MIC 1.5 μM) MRSA (MIC 3.0 μM) *Acinetobacter* sp. (MIC 3.0 μM)
Cembrene-C (**77**)	*S. trocheliophorum*	*C. albicans* (MIC 0.7 μM) *A. flavus* (MIC 0.7 μM)
Sarcophine (**78**)	*S. trocheliophorum*	*S. aureus* (MIC 9.4 μM) MRSA (MIC 9.4 μM) *Acinetobacter* sp. (MIC 9.4 μM)
Palustrol (**79**)	*S. trocheliophorum*	*S. aureus* (MIC 6.6 μM) MRSA (MIC 6.6 μM) *Acinetobacter* sp. (MIC 11.1 μM)
epi-Ilimaquinone (**80**)	*Hippospongia* sp.	MRSA (MIC 63 μg/mL) *S. aureus* (MIC 31 μg/mL) Vancomycin resistant *E. faecium* (MIC 16 μg/mL) Amphotericin-resistant *C. albicans* (MIC 125 μg/mL)
Penicillosides A (**81**)	*Penicillium* sp.	*C. albicans* (inhibition zone 23 mm at 100 μg/disk)
Penicillosides B (**82**)	*Penicillium* sp.	*S. aureus* (inhibition zone 19 mm at 100 μg/disk) *E. coli* (inhibition zone 20 mm at 100 μg/disk)
Ieodoglucomide A (**83**)	*Bacillus licheniformis*	*S. aureus* (MIC 8 μg/mL) *B. subtilis* (MIC 16 μg/mL) *B. cereus* (MIC 16 μg/mL) *S. typi* (MIC 16 μg/mL) *E. coli* (MIC 8 μg/mL) *P. aeruginosa* (MIC 8 μg/mL) *A. niger* (MIC 32 μg/mL) *C. albicans* (MIC 32 μg/mL)
Ieodoglucomide B (**84**)	*B. licheniformis*	*S. aureus* (MIC 8 μg/mL) *B. subtilis* (MIC 16μg/mL) *B. cereus* (MIC 8 μg/mL) *S. typi* (MIC 16 μg/mL) *E. coli* (MIC 16μg/mL) *P. aeruginosa* (MIC 8 μg/mL) *A. niger* (MIC 32 μg/mL) *C. albicans* (MIC 16 μg/mL)

Table 1. *Cont.*

Compound	Source	Activity Against Pathogen
Iedoglucomide C (**85**)	*Bacillus licheniformis*	*S. aureus* (MIC 0.03 µM) *B. subtilis* (MIC 0.03 µM) *B. cereus* (MIC 0.01 µM) *S. typi* (MIC 0.01 µM) *E. coli* (MIC 0.01 µM) *P. aeruginosa* (MIC 0.01 µM) *A. niger* (MIC 0.05 µM) *R. solani* (MIC 0.05 µM) *C. acutatum* (MIC 0.03 µM) *B. cenerea* (MIC 0.03 µM) *C. albicans* (MIC 0.03 µM)
Iedoglycoloipd (**86**)	*B. licheniformis*	*S. aureus* (MIC 0.03 µM) *B. subtilis* (MIC 0.05 µM) *B. cereus* (MIC 0.03 µM) *S. typi* (MIC 0.05 µM) *E. coli* (MIC 0.03 µM) *P. aeruginosa* (MIC 0.03 µM) *A. niger* (MIC 0.03 µM) *R. solani* (MIC 0.05 µM) *C. acutatum* (MIC 0.03 µM) *B. cenerea* (MIC 0.03 µM) *C. albicans* (MIC 0.05 µM)
Gageotetrin A (**87**)	*Bacillus subtilis*	*C. acutatum* (MIC 0.03 µM) *B. cinera.* (MIC 0.03 µM) *S. aureus* (MIC 0.03 µM) *B. subtilis* (MIC 0.03 µM)
Gageotetrin B (**88**)	*B. subtilis*	*C. acutatum* (MIC 0.01 µM) *B. cinera.* (MIC 0.01 µM) *S. aureus* (MIC 0.04 µM) *B. subtilis* (MIC 0.02 µM)
Gageotetrin C (**89**)	*B. subtilis*	*C. acutatum* (MIC 0.02 µM) *B. cinera* (MIC 0.01 µM) *S. aureus* (MIC 0.04 µM) *B. subtilis* (MIC 0.04 µM)
Lauramide diethanolamine (**90**)	*Streptomyces* sp.	*B. subtilis* (MIC 0.055 µg/mL) *E. coli* (MIC 0.055 µg/mL) *P. aeruginosa* (MIC 0.011 µg/mL) *S. aureus* (MIC 0.011 µg/mL) *S. cerevisiae* (MIC 0.022 µg/mL)
Linieodolide A (**91**)	*Bacillus* sp.	*B. subtilis* (MIC 8 µg/mL) *E. coli* (MIC 8 µg/mL) *S. cerevisiae* (MIC 32 µg/mL)
Linieodolide B (**92**)	*Bacillus* sp.	*B. subtilis* (MIC 64 µg/mL) *E. coli* (MIC 64 µg/mL) *S. cerevisiae* (MIC 128 µg/mL)
Dysiroid A (**93**)	*Dysidea* sp.	*S. aureus* ATCC 29213 (MIC 4 µg/mL) *S. aureus* ATCC 43300 (MIC 8 µg/mL) *E. faecalis* ATCC 29212 (MIC 4 µg/mL) *B. licheniformis* ATCC 10716 (MIC 16 µg/mL)
Dysiroid B (**94**)	*Dysidea* sp.	*S. aureus* ATCC 29213 (MIC 4 µg/mL) *S. aureus* ATCC 43300 (MIC 4 µg/mL) *E. faecalis* ATCC 29212 (MIC 4 µg/mL) *B. licheniformis* ATCC 10716 (MIC 8 µg/mL)
Halistanol sulfate A (**95**)	*Petromica ciocalyptoides*	*S. mutans* clinical isolate (MIC 15 µg/mL) *S. mutans* UA159 (MIC 15 µg/mL)
Peptides		
Rodriguesines A and B (**96** and **96a**)	*Didemnum* sp.	*S. mutans* clinical isolate (MIC 31 µg/mL) S. mutans UA159 (MIC 62 µg/mL)
Cyclo-(L-valyl-D-proline) (**97**)	*Rheinheimera japonica*	*S. aureus* (Zone of inhibition 17 mm at 0.5 mg/mL) *V. parahaemolyticus* (MIC 0.05 µg/mL) *V. vulnificus* (MIC 5 µg/mL) *M. luteus* (MIC 5 µg/mL)

Table 1. *Cont.*

Compound	Source	Activity Against Pathogen
Cyclo-(L-phenylalanyl-D-proline (**98**)	*R. japonica*	*S. aureus* (Zone of inhibition 15 mm at 0.5 mg/mL)
Cyclo-(S-Pro-R-Leu) (**99**)	*Haliclona oculata*	*V. parahaemolyticus* (MIC 0.5 μg/mL) *V. vulnificus* (MIC 5 μg/mL) *B. cereus* (MIC 0.05 μg/mL)
Isaridin G (**100**)	*Beauveria felina EN-135*	*E. coli* (MIC 64 μg/mL)
Desmethylisaridin G (**101**)	*B. felina EN-135*	*E. coli* (MIC 64 μg/mL)
Desmethylisaridin C1(**102**)	*B. felina EN-135*	*E. coli* (MIC 8 μg/mL)
Isaridin E (**103**)	*B. felina EN-135*	*E. coli* (MIC 16 μg/mL)
Desotamide (**104**)	*Streptomyces scopuliridis* SCSIO ZJ46	*S. pnuemoniae* (MIC 13 μg/mL) *S. aureus* (MIC 16 μg/mL) MRSE (MIC 32 μg/mL)
Desotamide B (**105**)	*S. scopuliridis SCSIO ZJ46*	*S. pnuemoniae* (MIC 13 μg/mL) *S. aureus* (MIC 16 μg/mL) MRSE (MIC 32 μg/mL)
Halogenated Compounds		
2-(2′,4′-dibromophenoxy)-3,5-dibromophenol (**106**)	*Dysidea granulosa*	MSSA (MIC 1 mg/L) *L. monocytogenes* (MIC 2 mg/L) *B. cereus* (MIC 0.1 mg/L) *C. diffiile* (MIC 4 mg/L) MRSA (MIC 0.1 mg/L) *Salmonella* sp. (MIC 1 mg/L) *E. coli* O157:H7 (MIC 8 mg/L) *Pseudomonas* (MIC 4 mg/L) *K. pneumoniae* (MIC 0.1 mg/L) *N. gonorrhoeae* (MIC 2 mg/L) *A. baumannii* (MIC 16 mg/L) *C. jejuni* (MIC 5 mg/L)
2-(2′,4′-dibromophenoxy)-3,4,5-tribromophenol (**107**)	*Dysidea granulosa*	*L. monocytogenes* (MIC 0.1 mg/L) *C. diffiile* (MIC 10 mg/L) MRSA (MIC 0.1 mg/L) *Salmonella* sp. (MIC 10 mg/L) *C. jejuni* (MIC 5 mg/L)
2-(2′,4′-dibromophenoxy)-4,6-dibromophenol (**108**)	*Dysidea* sp.	*L. monocytogenes* (MIC 1 mg/L) *B. cereus* (MIC 5 mg/L) *S. pneumoniae* (MIC 5 mg/L) *C. diffiile* (MIC 10 mg/L) MRSA (MIC 1 mg/L) *Salmonella* sp. (MIC 10 mg/L) *E. coli* O157:H7 (MIC 10 mg/L) *K. pneumoniae* (MIC 5 mg/L) *C. jejuni* (MIC 5 mg/L)
Aplysamine 8 (**109**)	*Pseudoceratina purpurea*	*E. coli* (MIC 125 μg/mL) *S. aureus* (MIC 31 μg/mL)
Sphaerodactylomelol (**113**)	*Sphaerococcus coronopifolius*	*S. aureus* (MIC 96 μM)
Bromosphaerol (**114**)	*Sphaerococcus coronopifolius*	*S. aureus* (MIC 22 μM)
12R-hydroxybromosphaerol (**115**)	*Sphaerococcus coronopifolius*	*S. aureus* (MIC 6 μM)
6-bromoindolyl-3-acetic acid (**116**)	*Pseudoalteromonas flavipulchra*	*V. anguillarum* (MIC 0.25 mg/mL)
Nagelamide X (**117**)	*Agelas* sp.	*S. aureus* (MIC 8 μg/mL) *M. luteus* (MIC 8 μg/mL) *A. niger* (IC_{50} 32 μg/mL) *T. mentagrophytes* (IC_{50} 16 μg/mL) *C. albicans* (IC_{50} 2 μg/mL)
Nagelamide Y (**118**)	*Agelas* sp.	*C. albicans* (IC_{50} 2 μg/mL)
Nagelamide Z (**119**)	*Agelas* sp.	*S. aureus* (MIC 16 μg/mL) *M. luteus* (MIC 8 μg/mL) *A. niger* (IC_{50} 4 μg/mL) *T. mentagrophytes* (IC_{50} 4 μg/mL) *C. albicans* (IC_{50} 0.25 μg/mL) *C. neoformans* (IC_{50} 2 μg/mL)

Table 1. *Cont.*

Compound	Source	Activity Against Pathogen
Ianthelliformisamine A (**120**)	*Suberea ianthelliformis*	*P. aeruginosa* (IC$_{50}$ 7 µM)
Ianthelliformisamine C (**122**)	*Suberea ianthelliformis*	*P. aeruginosa* (IC$_{50}$ 9 µM) *S. aureus* (IC$_{50}$ 4µM)
Polyketides		
N-acetyl-*N*-demethylmayamycin (**123**)	*Streptomyces* sp. 182SMLY	MRSA (MIC 20 µM)
Lindgomycin (**124**)	*Lindgomycetaceae strain*	*B. subtilis* (MIC 2.2 µM) *X. campestris* (MIC 17.8 µM) *S. epidermidis* (MIC 4.6 µM) *S. aureus* (MIC 2.7 µM) *S. aureus* (MRSA) (MIC 5.1 µM) *C. albicans* (MIC 5.7 µM) *S. tritici* (MIC 5.1 µM) *P. acnes* (MIC 4.7 µM)
Ascosetin (**125**)	*Lindgomycetaceae strain*	*B. subtilis* (MIC 3.4 µM) *X. campestris* (MIC 14.8 µM) *S. epidermidis* (MIC 6.3 µM) *S. aureus* (MIC 2.9 µM) *S. aureus* (MRSA) (MIC 3.2 µM) *C. albicans* (MIC 8 µM) *S. tritici* (MIC 10 µM) *P. acnes* (MIC 2.8 µM)
Manadodioxan E (**127**)	*Plakortis bergquistae*	*E. coli* (Zone of inhibition 16 mm at 10 µg/disk) *B. cereus* (Zone of inhibition 9 mm at 10 µg/disk)
2-(7-(2-Ethylbutyl)-2,3,4,4a,6,7-hexahydro-2-oxopyrano-[3,2b]-pyran-3-yl)-ethyl benzoate (**128**)	*B. subtilis* MTCC 10407	*V. parahemolyticus* (Zone of inhibition 11 mm at 10 µg/disk) *V. vulnificus* (Zone of inhibition 13 mm at 10 µg/disk) *A. hydrophilla* (Zone of inhibition 18 mm at 10 µg/disk)
2-((4Z)-2-ethyl-octahydro-6-oxo-3-((E)-pent-3-enylidene)-pyrano-[3,2b]-pyran-7-yl)-ethyl benzoate (**129**)	*B. subtilis* MTCC 10407	*V. parahemolyticus* (Zone of inhibition 11 mm at 10 µg/disk/mL) *V. vulnificus* (Zone of inhibition 14 mm at 10 µg/disk) *A. hydrophilla* (Zone of inhibition 15 mm at 10 µg/disk)
3-(methoxycarbonyl)-4-(5-(2-ethylbutyl)-5,6-dihydro-3-methyl-2H-pyran-2-yl)-butyl benzoate (**130**)	*Sargassum myriocystum*	*V. parahemolyticus* (Zone of inhibition 7 mm at 10 µg/disk) *V. vulnificus* (Zone of inhibition 7 mm at 10 µg/disk) *A. hydrophilla* (Zone of inhibition 8 mm at 10 µg/disk)
2-(8-butyl-3-ethyl-3,4,4a,5,6,8ahexahydro-2H-chromen-6-yl)-ethyl benzoate (**131**)	*Sargassum myriocystum*	*V. parahemolyticus* (Zone of inhibition 9 mm at 10 µg/disk) *V. vulnificus* (Zone of inhibition 8 mm at 10 µg/disk *A. hydrophilla* (Zone of inhibition 7 mm at 10 µg/disk)
Fradimycin A (**132**)	*Streptomyces fradiae*	*S. aureus* (MIC 6 µg/mL)
Fradimycin B (**133**)	*Streptomyces fradiae*	*S. aureus* (MIC 2 µg/mL)
Glycosylated Macrolactin A1 (**134**)	*Streptomyces* sp.	*B. subtilis* (MIC 0.027 µg/mL) *E. coli* (MIC 0.022 µg/mL) *P. aeruginosa* (MIC 0.055 µg/mL) *S. aureus* (MIC 0.055 µg/mL) *S. cerevisiae* (MIC 0.022 µg/mL)
Glycosylated Macrolactin B1 (**135**)	*Streptomyces* sp.	*B. subtilis* (MIC 0.055 µg/mL) *E. coli* (MIC 0.022 µg/mL) *P. aeruginosa* (MIC 0.055 µg/mL) *S. aureus* (MIC 0.055 µg/mL) *S. cerevisiae* (MIC 0.022 µg/mL)
Macrolactin X (**136**)	*Bacillus* sp.	*B. subtilis* (MIC 16 µg/mL) *E. coli* (MIC 16 µg/mL) *S. cerevisiae* (MIC 64 µg/mL)
Macrolactin Y (**137**)	*Bacillus* sp.	*B. subtilis* (MIC 16 µg/mL) *E. coli* (MIC 32 µg/mL) *S. cerevisiae* (MIC 64 µg/mL)
Macrolactin Z (**138**)	*Bacillus* sp.	*B. subtilis* (MIC 16 µg/mL) *E. coli* (MIC 32 µg/mL) *S. cerevisiae* (MIC 64 µg/mL)

Table 1. *Cont.*

Compound	Source	Activity Against Pathogen
Macrolactinic acid (**139**)	*Bacillus* sp.	*B. subtilis* (MIC 64 µg/mL) *E. coli* (MIC 32 µg/mL) *S. cerevisiae* (MIC 128 µg/mL)
Heronamycin A (**140**)	*Streptomyces* sp.	*B. subtilis* ATCC 6051 (MIC 18 µM) *B. subtilis* ATCC (MIC 14 µM)
Herbimycin A (**141**)	*Streptomyces* sp.	*B. subtilis* ATCC 6051 (MIC 5 µM) *B. subtilis* ATCC (MIC 5 µM) *C. albicans* ATCC 90028 (MIC 7 µM)
3-(Octahydro-9-isopropyl-2*H*-benzo[h]chromen-4-yl)-2-methylpropyl benzoate (**142**)	*Bacillus amyloliquefaciens*	*V. parahemolyticus* (Zone of inhibition 14 mm at 25 µg/disk) *V. vulnificus* (Zone of inhibition 16 mm at 25 µg/disk) *A. hydrophilla* (Zone of inhibition 18 mm at 25 µg/disk)
Methyl 8-(2-(benzoyloxy)ethyl)-hexahydro-4-((*E*)-pent-2-enyl)-2*H*-chromene-6-carboxylate (**143**)	*Bacillus amyloliquefaciens*	*V. parahemolyticus* (Zone of inhibition 12 mm at 25 µg/disk) *V. vulnificus* (Zone of inhibition 14 mm at 25 µg/disk) *A. hydrophilla* (Zone of inhibition 16 mm at 25 µg/disk)
3-*O*-(α-D-ribofuranosyl)questin (**144**)	*Eurotium cristatum*	*E. coli* (MIC 32 µg/mL)
Eurorubrin (**145**)	*Eurotium cristatum*	*E. coli* (MIC 64 µg/mL)
Isocoumarins		
Penicisimpins A (**146**)	*Penicillium simplicissimum* MA-332	*A. hydrophilia* (MIC > 64µg/mL) *E. coli* (MIC 4 µg/mL) *M. luteus* (MIC 8 µg/mL) *P. aeruginosa* (MIC 4 µg/mL) *V. alginolyticus* (MIC 8µg/mL) *V. harveyi* (MIC 4 µg/mL) *V. parahaemolyticus* (MIC 4 µg/mL) *C. gloeosprioides* (MIC 4 µg/mL)
Penicisimpins B (**147**)	*Penicillium simplicissimum* MA-332	*A. hydrophilia* (MIC 32 µg/mL) *E. coli* (MIC 32 µg/mL) *M. luteus* (MIC 64 µg/mL) *P. aeruginosa* (MIC 32 µg/mL) *V. alginolyticus* (MIC 32 µg/mL) *V. harveyi* (MIC 16 µg/mL) *V. parahaemolyticus* (MIC 32 µg/mL) *C. gloeosprioides* (MIC 16 µg/mL)
Penicisimpins C (**148**)	*Penicillium simplicissimum* MA-332	*A. hydrophilia* (MIC 16 µg/mL) *E. coli* (MIC 8 µg/mL) *M. luteus* (MIC 16 µg/mL) *P. aeruginosa* (MIC 8 µg/mL) *V. alginolyticus* (MIC 16 µg/mL) *V. harveyi* (MIC 8 µg/mL) *V. parahaemolyticus* (MIC 8 µg/mL) *C. gloeosprioides* (MIC 8 µg/mL
Citreoisocoumarin (**149**)	*Penicillium vinaceum*	*S. aureus* (Zone of inhibition 19 mm at 100 µg/disk)
Nucleosides		
Rocheicoside A (**150**)	*Streptomyces rochei* 06CM016	*E. coli* O157:H7 (MIC 16 µg/mL) MRSA DSM (MIC 8 µg/mL) *C. albicans* DSM (MIC 4 µg/mL)
Plicacetin (**151**)	*Streptomyces rochei* 06CM016	*E. coli* O157:H7 (MIC 8 µg/mL) MRSA DSM (MIC 4 µg/mL) *C. albicans* DSM (MIC 8 µg/mL)
Norplicacetin (**152**)	*Streptomyces rochei* 06CM016	*E. coli* O157:H7 (MIC 16 µg/mL) MRSA DSM (MIC 8 µg/mL) *C. albicans* DSM (MIC 4 µg/mL)
p-amino benzamido uracil (**153**)	*Streptomyces rochei* 06CM016	*E. coli* O157:H7 (MIC 16 µg/mL) MRSA DSM (MIC 16 µg/mL) *C. albicans* DSM (MIC 8 µg/mL)
Cytosamine (**154**)	*Streptomyces rochei* 06CM016	*E. coli* O157:H7 (MIC 16 µg/mL) MRSA DSM (MIC 16 µg/mL) *C. albicans* DSM (MIC 8 µg/mL)

Table 1. *Cont.*

Compound	Source	Activity Against Pathogen
Miscellaneous Compounds		
Terretrione A (**151**)	*Penicillium vinaceum*	*C. albicans* (Zone of inhibition 27 mm at 100 µg/disk)
Terretrione C (**152**)	*Penicillium* sp. CYE-87	*C. albicans* (MIC 32 µg/mL)
Terretrione D (**153**)	*Penicillium* sp. CYE-87	*C. albicans* (MIC 32 µg/mL)
α-Cyclopiazonic acid (**154**)	*Penicillium vinaceum*	*E. coli* (Zone of inhibition 20 mm at 100 µg/disk)
5-methoxydihydrosterigmato-cystin (**155**)	*Aspergillus versicolor*	*S. aureus* (MIC 13 µg/mL) *B. subtilis* (MIC 3 µg/mL)
Caerulomycin A (**156**)	*Actinoalloateichus cyanogriseus*	*C. albicans* (MIC 0.8–1.6 µg/mL) *C. albicans* CO9 (MIC 0.8–1.6 µg/mL) *C. glabrata* HO5^{FlucR} (MIC 0.4–0.8 µg/mL) *C. krusei* GO3^{FlucR} (MIC 0.8–1.6 µg/mL)
Trichodin A (**157**)	*Trichoderma* sp.	*B. subtilis* (IC$_{50}$ 27 µM) *S. epidermidis* (IC$_{50}$ 24 µM) *C. albicans* (IC$_{50}$ 25 µM)
Pyridoxatin (**158**)	*Trichoderma* sp.	*B. subtilis* (IC$_{50}$ 4 µM) *S. epidermidis* (IC$_{50}$ 4 µM) *S aureus* (MRSA), (IC$_{50}$ 4 µM) *C. albicans* (IC$_{50}$ 26 µM) *T. rubrum* (IC$_{50}$ 4 µM)
1-(10-Aminodecyl) Pyridinium (**159**)	*Amycolatopsis alba*	*B. subtilis* NCIM 2439 (MIC > 70 µg/mL) *B. pumilus* NCIM 2327 (MIC > 90 µg/mL) *S. aureus* NCIM 5021 (MIC > 160 µg/mL) *A. formicans* NCIM 2319 (MIC > 150 µg/mL) *E. coli* NCIM 2067 (MIC > 90 µg/mL)
Porric acid D (**160**)	*Alternaria* sp.	*S. aureus* (MIC 100 µg/mL)
Altenusin (**161**)	*Alternaria* sp.	*S. aureus* (MIC 25 µg/mL)
p-Hydroxybenzoic acid (**162**)	*Pseudoalteromonas flavipulchra*	*V. anguillarum* (MIC 1.25 mg/mL)
trans-Cinnamic acid (**163**)	*Pseudoalteromonas flavipulchra*	*V. anguillarum* (MIC 1.25 mg/mL)
N-hydroxybenzoisoxazolone (**164**)	*Pseudoalteromonas flavipulchra*	*V. anguillarum* (MIC 0.25 mg/mL)
Bacilosarcin B (**165**)	*Bacillus subtilis*	*S. aureus* (MIC 4 µM)
amicoumacin A (**167**)	*Bacillus subtilis*	*S. aureus* (MIC 19 µM) *B. subtillis* (MIC 19 µM)
Microluside A (**169**)	*Micrococcus* sp. EG45	*E. faecalis* JH212 (MIC 10 µM) *S. aureus* NCTC 8325 (MIC 13 µM)

4. Alkaloids

Alkaloids constitute the largest number of antimicrobial compounds reported from marine species (Figure 2). Polyoxygenated dihydropyrano[2,3-*c*]pyrrole-4,5-dione derivatives, pyranonigrin A (**16**) and F (**17**) were isolated and identified from *Penicillium brocae* MA-231, which is an endophytic fungus obtained from the fresh tissue of the marine mangrove plant *Avicennia marina*. These compounds possess a wide array of antimicrobial activities against a human-, aqua-, and plant-pathogens [34]. Indole diketopiperazine compounds (**18–21**) identified from the culture extract of *Eurotium cristatum* EN-220 (endophytic fungus species obtained from the marine alga *Sargassum thunbergii*) possess antimicrobial activities. While rubrumazine B (**18**) exhibited moderate activity (MIC 64 µg/mL) against the plant-pathogenic fungus *Magnaporthe grisea*, echinulin (**19**), dehydroechinulin (**20**) and variecolorin H (**21**) showed mild activity (MIC 256 µg/mL) against the human pathogen *S. aureus*. This particular trend in antimicrobial activity led to an assumption that indole diketopiperazine alkaloids formed by condensation of a tryptophan residue with a second amino acid such as L-alanine might have low relevance to the observed antimicrobial activity [35]. In a separate study by Du et al., a set of indole alkaloids including echinulin (**19**), cristatumin A (**22**), cristatumin D (**23**) and tardioxopiperazine A (**24**) were found to be active against *Escherichia coli* and *S. aureus* bacteria but were unable to inhibit the growth of the plant-pathogenic fungi, *Alternaria brassicae*, *Valsa mali*, *Physalospora obtuse*, *Alternaria* *solania*, and *Sclerotinia miyabeana*. The antibacterial activity of cristatumin A (**22**) appeared to be related to the serine residue in the 2,5-diketopiperazine moiety compared to that of neoechinulin A (**25**), which contained an alanine residue. The structural difference between isoechinulin A (**26**) and tardioxopiperazine A (**24**) was found at the C-8/C-9 position, where the single bond between C-8/C-9 in compound **24** was essential for its antibacterial activity [36].

N-alkylated hydantoin alkaloids, such as hemimycalins A (**27**) and B (**28**) and (Z)-5-(4-hydroxybenzylidene)imidazolidine-2,4-dione (**29**), isolated from the Red Sea sponge, *Hemimycale arabica* have previously demonstrated antimicrobial activity against *E. coli* and *C albicans* but were found to be inactive against *S. aureus* [37]. Peniciadametizine A (**30**), a dithiodiketopiperazine derivative possessing a unique spiro [furan-2,7'-pyrazino[1,2-b][1,2]oxazine] skeleton, and its highly oxygenated analogue, peniciadametizine B (**31**) were isolated from *Penicillium adametzioides* AS-53, a fungus obtained from an unidentified marine sponge collected in Wenchang, Hainan, China. These compounds exhibited selective activities against *A. brassicae* but were inactive against bacteria (*Aeromonas hydrophilia*, *Edwardsiella tarda*, *E. coli*, *S. aureus*, *Vibrio alginolyticus*, *V. anguillarum*, *V. parahaemolyticus*, and *V. harveyi*), and plant-pathogenic fungi (*Colletotrichum gloeosporioides*, *Fusarium graminearum*, and *Gaeumannomyces graminis*) [38]. The antimicrobial sulfide diketopiperazine derivatives, penicibrocazines A–E (**32–36**), were isolated from a culture extract of *Penicillium brocae* MA-231, an endophytic fungus obtained from the fresh tissue of the marine mangrove plant *Avicennia marina*. Penicibrocazines B–E (compounds **33–36** shown in Figure 2) have demonstrated a wide spectrum of activity against several human, aquatic and plant-pathogenic microbes including *S. aureus*, *Micrococcus luteus* and *Gaeumannomyces graminis* but were inactive against *Eromonas hydrophilia*, *E. coli*, *V. harveyi*, *V. parahaemolyticus* and the plant-pathogenic fungi, *A. brassicae*, *Colletotrichum gloeosporioides* and *Fusarium graminearum*. Penicibrocazine A (**32**) was inactive against this entire array of microorganisms. The double bonds at C-6 and C-6' increase the activity against *S. aureus* (**34** vs. **32**). In addition, a higher number of S-methyl groups is likely to strengthen their activity against *G. graminis* (**35** vs. **32**), while keto groups at C-5/5' (**36** vs. **34**) are responsible for the enhanced antimicrobial activity [39]. Crambescidin 800 (**37**), a pentacyclic guanidine alkaloid isolated from the sponge *Clathria cervicornis*, has expressed specific inhibitory activity against *Acinetobacter baumannii*, *Klebsiella pneumonia* and *P. aeruginosa* [40].

Xinghaiamine A (**38**), an alkaloid isolated from the marine-derived actinomycete *Streptomyces xinghaiensis* NRRL B24674, has exhibited a broad-spectrum of antibacterial activities against both Gram-negative (*Acinetobacter baumannii*, *P. aeruginosa* and *E. coli*) and Gram-positive pathogens (*S. aureus* and *B. subtilis*). The inhibition of these pathogenic bacteria and clinical MRSA isolates demonstrates the potential of compound **38** to be an effective antibiotic against multi-drug resistant pathogens, particularly *S. aureus* and *A. baumanii*. However, the molecule displayed no obvious antifungal activity against *C. albicans* when tested at concentrations up to 176.64 µM. The sulfoxide moiety present in Xinghaiamine A (**38**) is unusual among the metabolites produced by marine actinomycetes and confers compounds with a broad spectrum of biological activities, including potent antimicrobial activities [41]. Bioassay-directed fractionation performed on the ethyl acetate fraction of an organic extract from the Red Sea sponge *Hyrtios* species yielded the alkaloid compounds, hyrtioerectines D–F (**39–41**), which possessed antimicrobial activities against *C. albicans*, *S. aureus* and *P. aeruginosa* but not against *E. coli*. The relatively higher antimicrobial activity exerted by compounds **39** and **41** with respect to compound **40** could be attributed to the presence of diphenolic moieties in their structure. Amidation of the carboxylic moiety in compound **41** exerted a slight effect on activity when compared to compound **39** [42]. The diterpene alkaloids, ageloxime B (**42**) and (−)-ageloxime D (**43**), were isolated from the marine sponge *Agelas mauritiana*, isolated from Yongxing island in the South China Sea. Both compounds (**42** and **43**) were able to inhibit the growth of *C. neoformans*. Compound **42** also exhibited antibacterial activity against *S. aureus* and MRSA [43]. A manzamine alkaloid, i.e., zamamidine D (**44**), was isolated from an Okinawan *Amphimedon* sp. marine sponge. Compound **44** is the first manzamine alkaloid possessing a 2,2'-methylenebistryptamine unit as its aromatic moiety instead of a β-carboline unit, which affected growth of both bacteria and fungi [44]. Chemical investigation of the marine-sponge derived fungus *Penicillium adametzioides* AS-53 yielded the bisthiodiketopiperazine derivatives, adametizines A (**45**) and B (**46**), differing in the presence of a chlorine group at C-7 in compound **45** and hydroxyl in compound **46**. The presence of a chlorine atom at C-7 conferred compound **45** with better antibacterial activity than compound **46** [45].

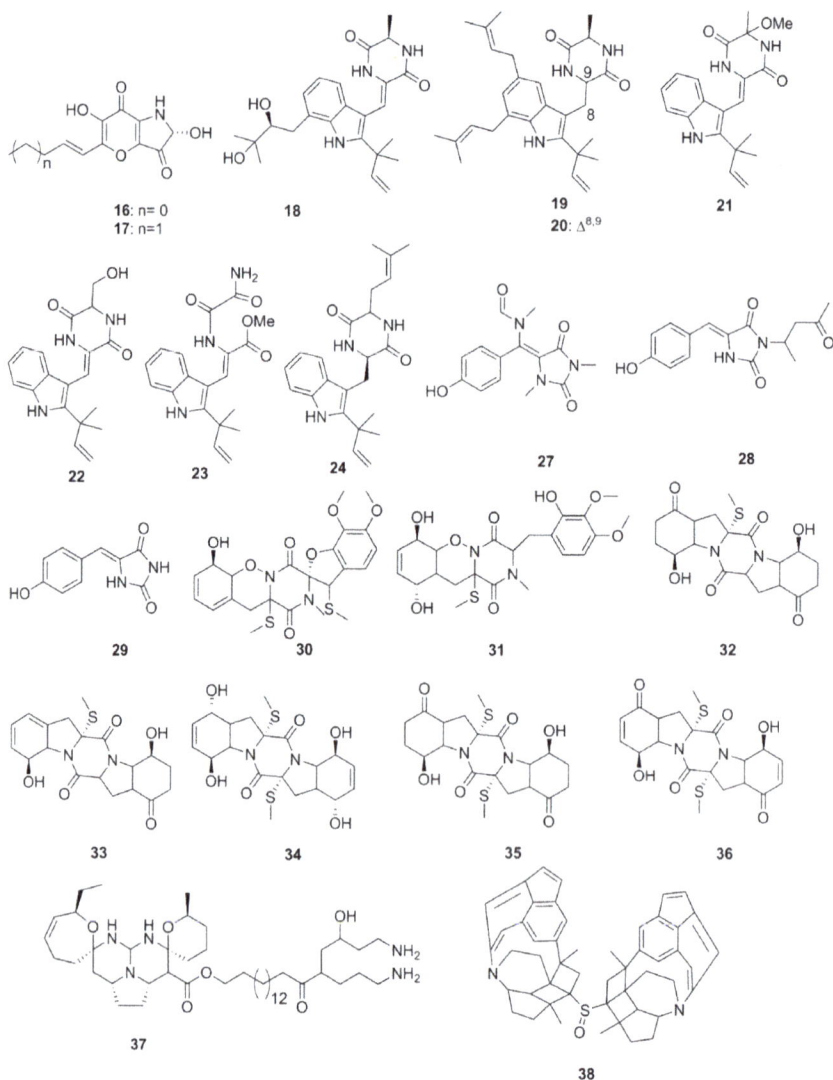

Figure 2. *Cont.*

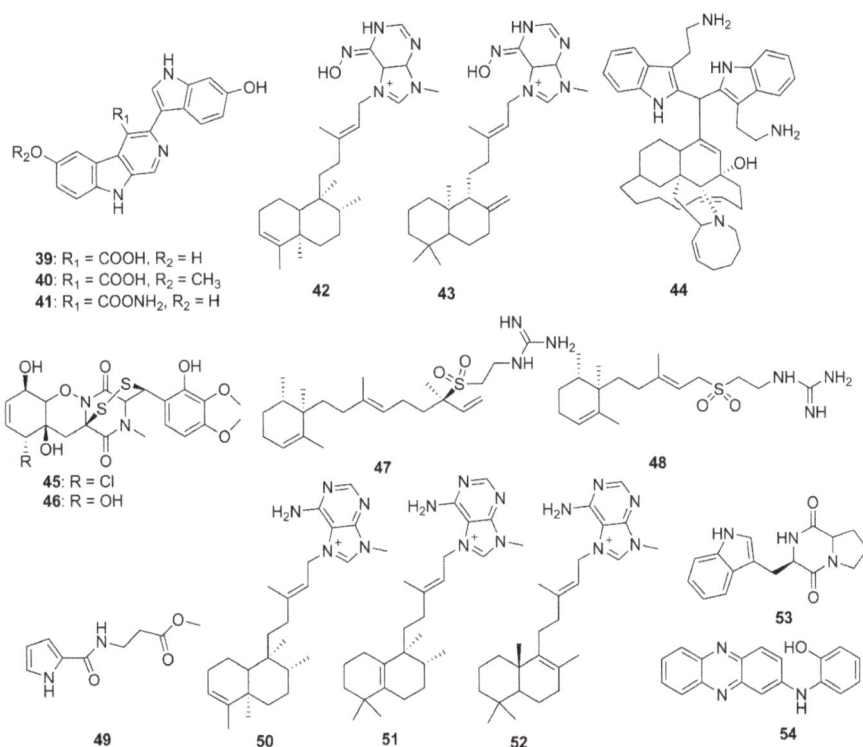

Figure 2. Antimicrobial alkaloids.

Diterpene alkaloids from the sponge *Agelas nakamurai* collected from the Xisha Islands in the South China Sea have demonstrated significant antimicrobial potential in vitro against *S. aureus*, *E. coli*, *Proteusbacillus vulgaris* and *C. albicans*. Iso-agelasidine B (**47**) and (−)-agelasidine C (**48**) exhibited pronounced antifungal activities (MIC 2.34 µg/mL) against *C. albicans* and weakly inhibited bacterial growth. In addition, diterpene alkaloids containing a 9-*N*-methyladeninium unit: iso-agelasine C (**49**), agelasine B (**50**), agelasine J (**51**) and nemoechine G (**52**) also possessed strong antifungal activities (MIC 0.6 µg/mL) against *C. albicans* and weak to moderate antibacterial activities [46]. Brevianamide F (**53**) was isolated from the marine-derived fungus, *Penicillium vinaceum*. This compound demonstrated antimicrobial activity against *S. aureus* and antifungal activity towards *C. albicans* [47]. A new secondary metabolite N-(2-hydroxyphenyl)-2-phenazinamine (**54**) was isolated from the saline culture broth of *Nocardia dassonvillei*, a marine actinomycete recovered from a sediment in the Arctic Ocean. This new compound has shown significant antifungal activity against *C. albicans*, with a MIC 64 µg/mL [48].

5. Terpenoids

As many as 26 different types of terpenoids have been found in marine species (Figure 3). A potent antimicrobial meroterpenoid, puupehenol (**55**) was isolated from the organic extract of a deep-water Hawaiian sponge *Dactylospongia* sp., along with puupehenone (**56**) which has been suggested to be a derivative of puupehenol. These compounds exhibited antibacterial activity towards the Gram-positive bacteria, *S. aureus* and *B. cereus*, but no inhibition was observed against *E. coli* and *P. aeruginosa* [49]. The sesquiterpenes, Penicibilaenes A (**57**) and B (**58**) isolated from *Penicillium bilaiae* MA-267, a fungus obtained from the rhizospheric soil of the mangrove plant *Lumnitzera racemose* possess a tricyclo[6.3.1.0^{1,5}]dodecane skeleton, which confers antimicrobial

activities to these compounds. Both of these molecules have demonstrated selective activity against
C. gloeosporioides. Compound **57** proved more active than **58** suggesting that the acetylation of
4-OH enhances the bioactivity of the compound [50]. Sesquiterpenoids including aspergillusene
A (**59**), (Z)-5-(Hydroxymenthyl)-2-(6′)-methylhept-2′-en-2′-yl)-phenol (**60**) and sydonic acid (**61**)
isolated from the sponge-associated fungus, *Aspergillus sydowii* ZSDS1-F6 displayed antimicrobial
activities against *Klebsiella pneumonia* and *Aeromonas hydrophila* [51]. Laurene-type sesquiterpenes,
12-hydroxy isolaurene (**62**), 8,11-dihydro-12-hydroxy isolaurene (**63**) and isolauraldehyde (**64**)
were isolated from the organic extract of the red alga *Laurencia obtuse* collected off the coast of
Jeddah in Saudia Arabia. These compounds exhibited potent activity against the Gram-positive
bacteria, *B. subtilis* and *S. aureus*, with compound **64** proving to be the most active (MIC 35 and
27 μg/mL, respectively). Compound **64** also significantly inhibited *C. albicans* (MIC of 70 μg/mL)
while, no significant activity against the Gram-negative bacterium *P. aeruginosa* was observed [52].
Napyradiomycins including Napyradiomycin 1 (**65**) and 2 (**66**), Napyradiomycin B2–B4 (**67–69**)
isolated from marine-derived, *Streptomyces* strains were found to be active against MRSA [53].
Antibacterial *N-N*-coupled indolo-sesquiterpene atropo-diastereomers, dixiamycin A (**70**), dixiamycin
B (**71**), oxiamycin (**72**), chloroxiamycin (**73**) and xiamycin A (**74**) were isolated from a marine-derived
actinomycete, *Streptomyces* sp. SCSIO 02999. Compounds **70** and **71** were identified as the first
examples of naturally occurring atropo-diastereomers containing an unusual *N-N*-coupled dimeric
indolo-sesquiterpene skeleton and a stereogenic *N-N* axis, whilst compound **72** was characterized to
contain an unusual seven-membered oxa-ring. Interestingly, the two dimeric compounds **70** and **71**
displayed better antibacterial activities than the monomers (compounds **72–74**) against four tested
strains [54]. Antimicrobial compounds with rare pyrane-based cembranoid structure, sarcotrocheliol
acetate (**75**) and sarcotrocheliol (**76**), along with cembranoid, cembrene-C (**77**), sarcophine (**78**), and
the aromadendrene sesquiterpenoid, palustrol (**79**) were isolated from the soft coral *Sarcophyton
trocheliophorum*. Compounds **75** and **76** displayed significant antibacterial activity, especially against
S. aureus, *Acinetobacter* sp., and MRSA with MICs ranging from 1.53 to 4.34 μM, while compound **77**
demonstrated antifungal activity against *Aspergillus flavus* and *C. albicans* with an MIC of 0.68 μM [55].
A sesquiterpenoid quinone, epi-ilimaquinone (**80**), isolated from the Fijian marine sponge *Hippospongia*
sp. was found to possess antibacterial activity against MRSA, wild-type *S. aureus* and VREF and
displayed antifungal activity against amphotericin-resistant *C. albicans* [56].

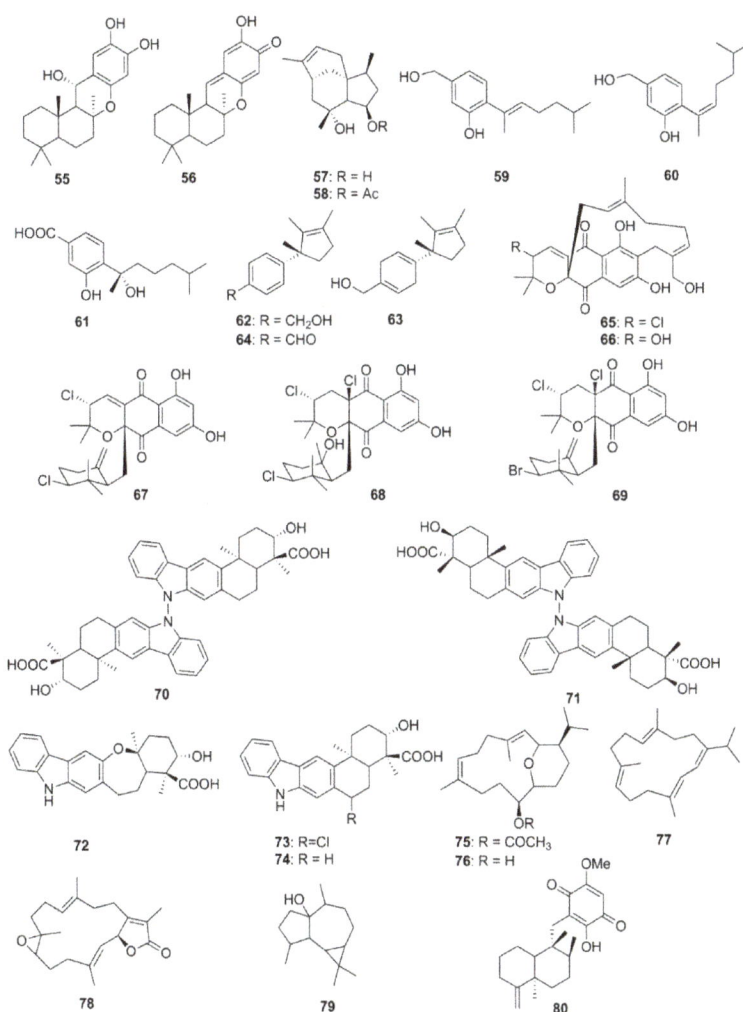

Figure 3. Antimicrobial terpenoids.

6. Lipids

In the past 5 years alone, 15 lipids possessing antibacterial activities were isolated from marine species (Figure 4). Two cerebrosides, penicillosides A (**81**) and B (**82**), were isolated from Red Sea marine-derived fungi and the fungus *Penicillium* species isolated from the tunicate *Didemnum* species. Penicilloside A (compound **81**) displayed antifungal activity towards *C. albicans* displaying an inhibition zone of 23 mm, while Penicilloside A (**82**) possessed antibacterial activities against *S. aureus* and *E. coli* displaying inhibition zones of 19 mm and 20 mm respectively, at 100 µg/disk concentration [57]. Glycolipopeptides-ieodoglucomides A–C (**83–85**), and a monoacyldiglycosyl glycerolipid-iedoglycolipid (**86**) isolated from the fermentation broth of the marine sediment-derived bacterium *Bacillus licheniformis* displayed promising antimicrobial activities against Gram-positive and Gram-negative bacteria. Moreover these two pencillosides were also active against the plant pathogenic fungi *Aspergillus niger*, *Rhizoctonia solani*, *Colletotrichum acutatum*, *B. cenerea* and *C. albicans* [58,59].

Gageotetrins A–C (**87–89**), fall under a unique class of linear lipopeptides (di- and tetrapeptides) isolated from a marine-derived *Bacillus* were found to be comparatively more active against fungi than to bacteria with MIC values of 0.02–0.04 μM. Moreover, compounds **88** and **89** at a concentration of 0.02 μM displayed potent motility inhibition and lytic activity against the late blight pathogen *Phytophthora capsici*, which causes enormous economic damage in cucumber, pepper, tomato and beans [60]. Lauramide diethanolamine (**90**) was isolated from the marine bacterial strain *Streptomyces* sp. (strain 06CH80). Although this compound (**90**) showed moderate antimicrobial activities against clinical pathogens, its chemical structure is particularly unusual containing a unique carbon skeleton which is different from any other existing antimicrobial agents. This unusual structure provides researchers with the exciting opportunity of exploring various chemical modifications of the compound with the aim of developing of potentially more efficient antimicrobial agents [61]. The unsaturated fatty acids, linieodolides A (**91**) and B (**92**), were isolated from the culture broth of a marine *Bacillus* and the mechanism of their antimicrobial activity was proposed through the inhibition of bacterial fatty acid synthesis [62]. Two highly acetylated steroids, dysiroid A (**93**) and dysiroid B (**94**), were isolated from the marine sponge, *Dysidea* sp. Compounds **93** and **94** showed potent activity against bacterial strains with MICs ranging from 4 to 8 μg/mL [63]. Halistanol sulfate A (**95**) was obtained from the sponge *Petromica ciocalyptoides* and determined to be an effective antibacterial compound against *Streptococcus mutans*. The compound inhibited *S. mutans* biofilm formation by down regulating the expression of the *gtfB*, *gtfC* and *gbpB* virulence genes. Compound **95** inhibited biofilm formation in two *S. mutans* strains at low MIC, but did not inhibit initial colonization by *S. sanguinis*. Such activity is highly desirable in preventative treatments that inhibits pathogenic bacteria via the disruption of biofilm formation without affecting the healthy normal microflora [64].

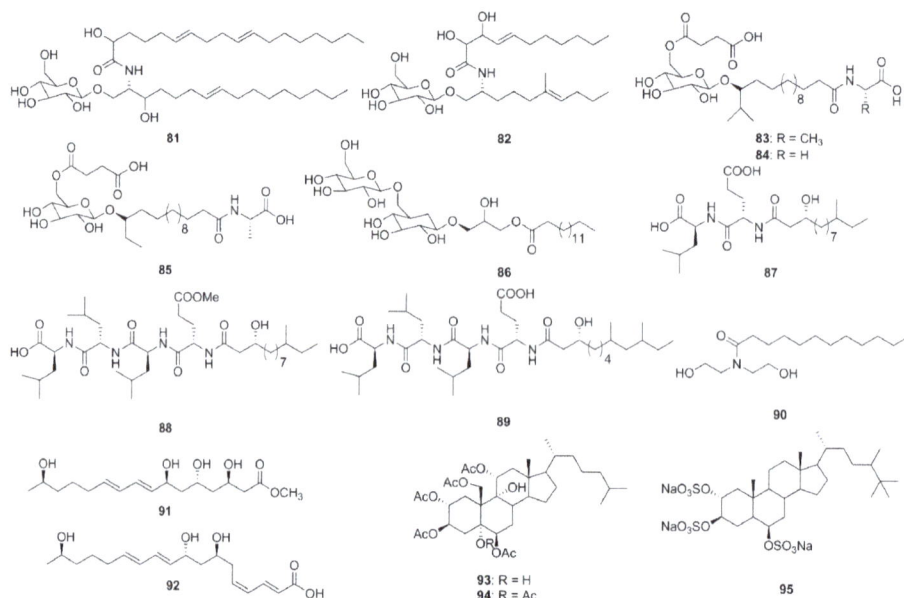

Figure 4. Antimicrobial lipids.

7. Peptides

Marine derived antimicrobial compounds of peptide origin are shown in Figure 5. Modified diketopiperazines, rodriguesines A and B (**96** and **96a**) isolated as an inseparable mixture from the ascidian *Didemnum* sp. were found to possess broad antimicrobial activities inhibiting both oral streptococci and pathogenic bacteria. These diketopiperazines are reported to modulate the LuxR-mediated quorum-sensing systems of Gram-negative and Gram-positive bacteria and are considered to influence cell-cell bacterial signaling, offering alternative ways to control biofilms by interfering with microbial communication [64]. Diketopiperazines, cyclo-(L-valyl-D-proline) (**97**) cyclo-(L-phenylalanyl-D-proline) (**98**), were isolated from a *Rheinheimera japonica* strain KMM 9513 collected from shores off of the Sea of Japan. These diketopiperazines inhibited the growth of *B. subtilis*, *Enterococcus faecium* and *S. aureus* but were inactive against *E. coli*, *S. epidermidis*, *Xanthomonas* sp. *pv. badrii* and *C. albicans* [65]. Compounds cyclo-(*S*-Proline-*R*-Leucine) (**99**) and (**97**) were isolated from *Bacillus megaterium* LC derived from the marine sponge *Haliclona oculata*. These compounds showed antimicrobial activity at MIC values ranging from 0.005 to 5 μg/mL against Gram-negative bacteria *V. vulnificus* and *V. parahaemolyticus*, together with the Gram-positive bacteria *B. cereus* and *M. luteus* [66]. Cyclohexadepsipeptides of the isaridin class including isaridin G (**100**), desmethylisaridin G (**101**), desmethylisaridin C1 (**102**) and isaridin E (**103**), were identified in the marine bryozoan-derived fungus *Beauveria felina* EN-135. These compounds possessed inhibitory activities against *E. coli* with MICs in the range of 8–64 μg/mL [67]. Desotamide (**104**) and desotamide B (**105**) are cyclic hexapeptides isolated from the marine microbe *Streptomyces scopuliridis* SCSIO ZJ46. When these compounds were explored for their antibacterial potential, notable antibacterial activities against strains of *S. pnuemoniae*, *S. aureus*, and methicillin-resistant *Staphylococcus epidermidis* (MRSE) were observed [68].

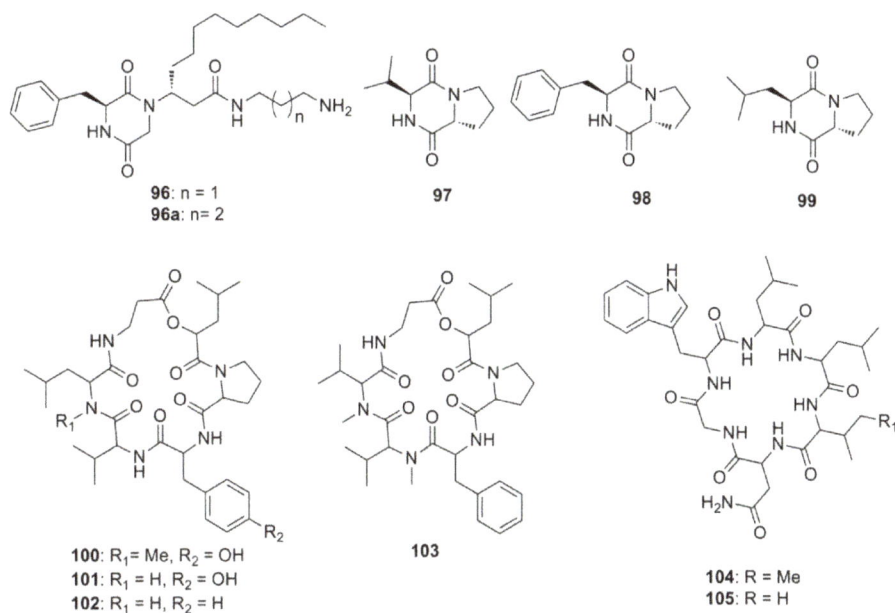

Figure 5. Antimicrobial peptides.

8. Halogenated Compounds

The chemical structures of halogenated compounds with antimicrobial properties isolated between 2012 and 2017 are presented in Figure 6. Compounds demonstrating a broad spectrum of antibacterial activities were polybrominated diphenyl ethers, 2-(2′,4′-dibromophenoxy)-3,5-dibromophenol (**106**) and 2-(2′,4′-dibromophenoxy)-3,4,5-tribromophenol (**107**) isolated from the marine sponge *Dysidea granulosa*; and 2-(2′,4′-dibromophenoxy)-4,6-dibromophenol (**108**) from *Dysidea* spp. These brominated ethers exhibited in vitro antibacterial activity against MRSA, methicillin-sensitive *Staphylococcus aureus* (MSSA), *E. coli* O157:H7, and *Salmonella*. Structurally compound **106** differed from compound **108** at a bromo-substituted position, and from compound **107** by containing an additional bromo group at the C-4 position. From the structure-activity relationships it was suggested that bromination and para-substitution decreases the antimicrobial activities of bromophenols, except against the human pathogen, *Listeria monocytogenes* [69]. Bromotyrosine-derived alkaloids were isolated by bioassay-guided fractionation of extracts from the Australian marine sponge *Pseudoceratina purpurea*. Aplysamine 8 (**109**) was not found to have any notable activity against *E. coli* or *S. aureus* while hexadellin (**110**), aplysamine 2 (**111**) and 16-debromoaplysamine 4 (**112**) displayed activity against *S. aureus* at concentrations ranging from 125–250 µg/mL [70]. The brominated diterpene, sphaerodactylomelol (**113**) which belongs to the rare dactylomelane family, bromosphaerol (**114**) and 12R-hydroxybromosphaerol (**115**) were isolated from cosmopolitan red algae, *Sphaerococcus coronopifolius*, collected in the Atlantic. These compounds inhibited the growth of *S. aureus* but were inactive against *E. coli* and *P. aeruginosa* [71]. A brominated compound, 6-bromoindolyl-3-acetic acid (**116**) displayed varied activities against both Gram-positive and Gram-negative bacteria: *V. harveyi*, *Photobacterium damselae* subsp. *damselae*. *A. hydrophila*, *S. aureus* and *B. subtilis* [72]. The bromopyrrole alkaloids, nagelamides X–Z (**117–119**), isolated from a marine sponge *Agelas* sp. exhibited antimicrobial activity against a range of bacteria and fungi. Compounds **117** and **118** are dimeric bromopyrrole alkaloids with a novel tricyclic skeleton, which consists of spiro-bonded tetrahydrobenzaminoimidazole and aminoimidazolidine moieties, while compound **118** is the first dimeric bromopyrrole alkaloid involving the C-8 position in dimerization [73]. The antibacterial bromotyrosine-derived metabolites, ianthelliformisamines A–C (**120–122**), were isolated from the marine sponge *Suberea ianthelliformis*. Compound **120** displayed selective inhibitory activity against *P. aeruginosa* while compound **121** showed little inhibition. The spermine moiety associated with compounds **120** and **122** appeared to be important for activity against *P. aeruginosa*, since replacement of spermine by spermidine in compound **121** reduced the activity significantly. Furthermore, the addition of an extra cinnamyl derivative in compound **122** to the terminal amine of the spermine chain decreased the antibacterial selectivity between *P. aeruginosa* and *S. aureus*; however the observed selectivity may be due to differential cell permeability between the Gram-negative and the Gram-positive bacteria [74].

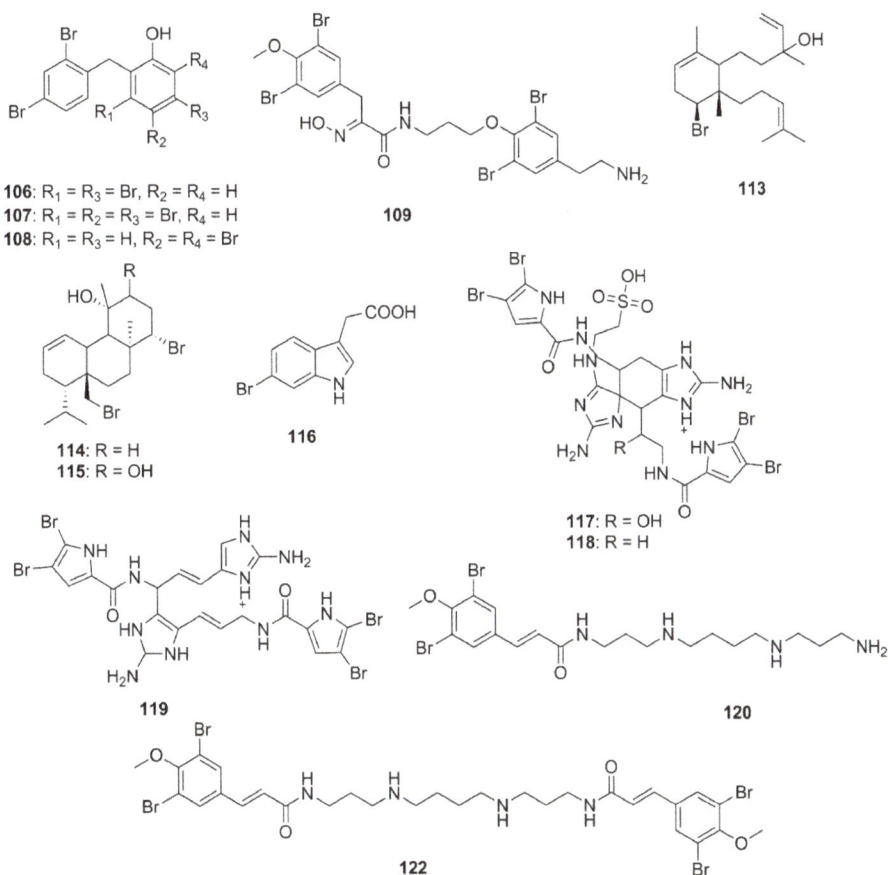

Figure 6. Antimicrobial halogenated compounds.

9. Polyketides

Antimicrobial marine natural products of polyketide origin are presented in Figure 7. Chemical investigations of cultures of the marine *Streptomyces* sp. 182SMLY led to the discovery of the antimicrobial polycyclic anthraquinone, *N*-acetyl-*N*-demethylmayamycin (**123**), which inhibited the growth of MRSA with a MIC 20 µM but displayed no inhibition against *E. coli* [75]. An unusual antibiotic polyketide with a new carbon skeleton, lindgomycin (**124**), and ascosetin (**125**) extracted from mycelia and culture broth of different *Lindgomycetaceae* strainswere two folds less active against the clinically relevant bacteria *Staphylococcus epidermidis*, *S. aureus*, MRSA, and *Propionibacterium acnes* than standard drug chloramphenicol. The antifungal activity of compound **124** and **125** was four times lower than nystatin against the human pathogenic yeast *C. albicans*. Moreover *Xanthomonas campestris* and *Septoria tritici*, were also inhibited by these metabolites but no inhibitory effects against Gram-negative bacteria was observed [76]. Among the polyketide endoperoxides, manadodioxans D (**126**) and E (**127**) isolated from the marine sponge *Plakortis bergquistae* in Indonesia, only manadodioxan E (**127**) displayed antimicrobial activities against bacteria, namely *E. coli* and *B. cereus*. The presence of a carbonyl group at C-13 position in compound **126** possibly has sequestered the antimicrobial activity. However, both compounds were found to be inactive against *C. albicans*, and *S. cerevisiae* [77]. Using antibacterial bioassay-guided fractionation, two *O*-heterocyclic compounds belonging to pyranyl benzoate analogues of polyketide origin 2-(7-(2-Ethylbutyl)-2,3,4,4*a*,6,7-hexahydro-2-oxopyrano-[3,2*b*]-pyran-3-yl)-ethyl benzoate (**128**) and 2-(4*Z*)-2-ethyl-octahydro-6-oxo-3-(*E*)-pent-3-enylidene-pyrano-[3,2*b*]-pyran-7-yl-ethyl benzoate (**129**), were isolated from the ethyl acetate extract of *B. subtilis* MTCC 10407. Two additional homologs, i.e., (3-(methoxycarbonyl)-4-(5-(2-ethylbutyl)-5,6-dihydro-3-methyl-2*H*-pyran-2-yl)-butylbenzoate) (**130**) and [2-(8-butyl-3-ethyl-3,4,4*a*,5,6,8*a*-hexahydro-2*H*-chromen-6-yl)-ethylbenzoate] (**131**), were also isolated from the ethyl acetate extract from the host seaweed *Sargassum myriocystum*. Although compounds **130** and **131** displayed weaker antibacterial activities than compounds **128** and **129** these four compounds possessed similar structures suggesting the ecological and metabolic symbiosis between seaweeds and bacteria. It was evident from the study that the presence of dihydro-methyl-2*H*-pyran-2-yl propanoate system was essential to impart the antibacterial activity. Tetrahydropyran-2-one moiety of the tetrahydropyrano-[3,2*b*]-pyran-2(3*H*)-one system of compound **128** might be cleaved by the metabolic pool of seaweeds to afford biologically active methyl 3-(dihydro-3-methyl-2*H*-pyranyl)-propanoate moiety of compound **130** (which was shown to have no significant antibacterial activity in intact form) [78].

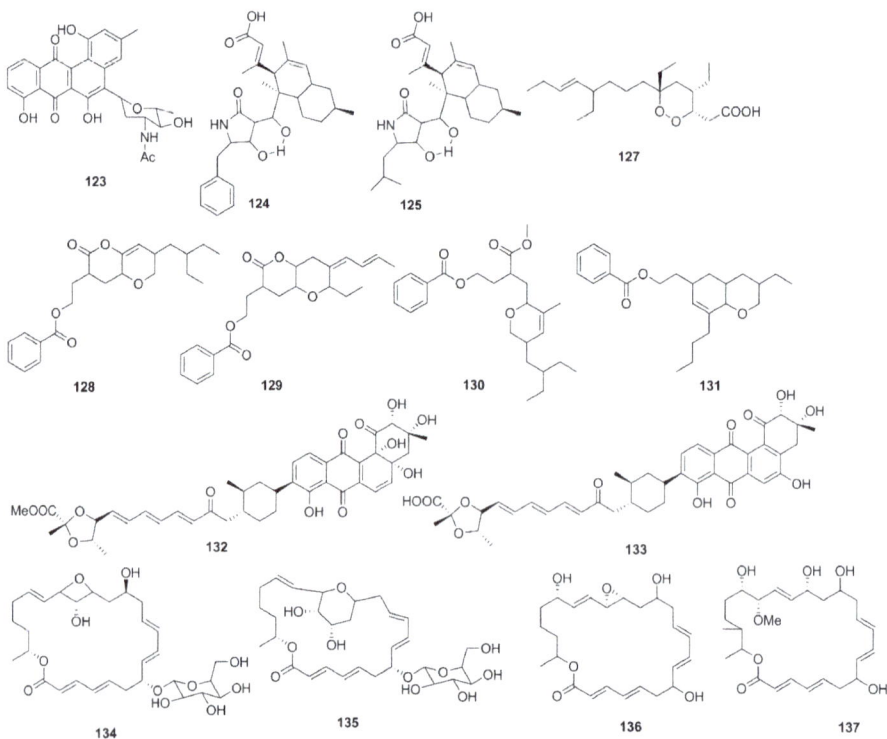

Figure 7. *Cont.*

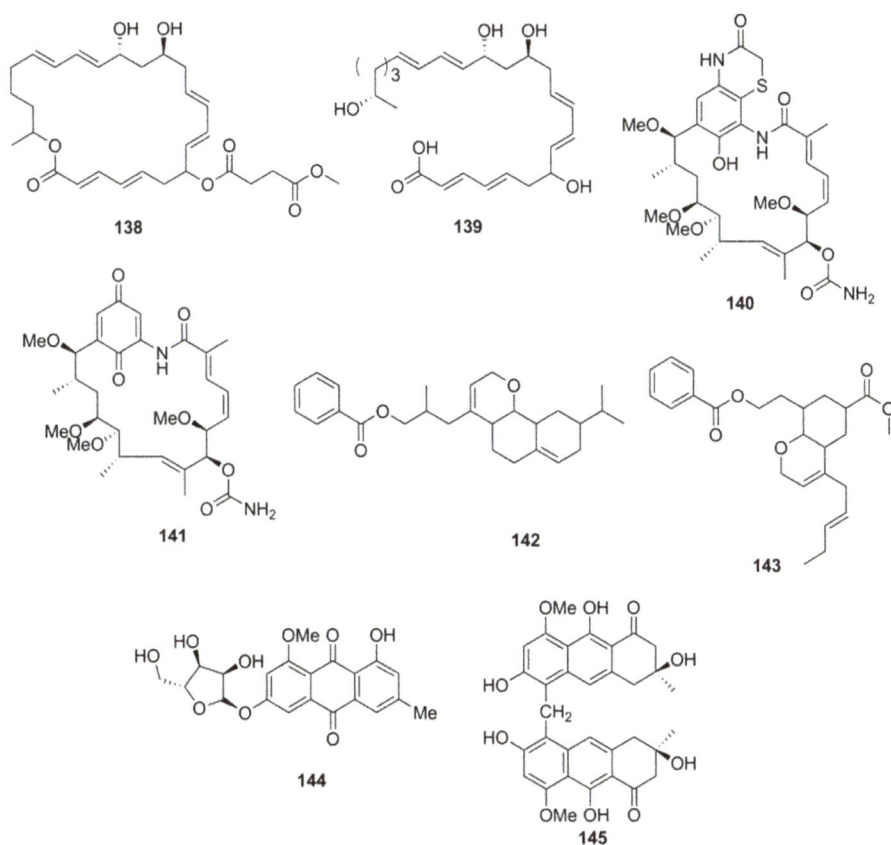

Figure 7. Antimicrobial polyketides.

Fradimycins A (**132**) and B (**133**) isolated from marine *Streptomyces fradiae* strain PTZ0025 displayed antimicrobial activity against *S. aureus* [79]. Macrolactins, polyene cyclic macrolactones possess a wide range of biological activities including antimicrobial, antiviral and anticancer. The majority of macrolactins are produced by *Bacillus* sp., whilst glycosylated macrolactins A1 (**134**) and B1 (**135**) are frequently isolated from marine *Streptomyces* species. The position of the hydroxyl group (OH) or the introduction of a keto group (C=O) in the macrolactone ring affects the antimicrobial activity of macrolactins, while the introduction of ester groups at the C-7 in macrolactins leads to better antibacterial activity. Compounds **134** and **135** were less active than their corresponding ether-containing macrolactins, probably due to the attachment of a sugar moiety at C-7 position of both compounds. However, glycosylated macrolactins are reported to have better polar solubility in comparison to the non-glycosylated macrolactins [61]. From the culture broth of a marine *Bacillus* sp., compounds including macrolactones (Macrolactin X–Z, **136–138**) and macrolactinic acid (**139**) were isolated. These metabolites displayed potential antimicrobial activity against microbial strains, *B. subtilis* (KCTC 1021), *E. coli* (KCTC 1923), and *S. cerevisiae* (KCTC 7913). Macrolactins exhibit their antibacterial activity by inhibiting peptide deformylase in a dose-dependent manner. The position of hydroxy groups in the macrolactone ring is also important for antimicrobial activity of macrolactins. The hydroxy group at C-15 in macrolactone ring increases the antimicrobial activity of macrolactins, whereas introduction of carbonyl group at C-15 decreases antimicrobial activity. The number of ring members and the presence of a hydroxy group at C-7 and C-9 position have no effect on the antibacterial activity [62].

Chemical analysis of a marine-derived *Streptomyces* sp. (CMB-M0392) isolated from sediment collected off Heron Island, Queensland, Australia, yielded the benzothiazine ansamycins, heronamycin A (**140**) and herbimycin A (**141**). Compound **140** showed antimicrobial activity against *B. subtilis* ATCC 6051 and 6633, but no activity against *S. aureus* ATCC 25923 and 9144, *E. coli* ATCC 11775, *P. aeruginosa* ATCC 10145 or *C. albicans* ATCC 90028 while compound **141** showed antimicrobial activity against *B. subtilis* ATCC 6051 and 6633 and *C. albicans* ATCC 90028 [80]. Antibacterial polyketide compounds were isolated from the heterotrophic bacterium, *Bacillus amyloliquefaciens* which is associated with the edible red seaweed *Laurenciae papillosa*. Bioactivity-guided techniques resulted in the isolation of 3-(octahydro-9-isopropyl-2*H*-benzo[*h*]chromen-4-yl)-2-methylpropyl benzoate (**142**) and methyl 8-(2-(benzoyloxy)-ethyl)-hexahydro-4-(*E*)-pent-2-enyl)-2*H*-chromene-6-carboxylate (**143**), compounds of polyketide origin which demonstrated activity against human opportunistic food pathogenic microbes. Compounds **142** and **143** demonstrated significant antibacterial activity (inhibitory zone diameter of greater than 18 mm against *V. vulnificus*, 25 μg on disk) against these pathogenic bacteria and lesser activity against *A. hydrophilla* (14–16 mm, 25 μg on disk) and *V. parahemolyticus* ATCC17802TM (inhibitory zone diameter of 12–14 mm, 25 μg on disk). In general antibacterial activities of compound **142** were greater than **143**. Various molecular descriptor variables, such as bulk, polarizability (electronic), and lipophilicity (octanol/water partition coefficient) were found to significantly influence the antibacterial activities of these compounds. Although there was no significant dissimilarities in polarizability of compounds **142** and **143**, the activity of the latter was lesser (inhibition zone diameter, 16 mm against *V. vulnificus*; 25 μg on disk) than that of the former (18 mm against *V. vulnificus*; 25 μg on disk) due to the lesser lipophilicity associated with **143** compared to **142**. The higher lipophilicity of compound **142** enabled the compound to effectively penetrate the intermembrane lipoprotein barrier to enter the receptor location, resulting in greater bioactivity against food pathogenic bacteria [81]. 3-*O*-(α-D-ribofuranosyl)questin (**144**) and eurorubrin (**145**) isolated from the fungal strain *Eurotium cristatum* EN-220, an endophyte obtained from the marine alga *Sargassum thunbergii* displayed antimicrobial activity against *E. coli* but did not inhibit the growth of *S. aureus*, *Physalospora obtuse*, *A. brassicae*, *Valsa mali*, *A. solania*, and *Sclerotinia miyabeana* [82].

10. Isocoumarins

Antimicrobial dihydroisocoumarin derivatives penicisimpins A–C (**146–148**) were reported from a rhizosphere-derived fungus, *Penicillium simplicissimum* MA-332 obtained from a marine mangrove plant *Bruguiera sexangula* var. rhynchopetala (Figure 8). These isocoumarins possess a broad-spectrum of antibacterial and antifungal activities. Among these three isocoumarins, penicisimpins A (**146**) exhibited the greatest activity against *E. coli*, *P. aeruginosa*, *V. parahaemolyticus*, *V. harveyi*, and *C. gloeosprioides* (each with MIC value of 4 μg/mL) while the other two congeners (**147** and **148**) were only moderately active against these pathogens. It was determined that the methyl group attached at C-7 was responsible for the enhanced activity observed in compound **146** relative to **147**, while the double bond at C-11 was responsible for the decreased activity (compound **146** vs. **148**) [83]. The marine-derived fungal species *Penicillium vinaceum* has been reported to produce citreoisocoumarin (**149**) that displayed activity against *S. aureus* (19 mm inhibition zone) [47].

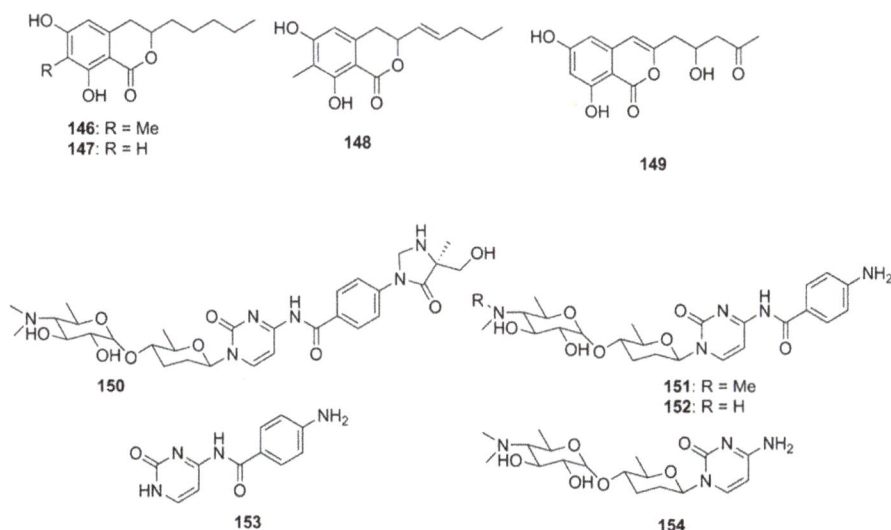

Figure 8. Antimicrobial isocoumarins (**146–149**) and nucleosides (**150–154**).

11. Nucleosides

Rocheicoside A (**150**), a nucleoside analogue possessing a novel 5-(hydroxymethyl)-5-methylimidazolidin-4-one substructure, and several other nucleosides such as plicacetin, norplicacetin, *p*-aminobenzanido uracil and cytosamine (**151–154**, Figure 8) have been isolated from the marine-derived actinomycete *Streptomyces rochei* 06CM016. These nucleosides have been described as potent antibiotics against microorganisms, including archaea, bacteria and eukarya. Rocheicoside A (**150**) displayed potential antimicrobial activity with MIC 4–16 µg/mL against a number of pathogens including *E. coli*, MRSA and *C. albicans* [84].

12. Miscellaneous Compounds

Several compounds which have not been classified under any of the above subheadings are discussed in this section and their chemical structures are illustrated in Figure 9. Terretrione A (**151**), terretrione C (**152**) and terretrione D (**153**) containing a 1,4-diazepane skeleton were isolated from an organic extract of the fungus *Penicillium* sp. CYE-87 derived from a tunicate-*Didemnum* sp. collected from the Suez Canal in Egypt. These compounds displayed antimicrobial activity against *C. albicans* but were inactive against *S. aureus* and *E. coli* [47,85]. Marine-derived fungi, *Penicillium vinaceum* species when investigated for antimicrobial compounds led to discovery of α-cyclopiazonic acid (**154**), which was active only against *E. coli* with an inhibition zone of 20 mm [47]. 5-methoxydihydrosterigmatocystin (**155**), a compound isolated from the marine-derived fungus, *Aspergillus versicolor* MF359, isolated from a marine sponge of *Hymeniacidon perleve* was active against *S. aureus* and *B. subtilis*. The compound did not display activity against MRSA and *P. aeruginosa* as the MIC was found to be >100 µg/mL against both the organisms [86]. Caerulomycin A (**156**), an antifungal compound isolated from marine actinomycetes *Actinoalloateichus cyanogriseus* showed potent activity against *Candida* isolates, *C. albicans* and *C. albicans* CO9, and two fluconazole resistant strains namely, *C. glabrata* HO5Fl and *C. krusei* GO3. The MICs of Caerulomycin A was in the range of 0.39–1.26 µg/mL. Furthermore, the MIC values obtained for Caerulomycin A against fluconazole resistant *C. glabrata* were comparable with the MIC values obtained for Amphotericin B [87]. Pyridones, trichodin A (**157**) and pyridoxatin (**158**), were extracted from both the mycelia and the culture broth of the marine fungus, *Trichoderma* sp. strain MF106 obtained from the Greenland Seas. Compounds **157** and **158** possess moderate antibiotic activities against the Gram-positive *B. subtilis*, *S. epidermidis*, MRSA and yeast, *C. albicans* but were inactive against *Trichophyton rubrum* [88]. A pyridinium compound, 1-(10-Aminodecyl) Pyridinium (**159**), isolated from the marine actinomycete, *Amycolatopsis alba* var. nov. DVR D4 demonstrated antimicrobial activities against Gram-positive and Gram-negative bacteria with MICs in the range of 70–160 µg/mL [89]. A dibenzofuran derivative porric acid D (**160**) and altenusin (**161**) were isolated from the methanol extract of the marine derived fungus, *Alternaria* sp., isolated from the Bohai Sea. These compounds were reported to have antibacterial activity against *S. aureus* [90]. Several small molecules including *p*-hydroxybenzoic acid (**162**), *trans*-cinnamic acid (**163**) and *N*-hydroxybenzoisoxazolone (**164**) were isolated from *Pseudoalteromonas flavipulchra* and showed antibacterial activity against *Vibrio anguillarum*. Greater growth inhibition was observed in *V. anguillarum* when a mixture of all three compounds was used compared to when each of the compounds was used individually, suggesting that they may act in a synergistic manner [72]. The antimicrobial compounds, bacilosarcin B (**165**) and C (**166**) and amicoumacin A (**167**) and B (**168**), were isolated from the culture broth of a marine-derived bacterium *B. subtilis*. The C-12 amide group of amicoumacin was found to be crucial for antibacterial activity against MRSA based on structural comparisons of amicoumacin A (**167**) and amicoumacin B (**168**). It was observed that only the compound amicoumacin A (**167**) with C-12′ amide groups exhibited antibacterial activities against *B. subtilis*, *S. aureus* and *L. hongkongensis*, which strongly supported the idea that the C-12′ amide group of amicoumcin acts as a pharmacophore in antibacterial activities. This conclusion was further supported by comparing antibacterial activities between compounds **167**/**168** and **165**/**166**. Compound **167** exhibited antibacterial activities against *B. subtilis*, *S. aureus* and *L. hongkongensis*, which were about six-fold higher than those of **168** [91]. Microluside A [4-(19-*p*-hydroxybenzoyloxy-*O*-β-D-cellobiosyl)-5-(30-*p*-hydroxybenzoyloxy-*O*-β-D-glucopyranosyl)xanthone] (**169**) is a unique *O*-glycosylated disubstituted xanthone isolated from the broth culture of *Micrococcus* sp. EG45 isolated from the Red Sea sponge *Spheciospongia vagabunda*. Compound **169** exhibited antibacterial activities against *E. faecalis* JH212, and *S. aureus* NCTC 8325 [92].

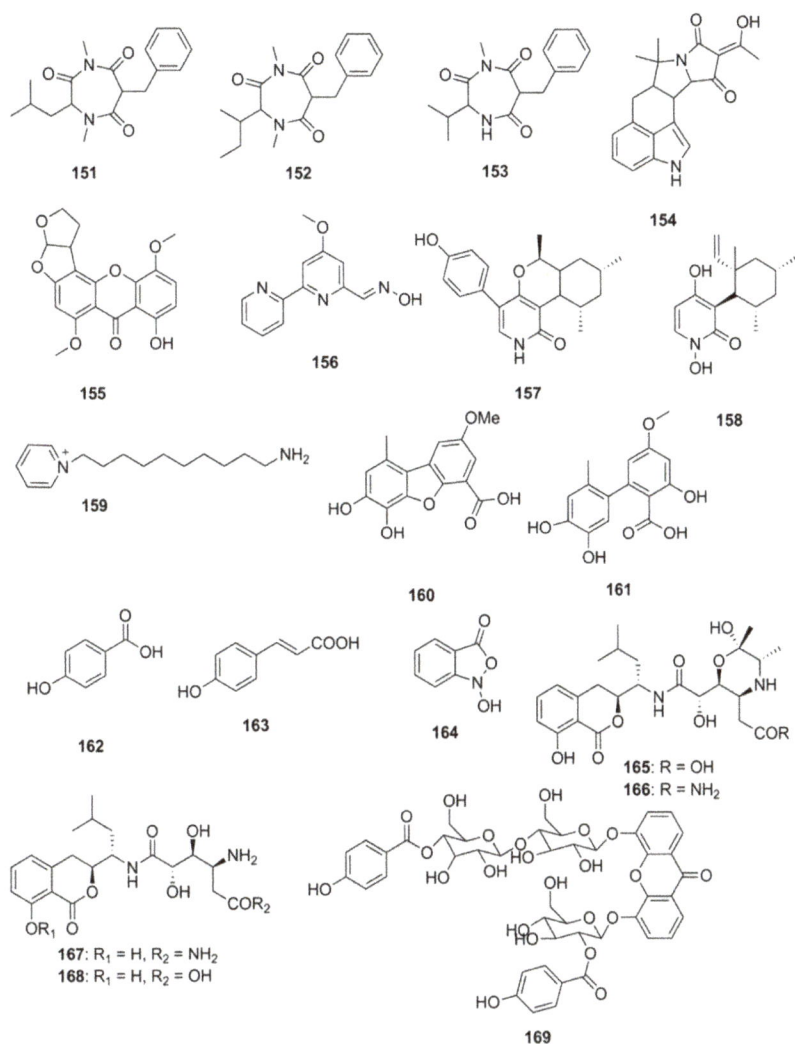

Figure 9. Antimicrobial miscellaneous compounds.

13. Synthetic Interventions

Although naturally occurring marine natural products are bestowed with interesting structural features (both chemical and stereochemical), some pharmacophoric modifications are still required to improve their biological efficacy [93]. In this respect, several synthetic chemists are engaged in tailoring these natural products to obtain new chemical entities by modifying their natural structures [94] (Figure 10). Recently, Sakata et al. identified isatin (**170**), an algicidal substance produced by the marine bacterium *Pseudomonas* sp. C55a-2 isolated from coastal sea water of Kagoshima Bay in Japan and targeted this compound for synthetic modification. This particular compound was chosen as many strains belonging to the genera *Pseudomonas*, *Alteromonas* and *Pseudoalteromonas* sp. have previously been reported to use chemical defences in the form of extracellular agents like isatin. With background knowledge of the antifungal activities of isatin, several structural modifications were made to the compound including bromination of the C-5 carbon of the isatin ring, altering the length of its alkyl chain and *N*-protections, resulting in the development of molecules with potentially better pharmacophoric features. Structure-activity relationships revealed that a bromine substitution at the C-5 carbon atom in isatin derivatives led to a decrease in antibacterial activity when compared structurally to the parent isatin molecule. The antibacterial activity remains unchanged for *N*-methyl and *N*-butyl isatin derivatives, however, the addition of a free NH group to the structure of these compounds results in a decrease in antibacterial activity. Hence, it can be deduced that the free NH moiety of isatin is necessary for its potent inhibitory activities against fouling bacteria. The antibacterial activity was found to be better in the class of compounds wherein acetonyl moiety is introduced as functionality, compared to more extended hydrophobic benzoyl, as well as electron donating ethoxy group. Compounds **171–173** from this group exhibited a greater inhibitory effect compared to the parent compound **170** and to the *N*-protected isatins **174–179** and the 5-bromoisatin derivatives **180–182**. The presence of a 3-acetonylidene group and a free NH moiety in compounds **171–173** were determined to be crucial structural elements responsible for enhancing the antibacterial activity of these compounds [95].

Bromopyrrole alkaloids are produced by marine sponges and possess an array of diverse biological activities, including antimicrobial and antineoplastic activities [96–98]. Many of these molecules are readily identifiable by a 4,5-dibromopyrrole ring contained within their structure, with oroidin (**183**) being the best characterized of these alkaloids noted for its antibiofilm activity [99] In 2012, Rane et al. synthesized marine bromopyrrole alkaloid derivatives containing 1,3,4-oxadiazole and thiazolidinone and evaluated their antimicrobial and antibiofilm properties. Compounds **184–188** are among these synthetic oroidin derivatives containing 1,3,4-oxadiazole and have exhibited antimicrobial activity against representative Gram-positive and Gram-negative bacteria [100,101]. These compounds demonstrated antibacterial activity comparable to ciprofloxacin® when used against *S. aureus* (MIC = 1.56 µg/mL). Further substitutions of 1-methyl-4,5-dibromopyrrole core with 4-thiazolidinone had been synthesized and tested for antibiofilm potentials against few Gram-positive bacteria. 4-thiazolidinone derivatives, compounds **189** and **190** showed antibiofilm activities (MIC = 0.78 µg/mL) 3-fold superior than those exerted following standard Vancomycin® use (MIC = 3.125 µg/mL), while activity of compounds **191–194** was 2-fold (MIC = 1.56 µg/mL) higher against *S. aureus* biofilm formation. Compounds **189–195** showed equal antibiofilm activity against *S. epidermidis* compared to standard Vancomycin (MIC = 3.125 µg/mL) [102].

In 2014, Zidar et al. isolated marine alkaloids, clathrodin (**196**) and oroidin (**183**), from sponges of the genus, *Agelas*, which possess significant antimicrobial activity against the bacterial strains *E. faecalis*, *S. aureus* and *E. coli* and *C. albicans*. The research group synthesized several derivatives using oroidin as a template. The most bioactive of all these derivatives was found to be 4-phenyl-2-aminoimidazole (**197**), which exhibited an MIC_{90} value of 12.5 µM against Gram-positive bacteria and 50 µM against *E. coli* [103].

A new marine-derived monoterpenoid compound, penicimonoterpene (+)-1 (**198**), was isolated from *Penicillium chrysogenum* QEN-24S in 2014 and had shown antifungal activity against *A. brassicae* and potent antibacterial activity against marine bacteria *Aeromonas hydrophila*, *V. harveyi*, and *V. parahaemolyticus*. This activity pattern encouraged chemists to develop a number of derivatives of compound **198**. Modifications focused on variation of the substituents at the C-8 position, the carbon-carbon double bond at the C-6/7 position, and carboxyl substituents at the C-1 position. Compounds **199–203** were synthesized according to these modifications and found to be particularly active against *F. graminearum*. It was determined that oxidation of the methyl to a hydroxymethyl group at the C-8 position or replacement of the methyl ester group at C-1 by an ethyl ester significantly increased the antifungal activity of these compounds. Compounds with a reduced double bond at C-6/7 also showed better inhibitory activities against *gloeosporioides* and *F. graminearum* except for those containing an aldehyde group [104].

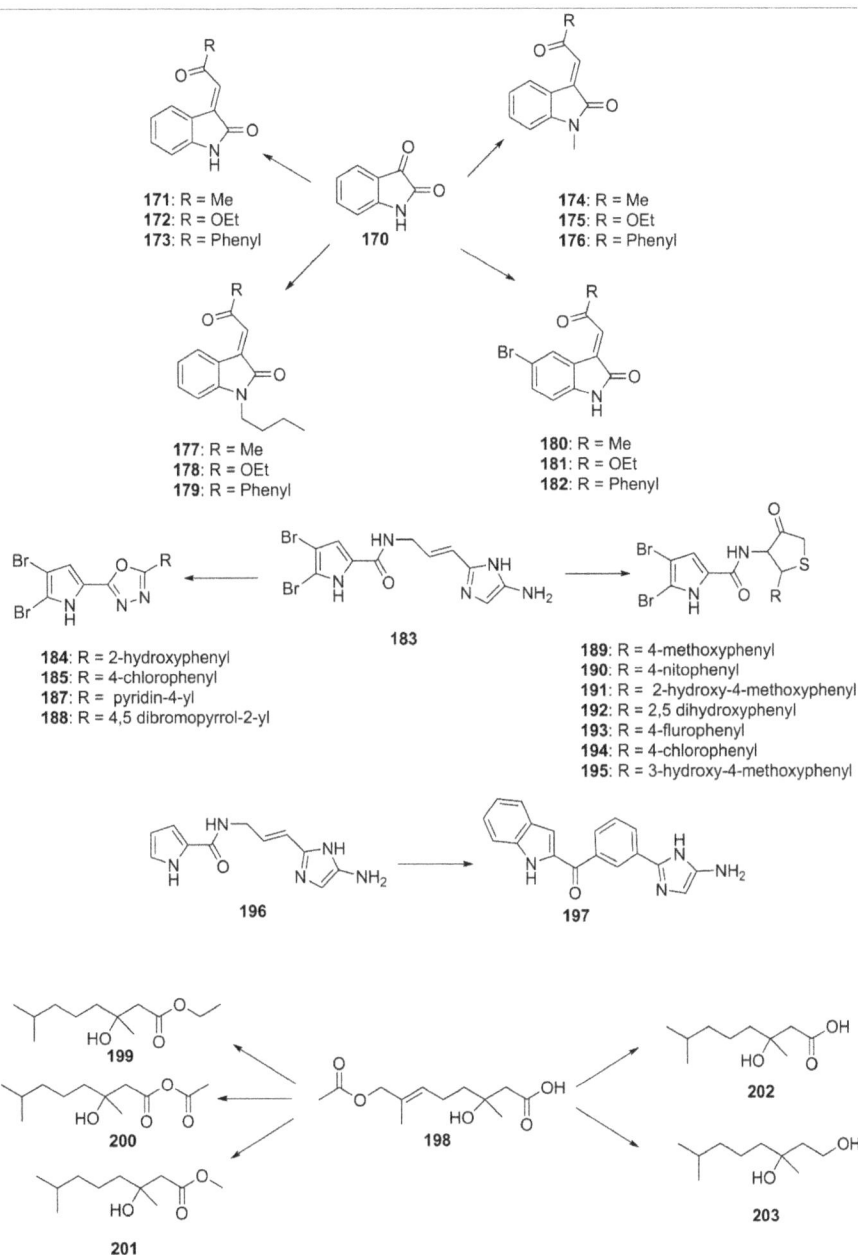

Figure 10. Semi-synthetic derivatives of MNPs.

14. Genome Mining of Marine Microorganisms—The Future of Antimicrobial Discovery

Antimicrobial compound discovery has traditionally mainly relied on bioassay-guided approaches involving the cultivation of microorganisms under a variety of growth conditions, the subsequent screening of culture extracts for bioactivity and chemical characterization of the compounds produced. Over time however, this approach has led to the frequent re-isolation of known compounds, resulting in a drastic decline in research efforts being undertaken by research groups and pharmaceutical companies [105], causing a significant deficit in the number of novel, natural products available for commercial and medicinal use. In this age of antimicrobial resistance, the demand for functionally diverse, unique antimicrobials has never been greater. Fortunately, advances in sequencing and '-omics' based technologies have revived the field of natural product discovery in recent years, owing in large part to the cost effectiveness and speed associated with next-generation sequencing. With more than 99,000 sequenced bacterial genomes currently publically available in the NCBI database [106] and sequence data from thousands of metagenome projects which can be accessed at the Genomes OnLine Database [107], researchers are now focusing their efforts on genome-guided investigations as a complementary approach to traditional bioactivity-guided methods in an effort to expedite the identification of 'talented' microbes, which are likely to possess the biosynthetic machinery typically associated with antimicrobial compound production.

Biosynthetic gene clusters (BGCs) are specialised groups of genes located in close proximity to each other in bacterial genomes and encode successive steps in the biosynthesis of natural products. Sequence-based detection, analysis and functional elucidation of these clusters are paramount to unlocking the true biosynthetic potential which resides within a microorganism. However functional elucidation of the products of these BGCs is not always easy, since most are either poorly expressed or not expressed at all under common laboratory culture conditions. Such was the case observed for the model antibiotic-producing actinobacterium, *Streptomyces coelicolor*, which, prior to having its entire genome sequenced in 2002 [108] was thought to contain BGCs responsible for the production of six distinct metabolites, previously identified by classic molecular genetic approaches [109]. However, upon inspection of the organism's complete genome sequence, 16 additional BGCs including BGCs potentially encoding nonribosomal peptide synthetases (NRPSs) and polyketide synthases (PKSs) were identified by bioinformatics-based predictions [110,111]. These clusters were deemed likely to encode enzymes involved in the synthesis of polyketides and nonribosomal peptides which are considered to be among the more valuable classes of microbial secondary metabolites from a biopharmaceutical perspective [112,113].

At the time of writing this review, data pertaining to over one million putative gene clusters currently resides in the Atlas of Biosynthetic gene Clusters which forms part of the Integrated Microbial Genomes component of the Joint Genome Institute (JGI IMG-ABC) [114]. With such a wealth of sequence information available, the challenges researchers now face primarily centre on how to effectively mine such a huge quantity of data in order to rapidly identify biosynthetically 'talented' microorganisms and furthermore, how to prioritize which BGCs to investigate among such a vast collection of uncharacterized clusters when targeting a desired anti-microbial bioactivity.

In this respect the Secondary Metabolite Bioinformatics Portal (SMBP), functions as a useful access point for investigators; containing website links to databases and tools used in genome mining and secondary metabolism research [115]. Automated tools such as antiSMASH [116] BAGEL [117] and PRISM [118] and databases such as Bactibase [119] ClusterMine360 [120] and MIBiG [121] represent just a few of the sophisticated in silico analytical tools which have quickly established themselves as the go-to resources for BGC identification, characterization and comparison. antiSMASH (antibiotics and Secondary Metabolite Analysis Shell) in particular is one of the most extensively used open-source BGC mining tools. Integrating and cross-linking several analysis tools including BLAST+, HMMer3 and FastTree, this computational platform not only facilitates the detection of known BGCs but can also detect unknown, BGC-like regions in genomes via ClusterFinder [122] an algorithm which uses Pfam domain, pattern based predictions to detect putative BGCs. ClusterFinder works on the premise that even the biosynthetic pathways for unknown compounds are likely to use the same broad families of enzymes for the catalysis of key reactions. The antiSMASH framework has been continuously developed [116,123] since its launch in 2010 [124]. antiSMASH version 4.0 was recently released in April 2017 [125] and contains improvements in prediction software for specialized secondary metabolites, including ribosomally synthesized and post-translationally modified peptides (RiPPs) and terpene products and has been expanded from BGC mining in bacteria and fungi to BGC mining in plants [126]. Another recently developed tool is EvoMining, which integrates evolutionary concepts related to the emergence of natural product biosynthesis into genome mining. It is a newly developed phylogenomics approach toward BGC analysis and has been successfully used in the retrieval of arseno-organic BGCs, which were previously unobtainable via antiSMASH analysis [127].

With respect to the specific identification of antibacterial compounds and the pathways involved in their biosynthesis, genome mining approaches have proven useful in the identification of these types of compounds from marine derived bacteria. Figure 11 highlights a number of antimicrobial compounds discovered via genome mining approaches. The gene cluster involved in the biosynthesis of heronamide F (**204**) was identified following genome scanning of the deep-sea derived *Streptomyces* sp. SCSIO 03032. Heronamides are polyketide macrolactams which belong to a class of potent antifungal metabolites that are produced by marine-derived actinomycetes [128]. Confirmation of the involvement of this cluster in the biosynthesis of heronamide F was achieved by the functional confirmation of one of the genes in the cluster, namely *herO*, encoding a cytochrome P450 by gene knockout experiments [129]. Another example resides with three cyclohexapeptides, destomides B–D (**205**–**207**) which were initially isolated from the marine microbe *Streptomyces scopuliridis* SCSIO ZJ46. Destomide B was found to display antibacterial activity against strains of *Streptococcus pneumonia*, *Staphylococcus aureus* and methicillin-resistant *Staphylococcus epidermidis* (MRSE) [68]. Genome mining identified a putative 39-kb desotamide *dsa* gene cluster in the *S. scopuliridis* strain, and was determined to contain three NRPS genes. Subsequent heterologous expression of the *dsa* gene cluster in the heterologous host *S. coelicolor* M1152 confirmed its involvement of the biosynthesis of desotamides [130]. Finally, another recent example involving ribosomally synthesized and post-translationally modified peptides (RiPPs), centres on the new cinnamycin-like lantiobiotic, mathermycin (**208**) which has recently been isolated from the marine-derived strain, *Marinactinospora thermotolerans* SCSIO 00652 and possesses antimicrobial activities towards *Bacillus subtilis* [131]. Genome mining of the strain facilitated the identification of the BGC for mathermycin, while subsequent expression of the gene cluster in the heterologous *Streptomyces lividans* host system allowed its antibacterial activity to be assessed.

Another popular approach in the identification of novel marine derived antimicrobials is to couple genomics with metabolomics, whereby genome mining together and metabolic profiling are used to identify novel natural products in marine microorganisms. Our group has previously employed this approach to identify bioactive compounds from *Streptomyces* sp. (SM8) isolated from the sponge *Haliclona simulans* which displayed antibacterial and antifungal activity [132]. Similarly, Paulus and co-workers employed both metabolomic and genomic profiling approaches on the marine *Streptomyces* sp. MP131-18 to identify a number of new biologically active compounds, including two new members of the bisindole pyrroles spirorindimycins, spiroindimicin E (**209**) and F (**210**). In addition, two new members of the α-pyrone lagunapyrone family, namely lagunapyrone D and E were also identified using similar approaches [133]. Another example of the successful use of this coupled approach resides in marformycins A-F (**211**–**216**), which display selective anti-microbial activity against *Micrococcus luteus*, *Propioniobacterium acnes* and *Propionobacterium granulosum*. The marformycin gene cluster was identified following genome scanning of the deep-sea sediment-derived *Streptomyces drozdowiczii* SCSIO 10141 strain. Confirmation of the involvement of this cluster in the biosynthesis of this group of cyclic peptides was achieved following in vivo inactivation studies coupled with metabolite identification [134].

Other techniques have also been developed to expedite the identification of BGCs, which include mining for the presence of self-resistance mechanisms within these gene clusters, allowing investigators to deduce the antibiotic compounds which an organism is likely to produce. Resistance mechanisms are characteristic traits associated with antibiotic producers, enabling an organism to avoid suicide from self-toxicity following the biosynthesis of its own molecule [135]. Resistance mechanisms include enzymes which degrade toxic compounds, efflux pumps for the effective removal of unwanted substances from the cell and target modification [136]. Wright and colleagues first used resistance as a discriminating criterion in 2013, demonstrating that organisms resistant to glycopeptide and ansamycin antibiotics are more likely to produce similar compounds [137]. Following this resistance based hypothesis increased the discovery rate of producers of the aforementioned antibacterial compounds by several orders of magnitude. Harnessing the success of this approach, Wright and colleagues further devised a method for isolating scaffold-specific antibacterial producers by taking advantage of the innate self-protection mechanisms employed by the producing organisms, isolating strains in the presence of a selective antibiotic [138]. In 2015, Moore and colleagues developed a target-directed genome mining method to identify BGCs [139]. As previously mentioned target modification is one of several resistance strategies employed by antibiotic producing bacteria and is effective in correlating an antibiotic to its mode of action. Since it is common for antibiotic producing bacteria to mutate and duplicate genes encoding proteins for resistance, Moore's group reasoned that identifying target-duplicated genes which are co-clustered with BGCs would provide valuable information pertaining to the molecular targets of the BGC products without any prior knowledge of the molecule synthesized. Since antibiotic targets are often the product of housekeeping genes, Moore and colleagues screened 86 *Salinispora* bacterial genomes for duplicated copies of housekeeping genes and related them to their presence in BGCs. Using this approach they successfully identified a duplicated fatty acid synthase in the direct vicinity of an orphan hybrid PKS-NPRS gene cluster and prioritized this for investigation. Following cloning, heterologous expression and mutational analysis, the authors linked the gene cluster to the biosynthesis of thiolactomycin (**217**) a previously characterized fatty acid synthase inhibitor, and to the production of a group of unusual thiotetronic acid natural products [139]. In 2016, Oakley and colleagues provided experimental validation for target-directed genome mining in a fungal BGC system [140]. The group identified a proteasome subunit-encoding gene within a gene cluster in *Aspergillus nidulans* and hypothesized that the cluster may be responsible for the production of a proteasome inhibitor. Following a number of molecular genetic based strategies, the investigators determined that the product of the cluster was indeed a proteasome inhibitor, fellutamide B (**218**). Recently, Johnston and co-workers used resistance-based mining to predict natural products with new modes of action [141]. The authors used a retrobiosynthetic algorithm to mine biosynthetic scaffolds and resistance determinants to identify structures with unknown modes of action. Using this approach, the investigators determined that the telomycin family of natural products from *Streptomyces canus* possess a new antibacterial mode of action which targets cardiolipin, a bacterial phospholipid.

Tracanna and co-workers recently suggested another strategy as a means of prioritizing BGCs which may encode novel antibiotics, based on synergistic interactions [142]. Natural products can occur as synergistic pairs in nature, as in the case of cephamycin and clavulanic acid, two compounds which are naturally produced by *Streptomyces clavuligerus*. The BGCs for these compounds are intertwined in a 'supercluster' configuration [143,144]. This genetic conformation inspired the production of the antibiotic, Augmentin®, which is comprised of a combination of the β-lactam antibiotic, amoxicillin with the β-lactamase inhibitor, clavulanic acid. Synergistic pairs of natural products are particularly attractive commodities in the ongoing attempt to tackle antimicrobial resistance, as it is more difficult for a pathogen to develop resistance to both compounds. They suggest that thorough analysis surrounding the evolutionary history of the genes associated with large, hybrid BGCs may provide a valuable insight into whether these BGCs are in fact 'superclusters', responsible for the production of a number of different compounds which may be synergistic.

It is important that in silico analytical tools keep pace with these genome-guided discovery strategies in order to expedite novel compound discovery. The Antibiotic Resistant Target Seeker (ARTS) is one such web tool which uses three criteria to detect known resistance genes as well as putative resistance house-keeping genes in actinobacterial genomes i.e., (i) duplication (ii) evidence for horizontal gene transfer (HGT) and (iii) localization within a BGC [145]. This inexpensive, computational analysis tool, facilitates high-throughput screening of bacterial genomes. Although the current focus of ARTS is on the analysis of actinobacterial genomes, the pipeline also works for other phyla and is being expanded to include reference sets for other taxa. Several other useful resources are available pertaining to the field of antimicrobial resistance, with the Comprehensive Antibiotic Resistance Database (CARD) [146] and Antibiotic Resistance Gene-ANNOTation (ARG-ANNOT) being the two most extensive databases [147].

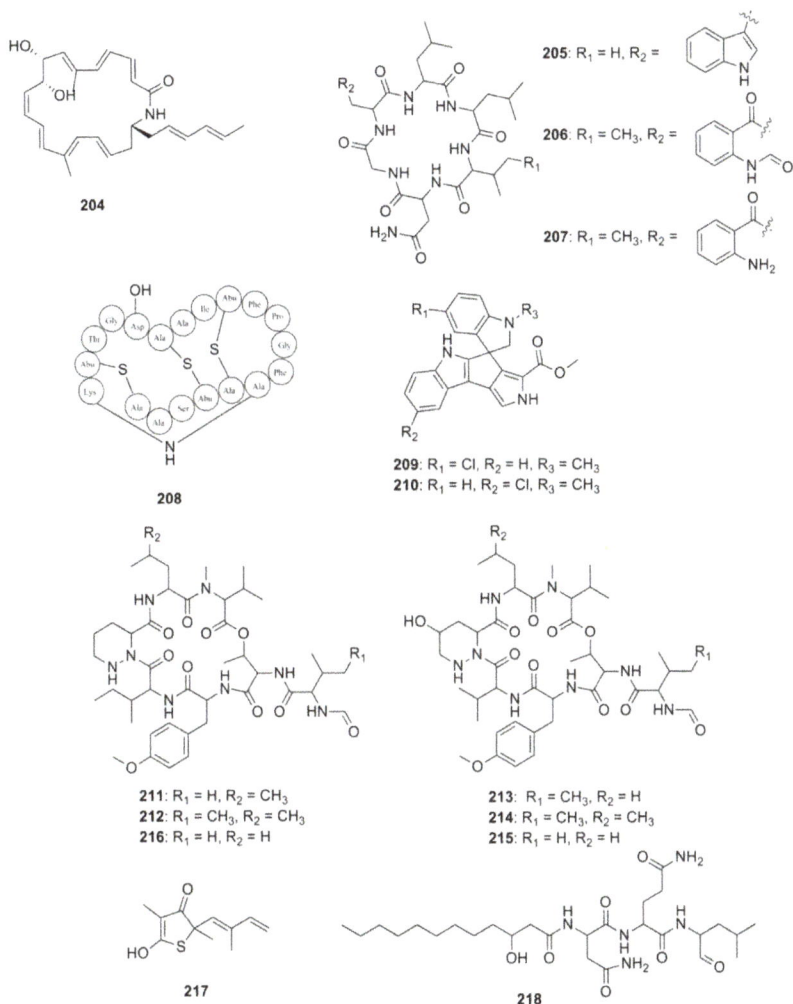

Figure 11. Antimicrobial compounds discovered via genome mining approaches.

15. Concluding Remarks and Future Prospects

The marine environment is home to a vast number of macro and microorganisms with untapped biosynthetic activities which are used to a large extent to ensure their survival in this diverse and often hostile habitat. This unique environment facilitates the biosynthesis of an array of secondary metabolites which act as chemical defenses and display a broad range of antimicrobial bioactivities. Nevertheless, despite extensive structural and stereo chemical diversity, only seven marine-derived metabolites have to date been approved as drugs, while 12 MNPs (or derivatives thereof) are currently in different phases of clinical trials.

As previously noted, none of the newly discovered marine natural products have as yet progressed to clinical trials, although a few of them are in preclinical studies. The slower pace of MNPs towards clinical trials is due to several factors that hinder their development as clinical agents. One of the major factors is the "continuous supply". Large quantities of a compound are required to carry out biological assays to determine the site of action, specific targets, selectivity of the compound and its cytotoxicity. Irrespective of the potential applications of a functionally promising compound, a significant challenge faced by researchers is that several hundred grams of the compound are required for preclinical development, and multi kilogram quantities required for clinical trials. This is often one of the major bottle-necks in the development of MNP for clinical applications. Synthetic chemists around the world however are continuing to develop synthetic and semi-synthetic strategies to help overcome the supply issue surrounding MNPs, in an effort to satisfy the requirements to help bring these molecules to the preclinical stage and eventual development for use in the commercial or medicinal arenas.

MNPs can be chemically modified with various biosteric structural units to develop 'drug-like' molecules. Also, developments in mariculture (farming the growth of the organism in its natural environment) and aquaculture (culturing an organism under artificial conditions) have been attempted in order to solve the problem of sustainable supply of macroorganisms. However, the unique and sometimes exclusive conditions of the sea make cultivation or maintenance of isolated samples still very challenging and often impossible.

The preclinical pipeline also demands elaborate, mechanistic and pharmacokinetic studies to develop tailored MNPs, which in itself is a hugely challenging but nevertheless exciting task. Regardless of these challenges, the preclinical pipeline continues to supply studies with several hundred novel bioactive marine compounds with the potential for use as therapeutics. From a global perspective, the marine pharmaceutical pipeline remains very active, and now appears to have sufficient momentum to deliver additional antimicrobial compounds to the marketplace in the near future. The efficiency of various marine compounds against pathogens is very encouraging and there is no doubt that their exploitation and application will continue to develop. Genome mining has ushered in a renaissance in the field of natural product discovery, providing new hope in the ongoing search for novel antimicrobial compounds. This strategy allows researchers to harness the true biosynthetic potential which resides within diverse groups of marine microorganisms and offers an invaluable insight into not only the biosynthetic, but also the evolutionary and defensive strategies these organisms employ in the marine environment. Collaborative endeavours involving marine natural products chemistry with organic chemistry, medicinal chemistry, pharmacology, biology, bioinformatics and associated disciplines will help to ensure and facilitate an increase in marine natural product reaching the market as antimicrobial therapeutics.

Acknowledgments: The authors wish to acknowledge the Irish Department of Agriculture, Food and the Marine for their funding to this project under the food institutional Research Measure (FIRM 11/F/009).

Author Contributions: Alka Choudhary, Lynn M. Naughton, Itxaso Montánchez, Alan D. W. Dobson and Dilip K. Rai contributed to the conception and writing of the manuscript. Alan D. W. Dobson and Dilip K. Rai edited the manuscript.

Conflicts of Interest: The authors declare no conflict of interest.

References

1.   O'Neill, J. Antimicrobial resistance: Tackling a crisis for the health and wealth of nations. *Rev. Antimicrob. Resist.* **2014**, 1–16. [CrossRef]
2.   Singer, A.C.; Shaw, H.; Rhodes, V.; Hart, A. Review of antimicrobial resistance in the environment and its relevance to environmental regulators. *Front. Microbiol.* **2016**, *7*, 1728. [CrossRef] [PubMed]
3.   Gyawali, R.; Ibrahim, S.A. Natural products as antimicrobial agents. *Food Control* **2014**, *46*, 412–429. [CrossRef]

4. Fenical, W. New pharmaceuticals from marine organisms. *Trends Biotechnol.* **1997**, *15*, 339–341. [CrossRef]
5. Shen, B. A new golden age of natural products drug discovery. *Cell* **2015**, *163*, 1297–1300. [CrossRef] [PubMed]
6. Montaser, R.; Luesch, H. Marine natural products: A new wave of drugs? *Future* **2011**, *3*, 1475–1489. [CrossRef] [PubMed]
7. Mayer, A.M.; Glaser, K.B.; Cuevas, C.; Jacobs, R.S.; Kem, W.; Little, R.D.; McIntosh, J.M.; Newman, D.J.; Potts, B.C.; Shuster, D.E. The odyssey of marine pharmaceuticals: A current pipeline perspective. *Trends Pharmacol. Sci.* **2010**, *31*, 255–265. [CrossRef] [PubMed]
8. Patridge, E.; Gareiss, P.; Kinch, M.S.; Hoyer, D. An analysis of fda-approved drugs: Natural products and their derivatives. *Drug Discov. Today* **2016**, *21*, 204–207. [CrossRef] [PubMed]
9. Newman, D.J.; Cragg, G.M. Natural products as sources of new drugs over the 30 years from 1981 to 2010. *J. Nat. Prod.* **2012**, *75*, 311–335. [CrossRef] [PubMed]
10. Carter, G.T. Natural products and pharma 2011: Strategic changes spur new opportunities. *Nat. Prod. Rep.* **2011**, *28*, 1783–1789. [CrossRef] [PubMed]
11. Newman, D.J.; Cragg, G.M. Natural products as sources of new drugs from 1981 to 2014. *J. Nat. Prod.* **2016**, *79*, 629–661. [CrossRef] [PubMed]
12. Margulis, L.; Chapman, M.J. *Kingdoms and Domains: An illustrated Guide to the Phyla of Life on Earth*; Elsevier Science, Marine Biological Laboratory: Woods Hole, MA, USA, 2009.
13. Blunt, J.W.; Copp, B.R.; Keyzers, R.A.; Munro, M.H.; Prinsep, M.R. Marine natural products. *Nat. Prod. Rep.* **2016**, *33*, 382–431. [CrossRef]
14. Mayer, A.M.; Nguyen, M.; Newman, D.J.; Glaser, K.B. The marine pharmacology and pharmaceuticals pipeline in 2015. *FASEB J.* **2016**, *30*, 932.7.
15. Mayer, A.M.; Nguyen, M.; Kalwajtys, P.; Kerns, H.; Newman, D.J.; Glaser, K.B. The marine pharmacology and pharmaceuticals pipeline in 2016. *FASEB J.* **2017**, *31*, 18.1.
16. Blunt, J.W.; Copp, B.R.; Keyzers, R.A.; Munro, M.; Prinsep, M.R. Marine natural products. *Nat. Prod. Rep.* **2017**, *34*, 235–294. [CrossRef] [PubMed]
17. Martins, A.; Vieira, H.; Gaspar, H.; Santos, S. Marketed marine natural products in the pharmaceutical and cosmeceutical industries: Tips for success. *Mar. Drugs* **2014**, *12*, 1066–1101. [CrossRef] [PubMed]
18. Gerwick, W.H.; Moore, B.S. Lessons from the past and charting the future of marine natural products drug discovery and chemical biology. *Chem. Biol.* **2012**, *19*, 85–98. [CrossRef] [PubMed]
19. Falaise, C.; François, C.; Travers, M.-A.; Morga, B.; Haure, J.; Tremblay, R.; Turcotte, F.; Pasetto, P.; Gastineau, R.; Hardivillier, Y. Antimicrobial compounds from eukaryotic microalgae against human pathogens and diseases in aquaculture. *Mar. Drugs* **2016**, *14*, 159. [CrossRef] [PubMed]
20. Newman, D.J.; Cragg, G.M. Drugs and drug candidates from marine sources: An assessment of the current "state of play". *Planta Med.* **2016**, *82*, 775–789. [CrossRef]
21. Mehbub, M.F.; Perkins, M.V.; Zhang, W.; Franco, C.M. New marine natural products from sponges (porifera) of the order dictyoceratida (2001 to 2012); a promising source for drug discovery, exploration and future prospects. *Biotechnol. Adv.* **2016**, *34*, 473–491. [CrossRef]
22. Deshmukh, S.K.; Agrawal, S.; Adholeya, A. The pharmacological potential of nonribosomal peptides from marine sponge and tunicates. *Front. Pharmacol.* **2016**, *7*, 333. [CrossRef]
23. Moreno-González, R.; Rodríguez-Mozaz, S.; Huerta, B.; Barceló, D.; León, V. Do pharmaceuticals bioaccumulate in marine molluscs and fish from a coastal lagoon? *Environ. Res.* **2016**, *146*, 282–298. [CrossRef] [PubMed]
24. Anjum, K.; Abbas, S.Q.; Shah, S.A.A.; Akhter, N.; Batool, S.; ul Hassan, S.S. Marine sponges as a drug treasure. *Biomol. Ther.* **2016**, *24*, 347–362. [CrossRef]
25. Putz, A.; Proksch, P. Chemical Defence in Marine Ecosystems. In *Functions and Biotechnology of Plant Secondary Metabolites*, 2nd ed.; Wink, M., Ed.; Wiley-Blackwell: Oxford, UK, 2010. [CrossRef]
26. Mayer, A.; Rodríguez, A.D.; Taglialatela-Scafati, O.; Fusetani, N. Marine pharmacology in 2009–2011: Marine compounds with antibacterial, antidiabetic, antifungal, anti-inflammatory, antiprotozoal, antituberculosis, and antiviral activities; affecting the immune and nervous systems, and other miscellaneous mechanisms of action. *Mar. Drugs* **2013**, *11*, 2510–2573.

27. Mayer, A.M.; Rodríguez, A.D.; Berlinck, R.G.; Fusetani, N. Marine pharmacology in 2007–2008: Marine compounds with antibacterial, anticoagulant, antifungal, anti-inflammatory, antimalarial, antiprotozoal, antituberculosis, and antiviral activities; affecting the immune and nervous system, and other miscellaneous mechanisms of action. *Comp. Biochem. Physiol. Part C Toxicol. Pharmacol.* **2011**, *153*, 191–222. [PubMed]
28. Huang, T.; Lin, S. Microbial natural products: A promising source for drug discovery. *J. Appl. Microbiol. Biochem.* **2017**, *2*, 1–3.
29. Pangestuti, R.; Kim, S.-K. Bioactive peptide of marine origin for the prevention and treatment of non-communicable diseases. *Mar. Drugs* **2017**, *15*, 67. [CrossRef] [PubMed]
30. Keffer, J.L.; Huecas, S.; Hammill, J.T.; Wipf, P.; Andreu, J.M.; Bewley, C.A. Chrysophaentins are competitive inhibitors of ftsz and inhibit Z-ring formation in live bacteria. *Bioorg. Med. Chem.* **2013**, *21*, 5673–5678. [CrossRef]
31. Plaza, A.; Keffer, J.L.; Bifulco, G.; Lloyd, J.R.; Bewley, C.A. Chrysophaentins A–H, antibacterial bisdiarylbutene macrocycles that inhibit the bacterial cell division protein ftsz. *J. Am. Chem. Soc.* **2010**, *132*, 9069–9077. [CrossRef]
32. Kwan, J.C.; Meickle, T.; Ladwa, D.; Teplitski, M.; Paul, V.; Luesch, H. Lyngbyoic acid, a "tagged" fatty acid from a marine cyanobacterium, disrupts quorum sensing in *Pseudomonas aeruginosa*. *Mol. BioSyst.* **2011**, *7*, 1205–1216. [CrossRef] [PubMed]
33. DiGirolamo, J.A.; Li, X.-C.; Jacob, M.R.; Clark, A.M.; Ferreira, D. Reversal of fluconazole resistance by sulfated sterols from the marine sponge *Topsentia* sp. *J. Nat. Prod.* **2009**, *72*, 1524–1528. [CrossRef] [PubMed]
34. Meng, L.-H.; Li, X.-M.; Liu, Y.; Wang, B.-G. Polyoxygenated dihydropyrano [2,3-c]pyrrole-4,5-dione derivatives from the marine mangrove-derived endophytic fungus *Penicillium brocae* MA-231 and their antimicrobial activity. *Chin. Chem. Lett.* **2015**, *26*, 610–612. [CrossRef]
35. Du, F.-Y.; Li, X.; Li, X.-M.; Zhu, L.-W.; Wang, B.-G. Indolediketopiperazine alkaloids from *Eurotium cristatum* EN-220, an endophytic fungus isolated from the marine alga *Sargassum thunbergii*. *Mar. Drugs* **2017**, *15*, 24. [CrossRef] [PubMed]
36. Du, F.-Y.; Li, X.-M.; Li, C.-S.; Shang, Z.; Wang, B.-G. Cristatumins A–D, new indole alkaloids from the marine-derived endophytic fungus *Eurotium cristatum* EN-220. *Bioorg. Med. Chem. Lett.* **2012**, *22*, 4650–4653. [CrossRef]
37. Youssef, D.T.; Shaala, L.A.; Alshali, K.Z. Bioactive hydantoin alkaloids from the red sea marine sponge *Hemimycale arabica*. *Mar. Drugs* **2015**, *13*, 6609–6619. [CrossRef] [PubMed]
38. Liu, Y.; Mándi, A.; Li, X.-M.; Meng, L.-H.; Kurtán, T.; Wang, B.-G. Peniciadametizine A, a dithiodiketopiperazine with a unique spiro [furan-2,7′-pyrazino [1,2-b][1,2] oxazine] skeleton, and a related analogue, peniciadametizine B, from the marine sponge-derived fungus *Penicillium adametzioides*. *Mar. Drugs* **2015**, *13*, 3640–3652. [CrossRef] [PubMed]
39. Meng, L.-H.; Zhang, P.; Li, X.-M.; Wang, B.-G. Penicibrocazines A–E, five new sulfide diketopiperazines from the marine-derived endophytic fungus *Penicillium brocae*. *Mar. Drugs* **2015**, *13*, 276–287. [CrossRef] [PubMed]
40. Sun, X.; Sun, S.; Ference, C.; Zhu, W.; Zhou, N.; Zhang, Y.; Zhou, K. A potent antimicrobial compound isolated from *Clathria cervicornis*. *Bioorg. Med. Chem. Lett.* **2015**, *25*, 67–69. [CrossRef]
41. Jiao, W.; Zhang, F.; Zhao, X.; Hu, J.; Suh, J.-W. A novel alkaloid from marine-derived actinomycete *Streptomyces xinghaiensis* with broad-spectrum antibacterial and cytotoxic activities. *PLoS ONE* **2013**, *8*, e75994. [CrossRef] [PubMed]
42. Youssef, D.T.; Shaala, L.A.; Asfour, H.Z. Bioactive compounds from the red sea marine sponge *Hyrtios species*. *Mar. Drugs* **2013**, *11*, 1061–1070. [CrossRef] [PubMed]
43. Yang, F.; Hamann, M.T.; Zou, Y.; Zhang, M.-Y.; Gong, X.-B.; Xiao, J.-R.; Chen, W.-S.; Lin, H.-W. Antimicrobial metabolites from the paracel islands sponge *Agelas mauritiana*. *J. Nat. Prod.* **2012**, *75*, 774–778. [CrossRef]
44. Kubota, T.; Nakamura, K.; Kurimoto, S.-I.; Sakai, K.; Fromont, J.; Gonoi, T.; Kobayashi, J.I. Zamamidine D, a manzamine alkaloid from an okinawan *Amphimedon* sp. Marine sponge. *J. Nat. Prod.* **2017**. [CrossRef] [PubMed]
45. Liu, Y.; Li, X.-M.; Meng, L.-H.; Jiang, W.-L.; Xu, G.-M.; Huang, C.-G.; Wang, B.-G. Bisthiodiketopiperazines and acorane sesquiterpenes produced by the marine-derived fungus *Penicillium adametzioides* AS-53 on different culture media. *J. Nat. Prod.* **2015**, *78*, 1294–1299. [CrossRef] [PubMed]
46. Chu, M.J.; Tang, X.L.; Qin, G.F.; Sun, Y.T.; Li, L.; de Voogd, N.J.; Li, P.L.; Li, G.Q. Pyrrole derivatives and diterpene alkaloids from the south china sea sponge *Agelas nakamurai*. *Chem. Biodivers.* **2017**. [CrossRef]

47. Asiri, I.A.; Badr, J.M.; Youssef, D.T. Penicillivinacine, antimigratory diketopiperazine alkaloid from the marine-derived fungus *Penicillium vinaceum*. *Phytochem. Lett.* **2015**, *13*, 53–58. [CrossRef]

48. Gao, X.; Lu, Y.; Xing, Y.; Ma, Y.; Lu, J.; Bao, W.; Wang, Y.; Xi, T. A novel anticancer and antifungus phenazine derivative from a marine actinomycete BM-17. *Microbiol. Res.* **2012**, *167*, 616–622. [CrossRef] [PubMed]

49. Hagiwara, K.; Garcia Hernandez, J.E.; Harper, M.K.; Carroll, A.; Motti, C.A.; Awaya, J.; Nguyen, H.-Y.; Wright, A.D. Puupehenol, a potent antioxidant antimicrobial meroterpenoid from a hawaiian deep-water *Dactylospongia* sp. Sponge. *J. Nat. Prod.* **2015**, *78*, 325–329. [CrossRef] [PubMed]

50. Meng, L.-H.; Li, X.-M.; Liu, Y.; Wang, B.-G. Penicibilaenes A and B, sesquiterpenes with a tricyclo [6.3. 1.01, 5] dodecane skeleton from the marine isolate of *Penicillium bilaiae* MA-267. *Org. Lett.* **2014**, *16*, 6052–6055. [CrossRef]

51. Wang, J.-F.; Lin, X.-P.; Qin, C.; Liao, S.-R.; Wan, J.-T.; Zhang, T.-Y.; Liu, J.; Fredimoses, M.; Chen, H.; Yang, B. Antimicrobial and antiviral sesquiterpenoids from sponge-associated fungus, *Aspergillus sydowii* ZSDS1-F6. *J. Antibiot.* **2014**, *67*, 581–583. [CrossRef] [PubMed]

52. Alarif, W.M.; Al-Lihaibi, S.S.; Ayyad, S.-E.N.; Abdel-Rhman, M.H.; Badria, F.A. Laurene-type sesquiterpenes from the red sea red alga *Laurencia obtusa* as potential antitumor—Antimicrobial agents. *Eur. J. Med. Chem.* **2012**, *55*, 462–466. [CrossRef] [PubMed]

53. Cheng, Y.B.; Jensen, P.R.; Fenical, W. Cytotoxic and antimicrobial napyradiomycins from two marine-derived *Streptomyces* strains. *Eur. J. Org. Chem.* **2013**, *2013*, 3751–3757. [CrossRef] [PubMed]

54. Zhang, Q.; Mándi, A.; Li, S.; Chen, Y.; Zhang, W.; Tian, X.; Zhang, H.; Li, H.; Zhang, W.; Zhang, S. N–N-coupled indolo-sesquiterpene atropo-diastereomers from a marine-derived actinomycete. *Eur. J. Org. Chem.* **2012**, *2012*, 5256–5262. [CrossRef]

55. Al-Footy, K.O.; Alarif, W.M.; Asiri, F.; Aly, M.M.; Ayyad, S.-E.N. Rare pyrane-based cembranoids from the red sea soft coral *Sarcophyton trocheliophorum* as potential antimicrobial—Antitumor agents. *Med. Chem. Res.* **2015**, *24*, 505–512. [CrossRef]

56. Kumar, R.; Subramani, R.; Aalbersberg, W. Three bioactive sesquiterpene quinones from the fijian marine sponge of the genus *Hippospongia*. *Nat. Prod. Res.* **2013**, *27*, 1488–1491. [CrossRef] [PubMed]

57. Murshid, S.S.; Badr, J.M.; Youssef, D.T. Penicillosides a and b: New cerebrosides from the marine-derived fungus *Penicillium* species. *Rev. Bras. Farmacogn.* **2016**, *26*, 29–33. [CrossRef]

58. Tareq, F.S.; Kim, J.H.; Lee, M.A.; Lee, H.-S.; Lee, Y.-J.; Lee, J.S.; Shin, H.J. Ieodoglucomides A and B from a marine-derived bacterium *Bacillus licheniformis*. *Org. Lett.* **2012**, *14*, 1464–1467. [CrossRef] [PubMed]

59. Tareq, F.S.; Lee, H.-S.; Lee, Y.-J.; Lee, J.S.; Shin, H.J. Ieodoglucomide C and ieodoglycolipid, new glycolipids from a marine-derived bacterium *Bacillus Licheniformis* 09IDYM23. *Lipids* **2015**, *50*, 513–519. [CrossRef] [PubMed]

60. Tareq, F.S.; Lee, M.A.; Lee, H.-S.; Lee, Y.-J.; Lee, J.S.; Hasan, C.M.; Islam, M.T.; Shin, H.J. Gageotetrins A–C, noncytotoxic antimicrobial linear lipopeptides from a marine bacterium *Bacillus subtilis*. *Org. Lett.* **2014**, *16*, 928–931. [CrossRef] [PubMed]

61. Mondol, M.A.M.; Shin, H.J. Antibacterial and antiyeast compounds from marine-derived bacteria. *Mar. Drugs* **2014**, *12*, 2913–2921. [CrossRef] [PubMed]

62. Mondol, M.A.M.; Tareq, F.S.; Kim, J.H.; Lee, M.A.; Lee, H.-S.; Lee, J.S.; Lee, Y.-J.; Shin, H.J. New antimicrobial compounds from a marine-derived *Bacillus* sp. *J. Antibiot.* **2013**, *66*, 89–95. [CrossRef]

63. Lu, Y.; Zhao, M. Two highly acetylated sterols from the marine sponge *Dysidea* sp. *Z. Naturforschung B* **2017**, *72*, 49–52. [CrossRef]

64. Lima, B.D.A.; de Lira, S.P.; Kossuga, M.H.; Gonçalves, R.B.; Berlinck, R.G.; Kamiya, R.U. Halistanol sulfate A and rodriguesines A and B are antimicrobial and antibiofilm agents against the cariogenic bacterium *Streptococcus mutans*. *Rev. Bras. Farmacogn.* **2014**, *24*, 651–659. [CrossRef]

65. Kalinovskaya, N.I.; Romanenko, L.A.; Kalinovsky, A.I. Antibacterial low-molecular-weight compounds produced by the marine bacterium *Rheinheimera japonica* KMM 9513T. *Anieke Van Leeuwenhoek* **2017**, *110*, 719–726. [CrossRef] [PubMed]

66. Cuong, P.V.; Cuc, N.T.K.; Quyen, V.T.; Binh, P.T.; Kiem, P.; Nam, N.H.; Dat, N.T. Antimicrobial constituents from the *Bacillus megaterium* IC isolated from marine sponge *Haliclona oculata*. *Nat. Prod. Sci.* **2015**, *21*, 202–205.

67. Du, F.-Y.; Zhang, P.; Li, X.-M.; Li, C.-S.; Cui, C.-M.; Wang, B.-G. Cyclohexadepsipeptides of the isaridin class from the marine-derived fungus *Beauveria felina* EN-135. *J. Nat. Prod.* **2014**, *77*, 1164–1169. [CrossRef] [PubMed]

68. Song, Y.; Li, Q.; Liu, X.; Chen, Y.; Zhang, Y.; Sun, A.; Zhang, W.; Zhang, J.; Ju, J. Cyclic hexapeptides from the deep south china sea-derived *Streptomyces scopuliridis* SCSIO ZJ46 active against pathogenic gram-positive bacteria. *J. Nat. Prod.* **2014**, *77*, 1937–1941. [CrossRef] [PubMed]

69. Sun, S.; Canning, C.B.; Bhargava, K.; Sun, X.; Zhu, W.; Zhou, N.; Zhang, Y.; Zhou, K. Polybrominated diphenyl ethers with potent and broad spectrum antimicrobial activity from the marine sponge *Dysidea*. *Bioorg. Med. Chem. Lett.* **2015**, *25*, 2181–2183. [CrossRef] [PubMed]

70. Gotsbacher, M.P.; Karuso, P. New antimicrobial bromotyrosine analogues from the sponge *Pseudoceratina purpurea* and its predator *Tylodina corticalis*. *Mar. Drugs* **2015**, *13*, 1389–1409. [CrossRef] [PubMed]

71. Rodrigues, D.; Alves, C.; Horta, A.; Pinteus, S.; Silva, J.; Culioli, G.; Thomas, O.P.; Pedrosa, R. Antitumor and antimicrobial potential of bromoditerpenes isolated from the red alga, *Sphaerococcus coronopifolius*. *Mar. Drugs* **2015**, *13*, 713–726. [CrossRef] [PubMed]

72. Yu, M.; Wang, J.; Tang, K.; Shi, X.; Wang, S.; Zhu, W.-M.; Zhang, X.-H. Purification and characterization of antibacterial compounds of *Pseudoalteromonas flavipulchra* JG1. *Microbiology* **2012**, *158*, 835–842. [CrossRef] [PubMed]

73. Tanaka, N.; Kusama, T.; Takahashi-Nakaguchi, A.; Gonoi, T.; Fromont, J.; Kobayashi, J.I. Nagelamides X–Z, dimeric bromopyrrole alkaloids from a marine sponge *agelas* sp. *Org. Lett.* **2013**, *15*, 3262–3265. [CrossRef] [PubMed]

74. Xu, M.; Davis, R.A.; Feng, Y.; Sykes, M.L.; Shelper, T.; Avery, V.M.; Camp, D.; Quinn, R.J. Ianthelliformisamines A–C, antibacterial bromotyrosine-derived metabolites from the marine sponge *Suberea ianthelliformis*. *J. Nat. Prod.* **2012**, *75*, 1001–1005. [CrossRef] [PubMed]

75. Liang, Y.; Xie, X.; Chen, L.; Yan, S.; Ye, X.; Anjum, K.; Huang, H.; Lian, X.; Zhang, Z. Bioactive polycyclic quinones from marine *streptomyces* sp. 182SMLY. *Mar. Drugs* **2016**, *14*, 10. [CrossRef] [PubMed]

76. Wu, B.; Wiese, J.; Labes, A.; Kramer, A.; Schmaljohann, R.; Imhoff, J.F. Lindgomycin, an unusual antibiotic polyketide from a marine fungus of the lindgomycetaceae. *Mar. Drugs* **2015**, *13*, 4617–4632. [CrossRef] [PubMed]

77. Gushiken, M.; Kagiyama, I.; Kato, H.; Kuwana, T.; Losung, F.; Mangindaan, R.E.; De Voogd, N.J.; Tsukamoto, S. Manadodioxans A–E: Polyketide endoperoxides from the marine sponge *Plakortis bergquistae*. *J. Nat. Med.* **2015**, *69*, 595–600. [CrossRef] [PubMed]

78. Chakraborty, K.; Thilakan, B.; Chakraborty, R.D.; Raola, V.K.; Joy, M. O-heterocyclic derivatives with antibacterial properties from marine bacterium *Bacillus subtilis* associated with seaweed, *Sargassum myriocystum*. *Appl. Microbiol. Biotechnol.* **2017**, *101*, 569–583. [CrossRef] [PubMed]

79. Xin, W.; Ye, X.; Yu, S.; Lian, X.-Y.; Zhang, Z. New capoamycin-type antibiotics and polyene acids from marine *Streptomyces fradiae* PTZ0025. *Mar. Drugs* **2012**, *10*, 2388–2402. [CrossRef] [PubMed]

80. Raju, R.; Piggott, A.M.; Khalil, Z.; Bernhardt, P.V.; Capon, R.J. Heronamycin a: A new benzothiazine ansamycin from an australian marine-derived *streptomyces* sp. *Tetrahedron Lett.* **2012**, *53*, 1063–1065. [CrossRef]

81. Chakraborty, K.; Thilakan, B.; Raola, V.K.; Joy, M. Antibacterial polyketides from *Bacillus amyloliquefaciens* associated with edible red seaweed *Laurenciae papillosa*. *Food Chem.* **2017**, *218*, 427–434. [CrossRef] [PubMed]

82. Du, F.Y.; Li, X.M.; Song, J.Y.; Li, C.S.; Wang, B.G. Anthraquinone derivatives and an orsellinic acid ester from the marine alga-derived endophytic fungus *Eurotium cristatum* EN-220. *Helv. Chim. Acta* **2014**, *97*, 973–978. [CrossRef]

83. Xu, R.; Li, X.-M.; Wang, B.-G. Penicisimpins A-C, three new dihydroisocoumarins from *Penicillium simplicissimum* MA-332, a marine fungus derived from the rhizosphere of the mangrove plant *Bruguiera sexangula* var. Rhynchopetala. *Phytochem. Lett.* **2016**, *17*, 114–118. [CrossRef]

84. Aksoy, S.Ç.; Uzel, A.; Bedir, E. Cytosine-type nucleosides from marine-derived *Streptomyces rochei* 06CM016. *J. Antibiot.* **2016**, *69*, 51–56. [CrossRef] [PubMed]

85. Shaala, L.A.; Youssef, D.T. Identification and bioactivity of compounds from the fungus *Penicillium* sp. CYE-87 isolated from a marine tunicate. *Mar. Drugs* **2015**, *13*, 1698–1709. [CrossRef] [PubMed]

86. Song, F.; Ren, B.; Chen, C.; Yu, K.; Liu, X.; Zhang, Y.; Yang, N.; He, H.; Liu, X.; Dai, H. Three new sterigmatocystin analogues from marine-derived fungus *Aspergillus versicolor* MF359. *Appl. Microbiol. Biotechnol.* **2014**, *98*, 3753–3758. [CrossRef] [PubMed]
87. Ambavane, V.; Tokdar, P.; Parab, R.; Sreekumar, E.; Mahajan, G.; Mishra, P.D.; D'Souza, L.; Ranadive, P. Caerulomycin A—An antifungal compound isolated from marine actinomycetes. *Adv. Microbiol.* **2014**, *4*, 567–578. [CrossRef]
88. Wu, B.; Oesker, V.; Wiese, J.; Schmaljohann, R.; Imhoff, J.F. Two new antibiotic pyridones produced by a marine fungus, *Trichoderma* sp. strain MF106. *Mar. Drugs* **2014**, *12*, 1208–1219. [CrossRef] [PubMed]
89. Dasari, V.R.R.K.; Muthyala, M.K.K.; Nikku, M.Y.; Donthireddy, S.R.R. Novel pyridinium compound from marine actinomycete, *Amycolatopsis alba* var. Nov. DVR D4 showing antimicrobial and cytotoxic activities in vitro. *Microbiol. Res.* **2012**, *167*, 346–351. [CrossRef] [PubMed]
90. Xu, X.; Zhao, S.; Wei, J.; Fang, N.; Yin, L.; Sun, J. Porric acid D from marine-derived fungus *Alternaria* sp. isolated from bohai sea. *Chem. Nat. Compd.* **2012**, *47*, 893–895. [CrossRef]
91. Li, Y.; Xu, Y.; Liu, L.; Han, Z.; Lai, P.Y.; Guo, X.; Zhang, X.; Lin, W.; Qian, P.-Y. Five new amicoumacins isolated from a marine-derived bacterium *Bacillus subtilis*. *Mar. Drugs* **2012**, *10*, 319–328. [CrossRef]
92. Eltamany, E.E.; Abdelmohsen, U.R.; Ibrahim, A.K.; Hassanean, H.A.; Hentschel, U.; Ahmed, S.A. New antibacterial xanthone from the marine sponge-derived *Micrococcus* sp. EG45. *Bioorg. Med. Chem. Lett.* **2014**, *24*, 4939–4942. [CrossRef] [PubMed]
93. Wright, G.D. Opportunities for natural products in 21 st century antibiotic discovery. *Nat. Prod. Rep.* **2017**, *34*, 694–701. [CrossRef] [PubMed]
94. Yao, H.; Liu, J.; Xu, S.; Zhu, Z.; Xu, J. The structural modification of natural products for novel drug discovery. *Expert Opin. Drug Discov.* **2017**, *12*, 121–140. [CrossRef] [PubMed]
95. Majik, M.S.; Rodrigues, C.; Mascarenhas, S.; D'Souza, L. Design and synthesis of marine natural product-based 1*H*-indole-2, 3-dione scaffold as a new antifouling/antibacterial agent against fouling bacteria. *Bioorg. Chem.* **2014**, *54*, 89–95. [CrossRef] [PubMed]
96. Xiong, S.; Pang, H.D.; Fan, J.; Ge, F.; Yang, X.X.; Liu, Q.Y.; Liao, X.J.; Xu, S.H. In vitro and in vivo antineoplastic activity of a novel bromopyrrole and its potential mechanism of action. *Br. J. Pharmacol.* **2010**, *159*, 909–918. [CrossRef] [PubMed]
97. Tasdemir, D.; Topaloglu, B.; Perozzo, R.; Brun, R.; O'Neill, R.; Carballeira, N.M.; Zhang, X.; Tonge, P.J.; Linden, A.; Rüedi, P. Marine natural products from the turkish sponge *Agelas oroides* that inhibit the enoyl reductases from *Plasmodium falciparum*, *Mycobacterium tuberculosis* and *Escherichia coli*. *Bioorg. Med. Chem.* **2007**, *15*, 6834–6845. [CrossRef] [PubMed]
98. Schillaci, D.; Petruso, S.; Cascioferro, S.; Raimondi, M.; Haagensen, J.A.J.; Molin, S. In vitro anti-gram-positive and antistaphylococcal biofilm activity of newly halogenated pyrroles related to pyrrolomycins. *Int. J. Antimicrob. Agents* **2008**, *31*, 380–382. [CrossRef] [PubMed]
99. Fattorusso, E.; Taglialatela-Scafati, O. *Modern Alkaloids: Structure, Isolation, Synthesis, and Biology*, 1st ed.; John Wiley & Sons: Chichester, UK, 2008; p. 271.
100. Rane, R.A.; Gutte, S.D.; Sahu, N.U. Synthesis and evaluation of novel 1, 3, 4-oxadiazole derivatives of marine bromopyrrole alkaloids as antimicrobial agent. *Bioorg. Med. Chem. Lett.* **2012**, *22*, 6429–6432. [CrossRef] [PubMed]
101. Rane, R.A.; Sahu, N.U.; Shah, C.P. Synthesis and antibiofilm activity of marine natural product-based 4-thiazolidinones derivatives. *Bioorg. Med. Chem. Lett.* **2012**, *22*, 7131–7134. [CrossRef] [PubMed]
102. Rane, R.A.; Sahu, N.U.; Shah, C.P.; Shah, N.K. Design, synthesis and antistaphylococcal activity of marine pyrrole alkaloid derivatives. *J. Enzym. Inhib. Med.* **2014**, *29*, 401–407. [CrossRef] [PubMed]
103. Zidar, N.; Montalvão, S.; Hodnik, Ž.; Nawrot, D.A.; Žula, A.; Ilaš, J.; Kikelj, D.; Tammela, P.; Mašič, L.P. Antimicrobial activity of the marine alkaloids, clathrodin and oroidin, and their synthetic analogues. *Mar. Drugs* **2014**, *12*, 940–963. [CrossRef] [PubMed]
104. Zhao, J.-C.; Li, X.-M.; Gloer, J.B.; Wang, B.-G. First total syntheses and antimicrobial evaluation of penicimonoterpene, a marine-derived monoterpenoid, and its various derivatives. *Mar. Drugs* **2014**, *12*, 3352–3370. [CrossRef] [PubMed]
105. Baker, D.D.; Chu, M.; Oza, U.; Rajgarhia, V. The value of natural products to future pharmaceutical discovery. *Nat. Prod. Rep.* **2007**, *24*, 1225–1244. [CrossRef] [PubMed]

106. NCBI Resource Coordinators. Database resources of the national center for biotechnology information. *Nucleic Acids Res.* **2017**, *45*, pp. D12–D17.

107. Genomes OnLine Database (GOLD). Available online: https://gold.Jgi.Doe.Gov/ (accessed on 20 July 2017).

108. Bentley, S.D.; Chater, K.F.; Cerdeno-Tarraga, A.M.; Challis, G.L.; Thomson, N.R.; James, K.D.; Harris, D.E.; Quail, M.A.; Kieser, H.; Harper, D.; et al. Complete genome sequence of the model actinomycete *Streptomyces coelicolor* A3(2). *Nature* **2002**, *417*, 141–147. [CrossRef] [PubMed]

109. Challis, G.L.; Ravel, J. Coelichelin. A new peptide siderophore encoded by the *Streptomyces coelicolor* genome: Structure prediction from the sequence of its non-ribosomal peptide synthetase. *FEMS Microbiol. Lett.* **2000**, *187*, 111–114. [CrossRef] [PubMed]

110. Challis, G.L. Exploitation of the *streptomyces coelicolor* A3(2) genome sequence for discovery of new natural products and biosynthetic pathways. *J. Ind. Microbiol. Biotechnol.* **2014**, *41*, 219–232. [CrossRef] [PubMed]

111. Pawlik, K.; Kotowska, M.; Chater, K.F.; Kuczek, K.; Takano, E. A cryptic type I polyketide synthase (CPK) gene cluster in *Streptomyces coelicolor* A3(2). *Arch. Microbiol.* **2007**, *187*, 87–99. [CrossRef] [PubMed]

112. Doroghazi, J.R.; Metcalf, W.W. Comparative genomics of actinomycetes with a focus on natural product biosynthetic genes. *BMC Genom.* **2013**, *14*, 611. [CrossRef] [PubMed]

113. Xiong, Z.Q.; Wang, J.F.; Hao, Y.Y.; Wang, Y. Recent advances in the discovery and development of marine microbial natural products. *Mar. Drugs* **2013**, *11*, 700–717. [CrossRef] [PubMed]

114. Atlas of Biosynthetic Gene Clusters. Available online: https://img.Jgi.Doe.Gov/abc (accessed on 20 July 2017).

115. The Secondary Metabolite Bioinformatics Forum. Available online: http://www.Secondarymetabolites.Org/ (accessed on 20 July 2017).

116. Weber, T.; Blin, K.; Duddela, S.; Krug, D.; Kim, H.U.; Bruccoleri, R.; Lee, S.Y.; Fischbach, M.A.; Muller, R.; Wohlleben, W.; et al. Antismash 3.0-a comprehensive resource for the genome mining of biosynthetic gene clusters. *Nucleic Acids Res.* **2015**, *43*, W237–W243. [CrossRef] [PubMed]

117. De Jong, A.; van Hijum, S.A.; Bijlsma, J.J.; Kok, J.; Kuipers, O.P. Bagel: A web-based bacteriocin genome mining tool. *Nucleic Acids Res.* **2006**, *34*, W273–W279. [CrossRef] [PubMed]

118. Skinnider, M.A.; Dejong, C.A.; Rees, P.N.; Johnston, C.W.; Li, H.; Webster, A.L.; Wyatt, M.A.; Magarvey, N.A. Genomes to natural products prediction informatics for secondary metabolomes (PRISM). *Nucleic Acids Res.* **2015**, *43*, 9645–9662. [CrossRef] [PubMed]

119. Hammami, R.; Zouhir, A.; Le Lay, C.; Ben Hamida, J.; Fliss, I. Bactibase second release: A database and tool platform for bacteriocin characterization. *BMC Microbiol.* **2010**, *10*, 22. [CrossRef]

120. Conway, K.R.; Boddy, C.N. Clustermine360: A database of microbial PKS/NRPS biosynthesis. *Nucleic Acids Res.* **2013**, *41*, D402–D407. [CrossRef] [PubMed]

121. Medema, M.H.; Kottmann, R.; Yilmaz, P.; Cummings, M.; Biggins, J.B.; Blin, K.; de Bruijn, I.; Chooi, Y.H.; Claesen, J.; Coates, R.C.; et al. Minimum information about a biosynthetic gene cluster. *Nat. Chem. Biol.* **2015**, *11*, 625–631. [CrossRef] [PubMed]

122. Cimermancic, P.; Medema, M.H.; Claesen, J.; Kurita, K.; Wieland Brown, L.C.; Mavrommatis, K.; Pati, A.; Godfrey, P.A.; Koehrsen, M.; Clardy, J.; et al. Insights into secondary metabolism from a global analysis of prokaryotic biosynthetic gene clusters. *Cell* **2014**, *158*, 412–421. [CrossRef] [PubMed]

123. Blin, K.; Medema, M.H.; Kazempour, D.; Fischbach, M.A.; Breitling, R.; Takano, E.; Weber, T. Antismash 2.0—A versatile platform for genome mining of secondary metabolite producers. *Nucleic Acids Res.* **2013**, *41*, W204–W212. [CrossRef]

124. Medema, M.H.; Blin, K.; Cimermancic, P.; de Jager, V.; Zakrzewski, P.; Fischbach, M.A.; Weber, T.; Takano, E.; Breitling, R. Antismash: Rapid identification, annotation and analysis of secondary metabolite biosynthesis gene clusters in bacterial and fungal genome sequences. *Nucleic Acids Res.* **2011**, *39*, W339–W346. [CrossRef] [PubMed]

125. Blin, K.; Wolf, T.; Chevrette, M.G.; Lu, X.; Schwalen, C.J.; Kautsar, S.A.; Suarez Duran, H.G.; de Los Santos, E.L.C.; Kim, H.U.; Nave, M.; et al. Antismash 4.0-improvements in chemistry prediction and gene cluster boundary identification. *Nucleic Acids Res.* **2017**, *45*, W36–W41. [CrossRef] [PubMed]

126. Kautsar, S.A.; Suarez Duran, H.G.; Blin, K.; Osbourn, A.; Medema, M.H. Plantismash: Automated identification, annotation and expression analysis of plant biosynthetic gene clusters. *Nucleic Acids Res.* **2017**, *45*, W55–W63. [CrossRef] [PubMed]

127. Cruz-Morales, P.; Kopp, J.F.; Martinez-Guerrero, C.; Yanez-Guerra, L.A.; Selem-Mojica, N.; Ramos-Aboites, H.; Feldmann, J.; Barona-Gomez, F. Phylogenomic analysis of natural products biosynthetic gene clusters allows discovery of arseno-organic metabolites in model *streptomycetes*. *Genome Biol. Evol.* **2016**, *8*, 1906–1916. [CrossRef] [PubMed]

128. Zhang, W.; Li, S.; Zhu, Y.; Chen, Y.; Chen, Y.; Zhang, H.; Zhang, G.; Tian, X.; Pan, Y.; Zhang, S.; et al. Heronamides D-F, polyketide macrolactams from the deep-sea-derived *streptomyces* sp. SCSIO 03032. *J. Nat. Prod.* **2014**, *77*, 388–391. [CrossRef] [PubMed]

129. Zhu, Y.; Zhang, W.; Chen, Y.; Yuan, C.; Zhang, H.; Zhang, G.; Ma, L.; Zhang, Q.; Tian, X.; Zhang, S.; et al. Characterization of heronamide biosynthesis reveals a tailoring hydroxylase and indicates migrated double bonds. *ChemBioChem* **2015**, *16*, 2086–2093. [CrossRef] [PubMed]

130. Li, Q.; Song, Y.; Qin, X.; Zhang, X.; Sun, A.; Ju, J. Identification of the biosynthetic gene cluster for the anti-infective desotamides and production of a new analogue in a heterologous host. *J. Nat. Prod.* **2015**, *78*, 944–948. [CrossRef] [PubMed]

131. Chen, E.; Chen, Q.; Chen, S.; Xu, B.; Ju, J.; Wang, H. Discovery and biosynthesis of mathermycin from marine-derived marinactinospora thermotolerans scsio 00652. *Appl. Environ. Microbiol.* **2017**. [CrossRef]

132. Viegelmann, C.; Margassery, L.M.; Kennedy, J.; Zhang, T.; O'Brien, C.; O'Gara, F.; Morrissey, J.P.; Dobson, A.D.; Edrada-Ebel, R. Metabolomic profiling and genomic study of a marine sponge-associated *streptomyces* sp. *Mar. Drugs* **2014**, *12*, 3323–3351. [CrossRef]

133. Paulus, C.; Rebets, Y.; Tokovenko, B.; Nadmid, S.; Terekhova, L.P.; Myronovskyi, M.; Zotchev, S.B.; Ruckert, C.; Braig, S.; Zahler, S.; et al. New natural products identified by combined genomics-metabolomics profiling of marine *streptomyces* sp. MP131-18. *Sci. Rep.* **2017**, *7*, 42382. [CrossRef] [PubMed]

134. Liu, J.; Wang, B.; Li, H.; Xie, Y.; Li, Q.; Qin, X.; Zhang, X.; Ju, J. Biosynthesis of the anti-infective marformycins featuring pre-NRPS assembly line N-formylation and O-methylation and post-assembly line C-hydroxylation chemistries. *Org. Lett.* **2015**, *17*, 1509–1512. [CrossRef] [PubMed]

135. D'Costa, V.M.; King, C.E.; Kalan, L.; Morar, M.; Sung, W.W.; Schwarz, C.; Froese, D.; Zazula, G.; Calmels, F.; Debruyne, R.; et al. Antibiotic resistance is ancient. *Nature* **2011**, *477*, 457–461. [CrossRef] [PubMed]

136. Cox, G.; Wright, G.D. Intrinsic antibiotic resistance: Mechanisms, origins, challenges and solutions. *Int. J. Med. Microbiol.* **2013**, *303*, 287–292. [CrossRef] [PubMed]

137. Thaker, M.N.; Wang, W.; Spanogiannopoulos, P.; Waglechner, N.; King, A.M.; Medina, R.; Wright, G.D. Identifying producers of antibacterial compounds by screening for antibiotic resistance. *Nat. Biotechnol.* **2013**, *31*, 922–927. [CrossRef] [PubMed]

138. Thaker, M.N.; Waglechner, N.; Wright, G.D. Antibiotic resistance-mediated isolation of scaffold-specific natural product producers. *Nat. Protoc.* **2014**, *9*, 1469–1479. [CrossRef] [PubMed]

139. Tang, X.; Li, J.; Millan-Aguinaga, N.; Zhang, J.J.; O'Neill, E.C.; Ugalde, J.A.; Jensen, P.R.; Mantovani, S.M.; Moore, B.S. Identification of thiotetronic acid antibiotic biosynthetic pathways by target-directed genome mining. *ACS Chem. Biol.* **2015**, *10*, 2841–2849. [CrossRef] [PubMed]

140. Yeh, H.H.; Ahuja, M.; Chiang, Y.M.; Oakley, C.E.; Moore, S.; Yoon, O.; Hajovsky, H.; Bok, J.W.; Keller, N.P.; Wang, C.C.; et al. Resistance gene-guided genome mining: Serial promoter exchanges in *Aspergillus nidulans* reveal the biosynthetic pathway for fellutamide B, a proteasome inhibitor. *ACS Chem. Biol.* **2016**, *11*, 2275–2284. [CrossRef] [PubMed]

141. Johnston, C.W.; Skinnider, M.A.; Dejong, C.A.; Rees, P.N.; Chen, G.M.; Walker, C.G.; French, S.; Brown, E.D.; Berdy, J.; Liu, D.Y.; et al. Assembly and clustering of natural antibiotics guides target identification. *Nat. Chem. Biol.* **2016**, *12*, 233–239. [CrossRef] [PubMed]

142. Tracanna, V.; de Jong, A.; Medema, M.H.; Kuipers, O.P. Mining prokaryotes for antimicrobial compounds: From diversity to function. *FEMS Microbiol. Rev.* **2017**, *41*, 417–429. [CrossRef] [PubMed]

143. Ward, J.M.; Hodgson, J.E. The biosynthetic genes for clavulanic acid and cephamycin production occur as a 'super-cluster' in three streptomyces. *FEMS Microbiol. Lett.* **1993**, *110*, 239–242. [CrossRef] [PubMed]

144. Medema, M.H.; Trefzer, A.; Kovalchuk, A.; van den Berg, M.; Muller, U.; Heijne, W.; Wu, L.; Alam, M.T.; Ronning, C.M.; Nierman, W.C.; et al. The sequence of a 1.8-MB bacterial linear plasmid reveals a rich evolutionary reservoir of secondary metabolic pathways. *Genome Biol. Evol.* **2010**, *2*, 212–224. [CrossRef]

145. Alanjary, M.; Kronmiller, B.; Adamek, M.; Blin, K.; Weber, T.; Huson, D.; Philmus, B.; Ziemert, N. The antibiotic resistant target seeker (ARTS), an exploration engine for antibiotic cluster prioritization and novel drug target discovery. *Nucleic Acids Res.* **2017**, *45*, W42–W48. [CrossRef] [PubMed]

146. Jia, B.; Raphenya, A.R.; Alcock, B.; Waglechner, N.; Guo, P.; Tsang, K.K.; Lago, B.A.; Dave, B.M.; Pereira, S.; Sharma, A.N.; et al. Card 2017: Expansion and model-centric curation of the comprehensive antibiotic resistance database. *Nucleic Acids Res.* **2017**, *45*, D566–D573. [CrossRef] [PubMed]
147. Gupta, S.K.; Padmanabhan, B.R.; Diene, S.M.; Lopez-Rojas, R.; Kempf, M.; Landraud, L.; Rolain, J.M. Arg-annot. A new bioinformatic tool to discover antibiotic resistance genes in bacterial genomes. *Antimicrob. Agents Chemother.* **2014**, *58*, 212–220. [CrossRef]

marine drugs

MDPI

Review

Biogenetic Relationships of Bioactive Sponge Merotriterpenoids

Thomas E. Smith [1,2,*]

[1] Department of Medicinal Chemistry, University of Utah, 30 S. 2000 E., Salt Lake City, UT 84112, USA
[2] Department of Integrative Biology, University of Texas at Austin, 2506 Speedway, NMS 4.216 stop C0930, Austin, TX 78712, USA

Received: 25 July 2017; Accepted: 7 September 2017; Published: 10 September 2017

Abstract: Hydroquinone meroterpenoids, especially those derived from marine sponges, display a wide range of biological activities. However, use of these compounds is limited by their inaccessibility; there is no sustainable supply of these compounds. Furthermore, our knowledge of their metabolic origin remains completely unstudied. In this review, an in depth structural analysis of sponge merotriterpenoids, including the adociasulfate family of kinesin motor protein inhibitors, provides insight into their biosynthesis. Several key structural features provide clues to the relationships between compounds. All adociasulfates appear to be derived from only four different hydroquinone hexaprenyl diphosphate precursors, each varying in the number and position of epoxidations. Proton-initiated cyclization of these precursors can lead to all carbon skeletons observed amongst sponge merotriterpenoids. Consideration of the enzymes involved in the proposed biosynthetic route suggests a bacterial source, and a hypothetical gene cluster was constructed that may facilitate discovery of the authentic pathway from the sponge metagenome. A similar rationale can be extended to other sponge meroterpenoids, for which no biosynthetic pathways have yet been identified.

Keywords: sponge; meroterpenoid; marine natural product; medicinal chemistry; biosynthesis; drug discovery

---

1. Introduction

Meroterpenes have long been recognized for their diverse biological activities. In particular, hydroquinone meroterpenes are interesting because of their potential for redox chemistry and wide distribution in nature [1,2]. Marine sponges represent a prolific source of hydroquinone meroterpenoids, some of which exhibit unique activities that cannot be substituted for using alternative compounds. The diversity of structures and activities of sponge hydroquinone meroterpenoids have been thoroughly reviewed by Menna et al. [1]. This review focuses on the merotriterpenoids, including the adociasulfates (Figure 1). This family includes several unique carbon skeletons, its members are frequently sulfated, and it also encompasses a wide variety of biological activities. The toxicols (**17–19**) and shaagrockol C (**22**) inhibit the DNA polymerase function of HIV-1 reverse transcriptase [3]. Akaterpin (**25**) inhibits hydrolysis of phosphatidylinositol by phospholipase C, a key step in eukaryotic signaling pathways by its production of diacylglycerol and inositol triphosphate [4]. Indoleamine 2,3-dioxygenase, whose activity mediates T-cell activation and whose overexpression in cancer may prevent tumor rejection, is inhibited by halicloic acids A and B (**15**, **16**) [5]. Some of these compounds also display weak antimicrobial activities [6,7]. The adociasulfate family has been shown to inhibit H⁺-ATPases and kinesin motor proteins [8–11]. Inhibition of kinesin by adociasulfate-2 (**2**) involves competition with microtubules for binding [11,12]. This mode of kinesin inhibition is known for only two other compounds, rose bengal lactone and the polyoxometalate NSC 622124, which both display characteristic features of nonspecific inhibition, including aggregate formation, indiscriminate binding to positively charged protein surfaces, and inhibition of a variety of enzyme activities [13–17]. Thus, sponge hydroquinone merotriterpenoids display a variety of therapeutically interesting activities, including some with unique mechanisms of action.

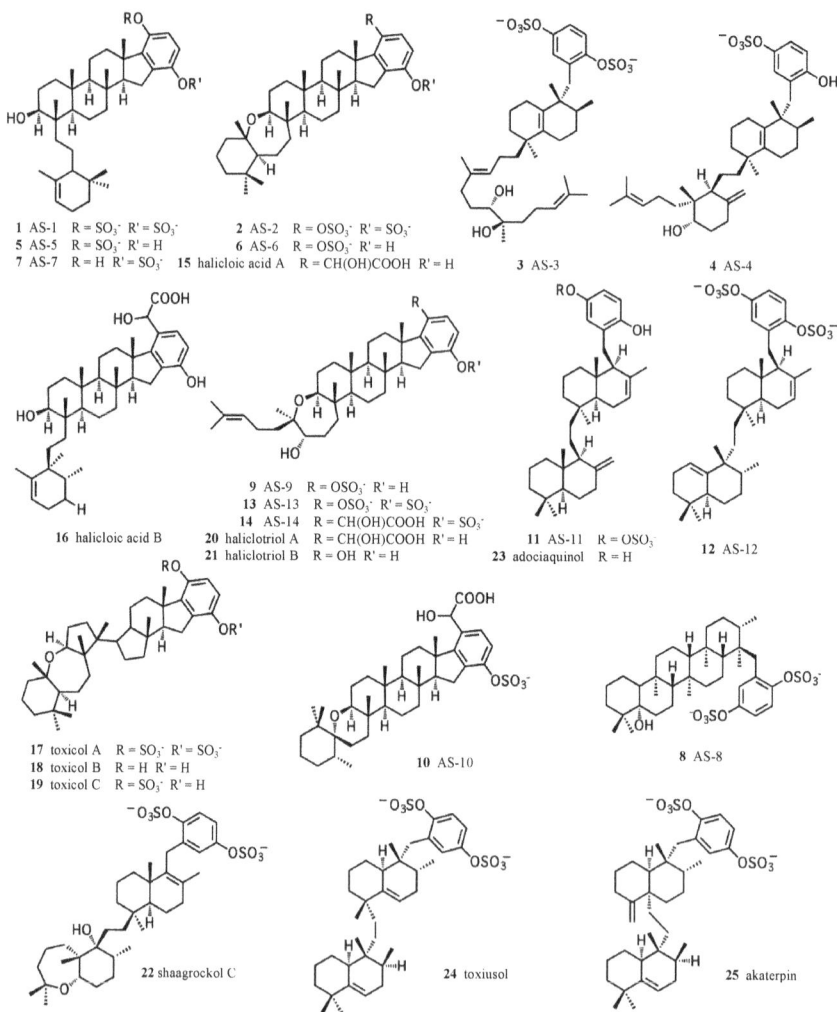

Figure 1. Chemical structures of sponge hydroquinone merotriterpenoids. The adociasulfates and related compounds are derived from sponges of the family *Chalinidae*, with the exception of akaterpin, isolated from *Callyspongia sp.*

The adociasulfates in particular, with their unique mechanism of action, are not only interesting from a medicinal perspective, but have great potential as tools for studying the function of kinesins in cell biology. Neurons rely on precise intracellular organization and transport to function, as different cellular regions have very distinct roles in responding to and relaying signals. Use of **2** revealed a role for kinesin motor proteins in the transport of cytoskeletal filaments within axons [18], and in intracellular spatio-temporal control of gene expression via transport of synapse-specific mRNAs [19]. Kinesins are also involved in reconstructing the nucleus after cell division. Formation of nuclear pore complexes (NPCs) in *Xenopus laevis* eggs was inhibited by **2**, but not the double membrane of the nuclear envelope (NE), indicating the existence of a distinct vesicle population for delivering NPCs that utilize kinesin-guided microtubule transport [20]. Developmental processes have also been probed using **2**. Asymmetric, kinesin-dependent shuttling of cargo was shown to occur very early in the development of frog and chick embryos, suggesting a cytoskeletal role in establishing left-right asymmetry [21]. Treatment of early embryos with **2** led to disruption of this asymmetry. Finally, adociasulfates have been used to interrogate kinesin function directly. The kinesin microtubule binding site was mapped based on binding experiments with **2** [22], and, more recently, adociasulfates were shown to display affinity for non-kinesin microtubule binding sites, indicating their potential as probes of other microtubule-binding proteins [12].

Despite their useful biological activities, sponge meroterpenoids are often unobtainable due to a lack of practical chemical syntheses and the difficulties associated with obtaining material from biological sources [23–26]. For this reason, studies using these compounds in biological applications are scarce and relatively infrequent. For example, with the exception of the most recent study, all of the studies described above obtained **2** from the authors of its original publication [11]. Thus, there is a need for a sustainable means of producing such compounds in order to make full use of their potential. This could be accomplished using a biosynthetic approach. However, there is a lack of knowledge with regard to meroterpenoid biosynthesis in marine invertebrates. No pathways for such compounds have been described despite hundreds of known compounds [1]. The characterization of one meroterpenoid pathway could reveal other the existence of other pathways, as sponge-derived hydroquinone meroterpenoids share many overlapping structural features that suggest common metabolic origins. To this end, adociasulfates provide an excellent starting point because a relatively simple biosynthetic hypothesis can be derived from a limited number of precursors (Figure 2). In fact, it is conceivable that all sponge triterpene hydroquinones are derived from a single parent pathway. The purpose of this review is to draw attention to the structural relationships between compounds and show that a thorough analysis of these relationships can reveal clues to their biosynthetic origin. Below, I discuss the features that unify the adociasulfates and other merotriterpenoids, make a case for the enzymes that are likely to be involved in their construction, and establish a biogenetic hypothesis. This analysis results in a hypothetical, bacteria-derived adociasulfate pathway.

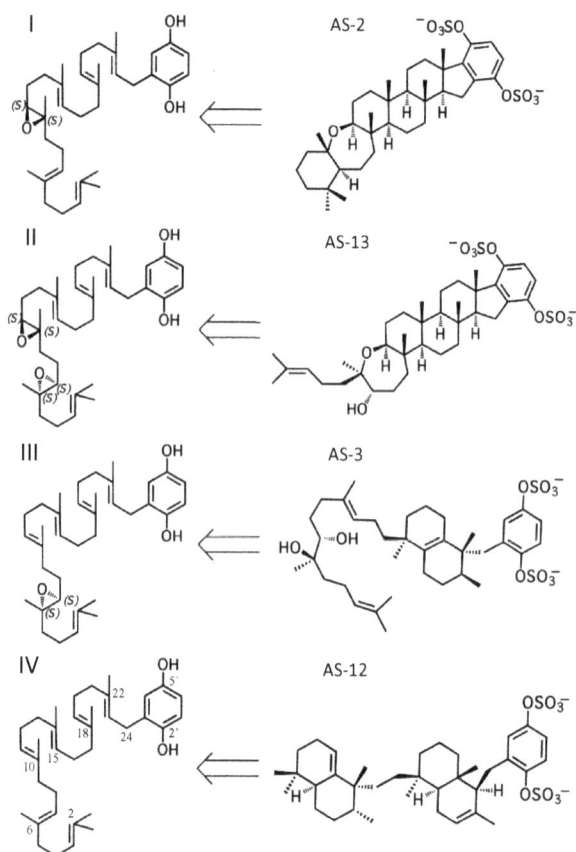

Figure 2. The four, putative hydroquinone merotriterpenoid biosynthetic classes. Sponge merotriterpenoids can be divided into four groups by the number and position of epoxidations of the linear hexaprenoid precursor (left side). Representative adociasulfates of each major group are shown (right side).

2. A proposed biosynthetic route for sponge hydroquinone merotriterpenoids

A defining feature of the adociasulfates is that the arrangement of methyl groups implies a linear triterpene-diphosphate precursor, as opposed to squalene. Prenyl diphosphates are typically formed by a head-to-tail condensation of isopentenyl diphosphate (IPP) with either dimethylallyl diphosphate (DMAPP) or the product of a previous such condensation, yielding linear terpenes extended by five carbons. Squalene, however, is made by the tail-to-tail condensation of two C_{15} farnesyl-diphosphates (FPP) to produce a symmetrical triterpene. The consequences of this are twofold. First, without the diphosphate, squalene is no longer activated for prenyl transfer to a hydrobenzoquinoid substrate. Second, cyclized derivatives of squalene display a characteristic arrangement of methyls that is not observed for adociasulfates or any other hydroquinone meroterpenoids. Linear meroterpenoids have been reported from sponges before, though not from sponges that produce adociasulfates [1]. Nonetheless, there is a precedent for prenyl transfer of linear triterpenes to quinones, resembling ubiquinone biosynthesis, while there is none for the equivalent transfer of squalene.

All sponge merotriterpenoids can potentially be derived from a common series of linear precursors (Figure 2). These universal precursors, the products of aromatic prenylation by hexaprenyl diphosphate, would then be cyclized via a proton-initiated (type II), carbocation-mediated cyclization cascade. Most adociasulfates are hydroxylated at one (e.g., **1**, **2**) or two (e.g., **13**) carbons at positions corresponding to alkenes in hexaprenyl diphosphate. This suggests that epoxidation of the linear substrate occurs prior to cyclization. The number and position of epoxides in the cyclization substrate provides a convenient way to group biosynthetically related sponge merotriterpenoids. Thus, group I precursors are epoxidized at position 10,11, group II at both positions 6,7 and 10,11, and group III at position 6,7, while group IV compounds are not epoxidized. All proposed cyclization schemes described herein are based on (*S,S*) epoxide configurations, as predicted from the configurations of the putative epoxide-derived hydroxy carbons present in group I, II, and III adociasulfates. A variety of skeletons resulting from multiple cyclization events of these precursors is shown in Figure 3.

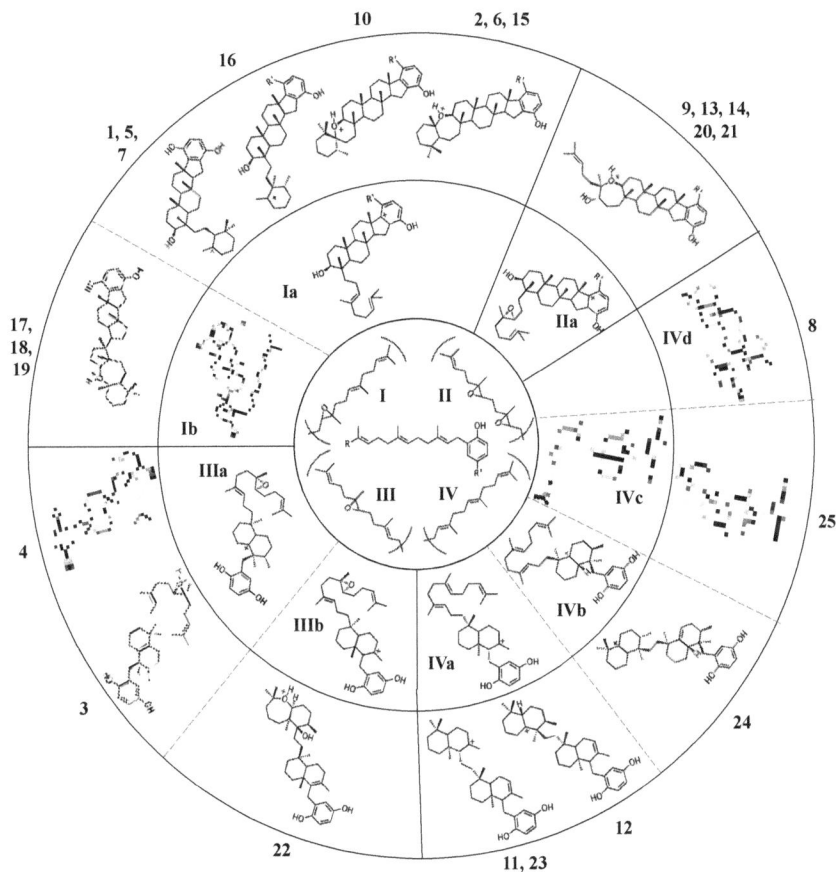

Figure 3. Diverse merotriterpenoid skeletons can be derived from a small number of related precursors via different cyclization routes. The innermost circle shows the four putative precursor molecules. Carbocation products of a single cyclization event are shown in the middle ring. Carbocation products of a second cyclization event are shown in the outermost ring. Numeric designations of final products lie outside of the circle. The number and complexity of structures expands outwards from the simple precursors in the center. The R group of the center linear precursor is substituted with one of the four groups in shown in parentheses, while the R′ group denotes a hydroxyl group for most compounds, or a glycolic acid moiety for some group I and II compounds.

2.1. Group I compounds

The simplest hypothetical cyclization schemes involve the group I meroterpenoids. Compounds in this group likely undergo two independent cyclization cascades and exhibit few rearrangements. The initial cyclization of **1**, **2**, **5**, **6**, **7**, and halicloic acid A (**15**) would be identical for each compound, with epoxide opening to form a hydroxyl group at C11, establishing the sterol-like, four-ring system with ring D fused to the hydrobenzoquinone moiety (Figure 4A) [5,9,10]. The resulting carbocation would then be quenched by proton abstraction, restoring aromaticity. A second proton-initiated cyclization of the remaining two olefins would produce a fifth ring and a carbocation at position C6. Here, **1**, **5**, and **7** would differ from **2**, **6**, and **15** in the manner of base abstraction. In the former group, deprotonation would occur at C5 to introduce a new double bond, leaving the fifth ring independent of the core. In the latter group, a sixth, seven-membered ring would be formed by attack of the C11 hydroxyl on the C6 carbocation. Proton abstraction would then occur at the cyclic ether oxygen. AS-10 (**10**) could be obtained from the same initial cyclization, but would involve a hydride shift in the second cyclization event, placing the carbocation on C7 instead of C6 and resulting in a six-membered heterocycle (Figure 4B) [8]. The 3D structure of **2** would be flat relative to **10**, whose terminal ring would be twisted perpendicular to the plane of the core ring system. Halicloic acid B (**16**) resembles **10**, but the second cyclization event would involve an additional rearrangement: a methyl transfer following the hydride shift, placing the carbocation on C2 (Figure 4C) [5]. Deprotonation at C3 would then yield a tri-substituted alkene. A glycolic acid moiety substitutes for the 5' hydroxyl in **10**, **15**, and **16**, suggesting an alternative aromatic prenyl acceptor to hydrobenzoquinone may be used. The final group I terpenes, toxicols A-C (**17–19**), likely undergo a unique cyclization that could occur in two different ways. In the first, an alkyl shift would condense the initial six-membered ring into a five-membered ring, resulting in an unstable secondary carbocation at C15 (Figure 4(Di)) [7]. Cyclization would then continue with subsequent attack on the C15 carbocation by C19. In the second, the initial epoxide opening would involve a direct attack by the 14,15-olefin on C10, which would be sterically hindered by the two methyls of C10 and C14 (Figure 4(Dii)). A second cyclization step and proton abstraction would result in the final product, with two independent ring systems. Finally, adociasulfates and related meroterpenoids would be sulfated at either, none, or both hydrobenzoquinone hydroxyls, while 5' glycolic acids appear not to be modified further.

2.2. Group II compounds

Adociasulfates and related meroterpenoids of group II are likely derived from a diepoxy precursor (Figure 5). Three of five members of this group exhibit a 5' glycolic acid substitution akin to **10**, **15**, and **16** [6,12,27]. The first cyclization event of **9** may mirror that of **2** from group I, with the opening the 10,11-epoxide and establishment of the adociasulfate core. The second cyclization would then involve the opening of the 6,7-epoxide by back-side attack of the C11 hydroxyl at the more-substituted C6 position in a typical acid-catalyzed epoxide opening. This would result in the formation of a seven-membered ring and an inversion of C6 stereochemistry. Assuming a *pro*-chair conformation would position C6 into a pro-(*R*) configuration relative to the C11 hydroxyl attack, resulting in the axial-oriented terminal olefin. For group I compounds, the lack of the 6,7-epoxide likely allows for inclusion of the 2,3-terminal alkene in the second cyclization event (Figure 4A), whereas all group II compounds display a free terminal olefin. This may reflect an enzymatic preference for protonation of epoxides over alkenes, resulting in early termination of the cyclization cascade.

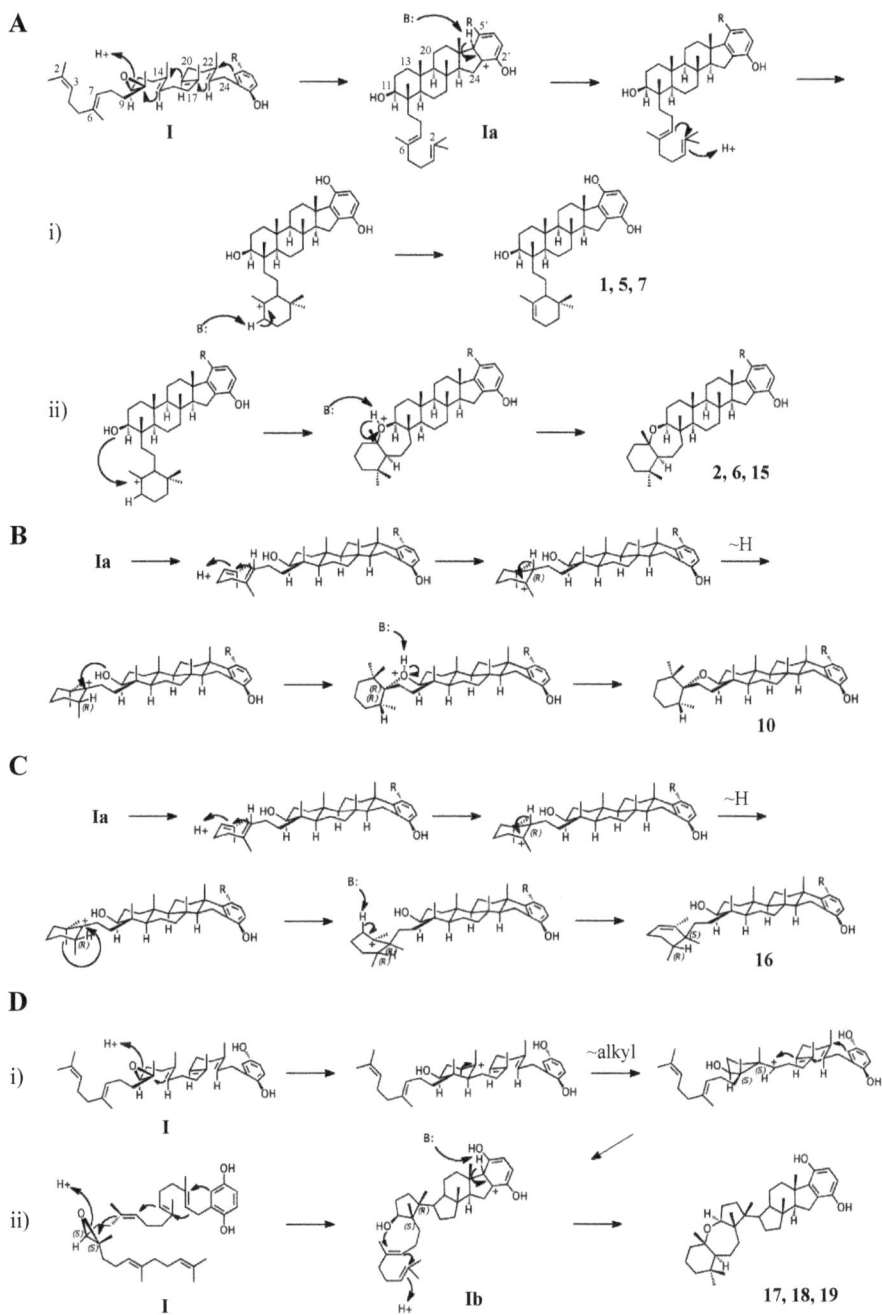

Figure 4. Putative cyclization routes of group I sponge merotriterpenoids, derived from 10,11-epoxyhexaprenyl diphosphate, including: (**A**) **1**, **2**, **5**, **6**, **7**, halicloic acid A (**15**), (**B**) **10**, (**C**) halicloic acid B (**16**), and (**D**) toxicols A-C (**17–19**).

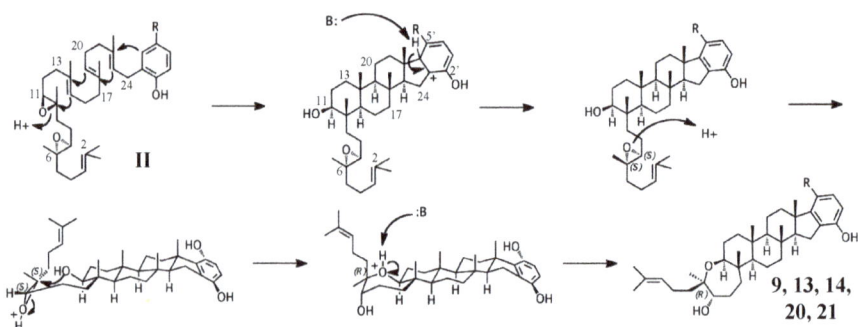

Figure 5. Putative cyclization routes of group II sponge merotriterpenoids, derived from 6,7-10,11-diepoxyhexaprenyl diphosphate, including: **9**, **13**, **14**, and haliclotriols A (**20**) and B (**21**).

2.3. Group III compounds

Group III merotriterpenoids are likely derived from a 6,7-epoxy precursor. This group is characterized by a lack of fusion to the aromatic ring and quenching by water. In the proposed cyclization of **3** and **4**, initiation by protonation would result in a bicyclic drimane-like skeleton that undergoes rearrangement before deprotonation by an active-site base, yielding a highly stable tetra-substituted double bond and unique configurations of methylated carbons (Figure 6A) [9]. The first cyclization event of **3** and **4**, involving the 14,15-, 18,19-, and 22,23-olefins, likely involves prearrangement of the substrate in a chair-chair orientation, placing the remaining linear terpene chain in a pre-equatorial position. For **4**, protonation of the 6,7-epoxide would initiate the second cyclization event involving the 10,11-olefin (Figure 6(Ai)), while hydrolysis of the epoxide would lead to **3** (Figure 6(Aii)). The first cyclization event of shaagrockol C (**22**) would also produce a bicyclic system, though deprotonation would occur prior to any rearrangement, yielding a tetra-substituted alkene (Figure 6B) [28]. The second cyclization would be similar to that of **4**. Prearrangement of the remaining linear portion of the substrate in a boat conformation, followed by hydride transfer from the C11 axial hydrogen to the C10 carbocation would allow for the (R) configuration at C10, as opposed to the (S) configuration that would result from a chair prearrangement and analogous hydride shift. Water would attack the C11 carbocation with inversion of stereochemistry. The net result of this dramatically different cyclization route is that the newly formed ring of **22** would incorporate an axial hydroxyl group in place of a proton at C11. Thus, **22** and **4** display the same relative configuration about C11, despite differing absolute configurations. Finally, the C7 hydroxyl would initiate a final cyclization with the 2,3-alkene, forming a 7-membered terminal heterocycle. Shaagrockol B, isolated together with **22**, is the oxidation product of **22** about the 22,23-alkene and is likely not enzymatic in origin [28].

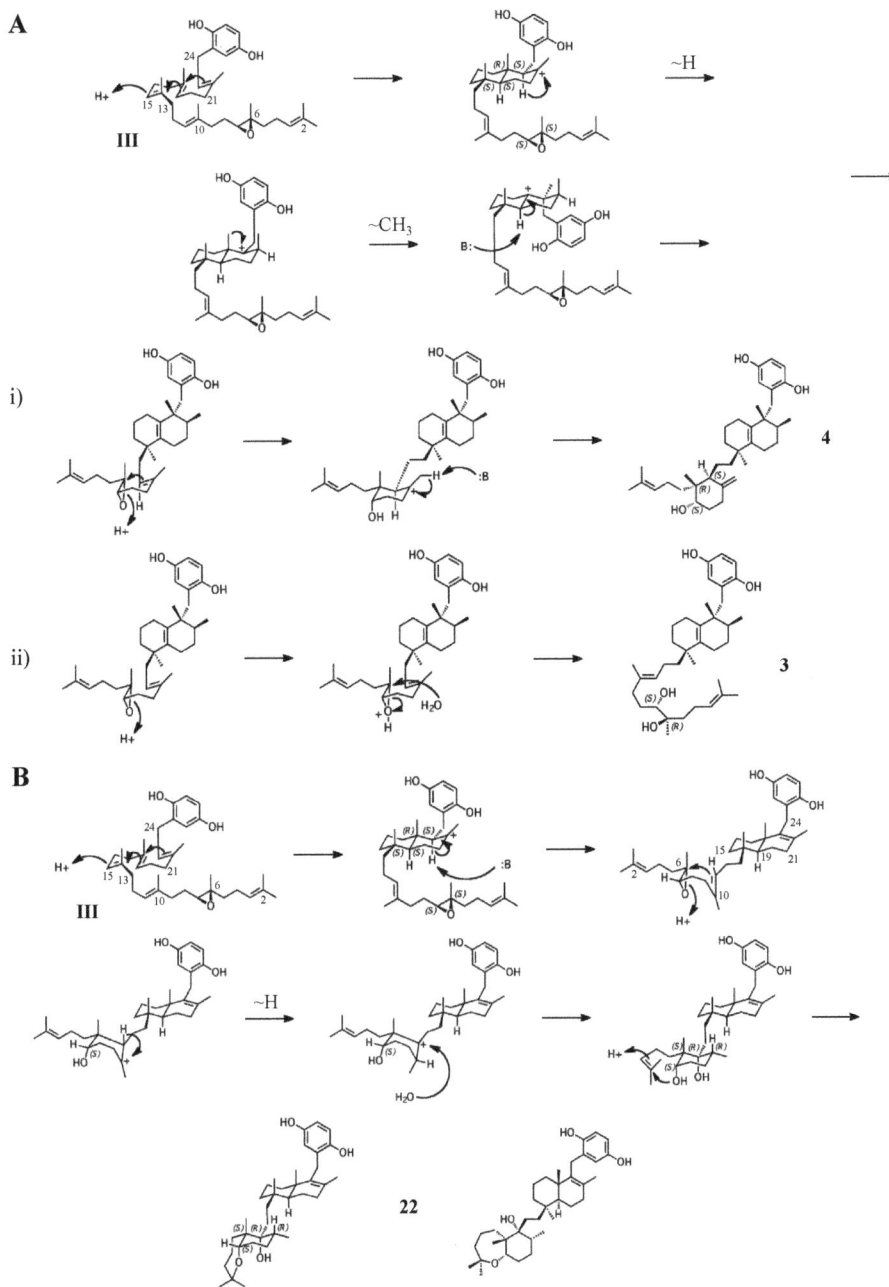

Figure 6. Putative cyclization routes of group III sponge merotriterpenoids, derived from 6,7-epoxyhexaprenyl diphosphate, including: (**A**) **3**, **4**, (**B**) and shaagrockol C (**22**).

2.4. Group IV compounds

The remaining six known sponge merotriterpenoids of group IV are likely derived from a substrate lacking epoxidation. The majority of these compounds undergo complex cyclizations followed by rearrangements, as evidenced by their atypical methyl positions. Like the group III compounds, none of the group IV members exhibit fused rings with the aromatic moiety, suggesting that aromatic ring fusion requires the presence of the 10,11-epoxide. Another common feature between groups III and IV is the absence of 5′ glycolic acid substitution. For **11**, **12**, and adociaquinol (**23**), the proposed initial cyclization would, like the group III compounds, yield a two-ring system, but would differ from these in the prearrangement of the substrate in a boat-chair conformation, placing the linear terpene chain in the less favorable axial position. (Figure 7A) [29]. Due to the absence of the 6,7-epoxide, the second cyclization event of **11**, **12**, and **23** would include the terminal olefin that was excluded by the group III compounds. The second cyclization event of **11** and **23** likely resembles the initial cyclization event of **3** and **4**, involving preorganization of the substrate in the chair-chair orientation that places the terpene chain in the more stable equatorial position (Figure 7(Ai)). Deprotonation at the C10 methyl would introduce the exocyclic alkene. The second cyclization event of **12** would involve a chair-boat conformation, placing the ring system established in the first cyclization event in the axial position, with deprotonation at C5 following both a hydride and methyl shift to form the trisubstituted alkene (Figure 7(Aii)). Cyclization of the initial bicyclic ring system of toxiusol (**24**) likely involves the chair-boat conformation, placing the hydroquinone in the axial position (Figure 7B) [7,29]. Two hydride shifts and a methyl transfer would occur prior to deprotonation to complete first cyclization. The second cyclization event of **23** would occur via the chair-chair conformation similar to **11** and **23**, but a series of hydride transfers would place the trisubstituted alkene on the opposite ring relative to **12**. The cyclization of akaterpin (**25**) likely follows a similar cyclization scheme as **24** but would involve an alkyl shift during the first event, relocating the remaining linear isoprene chain from C14 to the bridgehead carbon, C19 (Figure 7C) [4]. The final sponge merotriterpenoid, **8**, can be reached with a single proton-initiated cascade followed by extensive rearrangement. The substrate is likely prearranged in the antipodal conformation, the opposite orientation of the group I and II cyclizations, such that the end result appears structurally distinct from the sterol-like adociasulfate core of group I and II meroterpenoids (Figure 7D) [10]. In total, five hydride shifts and four methyl shifts would need to occur before an attack by water at the bridgehead carbon C7.

From this model of the origin of adociasulfates, it should be clear that all sponge merotriterpenoids of the hydrobenzoquinone family are related biosynthetically. In each adociasulfate discovery reported, mixtures of compounds from multiple groups were identified, suggesting a common synthetic route that is independent of the epoxidation state of the substrate [7–9,12,27,29]. Of this class of compounds, all but one member has been isolated from sponges within the family *Chalinidae*. The exception is **25**, which was reportedly discovered from *Callyspongia sp* [4]. Though *Callyspongia* is a member of the same order as *Chalinidae* (order *Haplosclerida*), *Callyspongia* is far enough removed in this case to be considered unrelated (Mary Kay Harper, personal communication, 2016). Thus, these compounds can be used as taxonomic identifiers, potentially due to a shared biosynthetic pathway.

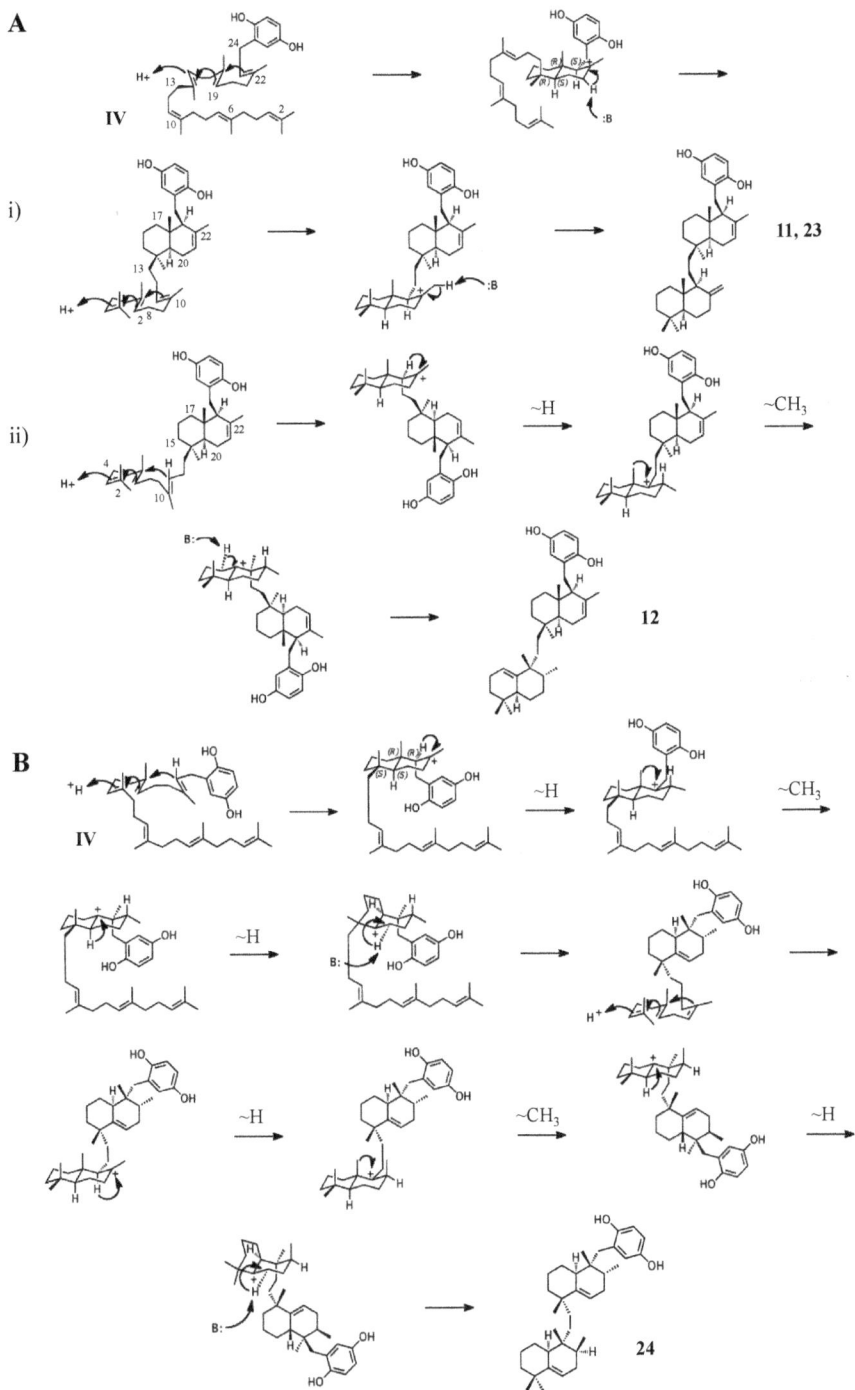

Figure 7. *Cont.*

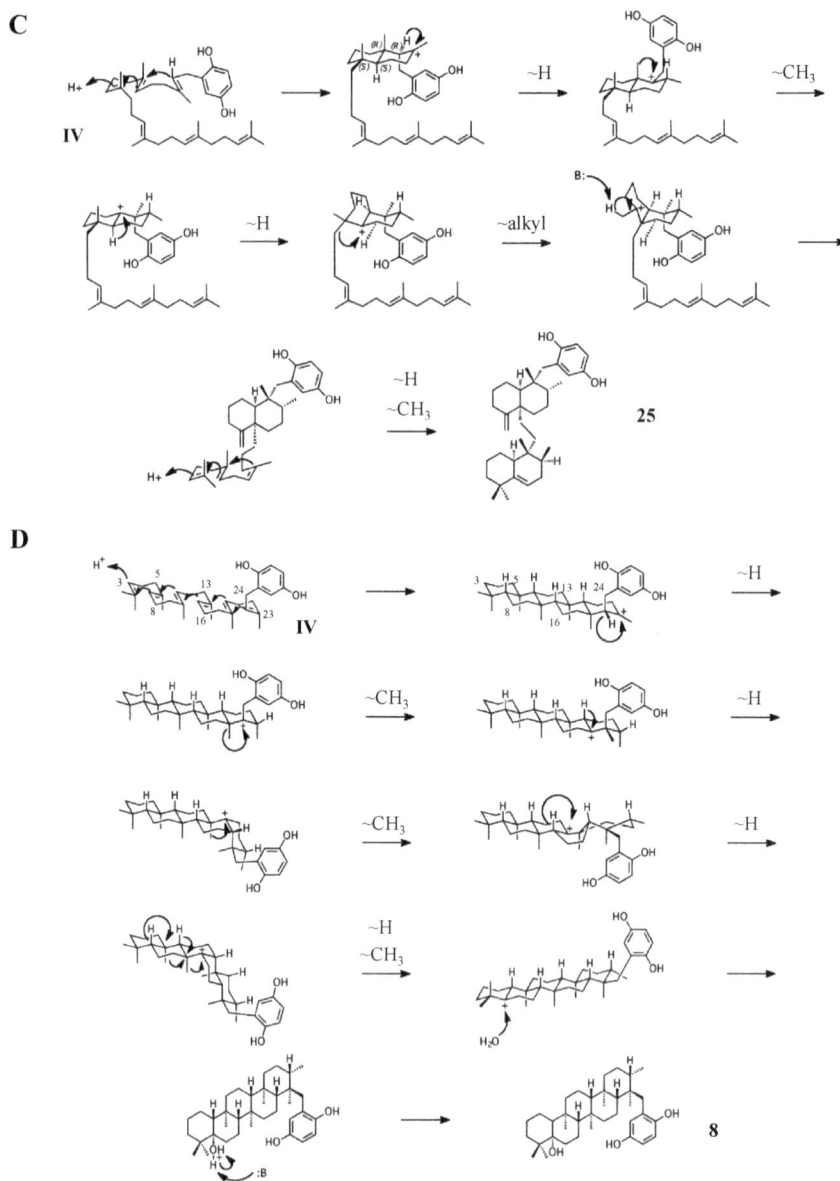

Figure 7. Putative cyclization routes of group IV sponge merotriterpenoids, derived from hexaprenyl diphosphate, including: (**A**) **11**, **12**, adociaquinol (**23**), (**B**) toxiusol (**24**), (**C**) akaterpin (**25**), and (**D**) **8**.

3. Considerations of the enzymatic origin of sponge merotriterpenoids

Only a few key biosynthetic steps are required for all four groups of merotriterpenoids described above: aromatic prenylation, proton-initiated cyclization, and sulfation. Epoxidation also occurs for the majority of these compounds, with the exception of group IV. The potential enzyme families responsible for these key steps of adociasulfate construction are discussed in this section. The source of the terpene and benzoquinone precursors is also considered, as these metabolites can be derived from multiple routes and the enzymes involved in their synthesis may be components of an adociasulfate biosynthetic gene cluster. In addition to the enzymatic origins of sponge merotriterpenoids, the identity of the producing organism is taken into account, as this will dramatically affect the genetic organization of the pathway.

3.1. Origin of precursors

The majority of the adociasulfate structure is constructed of five-carbon isoprene units. There are two known biosynthetic pathways for isoprene production: the mevalonate (MEV) pathway, which provides the precursors for steroids in eukaryotes but is also present in some bacteria, and the 1-deoxy-D-xylulose-5 phosphate (MEP/non-mevalonate) pathway unique to plants, bacteria, and some parasites. Both of these are considered primary metabolic pathways. It is possible that the adociasulfate pathway draws IPP directly from an endogenous metabolite pool and lacks any dedicated genes for IPP/DMAPP synthesis. However, the producing organism's native isoprene source does not necessarily imply that pathway's involvement in secondary metabolism. Bacteria normally lacking the MEV pathway are known to incorporate horizontally acquired MEV pathway genes into meroterpenoid biosynthetic clusters as a pathway-specific source of IPP/DMAPP [30–39]. Some MEP pathway bacteria contain duplications of MEP genes in secondary metabolite clusters [40,41]. The role of these seemingly redundant genes may be to enhance production of precursor metabolites or to establish regulation of early steps in the pathway. Thus, copies of MEV or MEP pathway genes might be involved in meroterpenoid production. However, as there is no evidence to suggest one isoprene pathway being involved over the other, adociasulfate pathway identification should focus on the biosynthetic steps unique to merotriterpenoids. The presence of isoprene pathway elements should be considered a secondary indication of a terpene pathway.

The adociasulfate prenyl donor, consisting of six isoprene units, is almost certainly a product of a *trans* isoprenyl diphosphate synthase. Isoprenyl diphosphate synthases are soluble, Mg^{2+}-dependent prenyltransferases (PTases) mechanistically related to aromatic UbiA-like PTases [42,43]. These enzymes are responsible for producing prenyl diphosphates of different lengths for various biological functions, including polyprenyl diphosphates of 30–50 carbons used in ubiquinone and menaquinone biosynthesis, and the FPP used to make squalene in steroid biosynthesis. Isoprenyl diphosphate synthases are sometimes components of meroterpenoid gene clusters [30–35,38,40]. Their inclusion in secondary metabolite pathways may reflect a selection mechanism for a particular length polyprenyl substrate, establishing a distinct substrate pool for meroterpenoid biosynthesis separate from the endogenous IPP pool. However, native isoprenyl diphosphate synthases are likely capable of providing the prenyl substrate for secondary metabolism.

Like prenyl diphosphates, quinones can also derived from primary metabolic pathways like the phenylalanine/tyrosine pathway, from which hydroquinone and 4-hydroxyphenylacetate (4HPA), a potential pre-hydroxylation precursor of the 5'-glycolic acid substituted adociasulfates, can be derived (Figure 8). 4HPA may be derived from 4-hydroxyphenylpyruvate (4HPP), a product of tyrosine degradation. Oxidative decarboxylation, such as that catalyzed by 4-hydroxyphenylpyruvate (4HPP) dioxygenase, an Fe^{2+}-dependent internal ketoacid dioxygenase, could be used to generate 4HPA from 4HPP [44]. Alternatively, 4HPA could potentially be obtained from 4HPP via 4HPA decarboxylase, such as the enzyme of *Clostridium difficile* that produces *p*-cresol from 4HPA and is a member of the glycyl radical enzymes (GRE) of the radical-SAM superfamily [45]. In a less direct route, decarboxylation of 4HPP to the aldehyde with subsequent oxidation to 4HPA by either an aldehyde dehydrogenase (ALDH) or an aldehyde oxidase (AOX) could be possible [46,47]. Both the NAD(P)$^+$-dependent ALDHs and flavin-dependent molybdenum/tungsten AOXs are described as broad-substrate and are largely uncharacterized. Subsequent hydroxylation of the 4HPA acyl side-chain could be carried out by an α-ketoglutarate-dependent Fe^{2+} enzyme or a cytochrome P450 (P450) [48,49]. 4HPA could also enter into the homogentisate pathway, where hydroquinone could be obtained from homogentisate in a few enzymatic steps [50,51]. Hydroquinone could be derived from gentisate by decarboxylation, potentially requiring a nonoxidative decarboxylase like 5-carboxyvanillate or γ-resorcylate decarboxylase, both members of the ACMSD decarboxylase family [52–54]. Oxidative decarboxylation of aromatic substrates can also be carried out by flavin monooxygenases (FMOs) [55]. Though it is unclear whether tyrosine metabolism factors into meroterpenoid biosynthesis, enzymes similar to these are capable of supplying the prenyl acceptor.

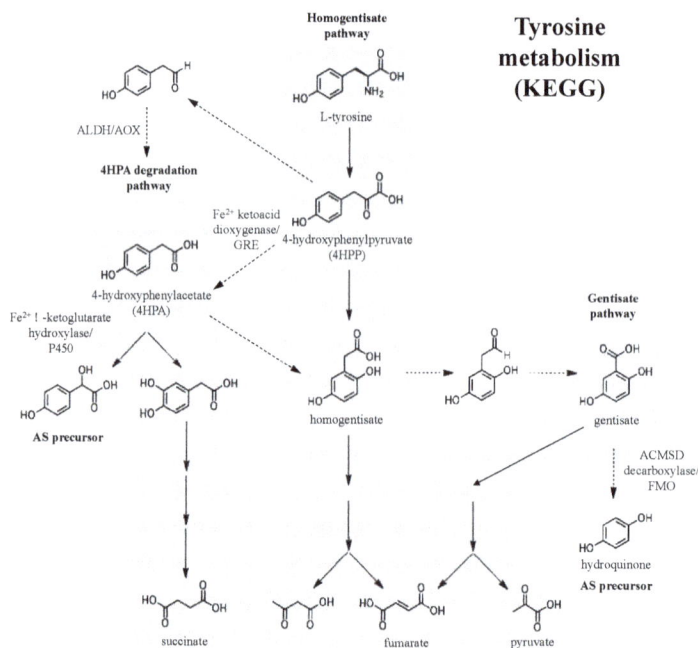

Figure 8. Potential biosynthetic origins of the aromatic adociasulfate prenyl acceptor from tyrosine metabolism pathways, as mapped by the Kyoto Encyclopedia of Genes and Genomes (KEGG). Dashed arrows indicate possible or uncharacterized enzymatic transformations.

The majority of meroterpenoid pathways contain genes responsible for providing or modifying existing aromatic precursors, but these genes represent a variety of distinct biosynthetic routes. Hydroquinone prenyl acceptors of known meroterpenoid pathways are derived primarily from polyketides [31,32,34,38,56–60], but can also be derived from tyrosine [41,61], and from the carbohydrate sedoheptulose 7-phosphate [33]. Another possibility is that the prenyl acceptor is extensively modified after the initial prenylation event, as is the case in ubiquinone synthesis. 4-hydroxybenzoate (4HB) and homogentisate, similar in structure to 4HPA and hydroquinone, are known prenyl acceptors in the ubiquinone and plastoquinone/tocopherol pathways, respectively [62,63]. Prenyl-4HB/homogentisate could be decarboxylated and then hydroxylated to generate the precursor of adociasulfate cyclization. From the examples described here, merotriterpenoids are likely to include specific genes devoted to hydroquinone synthesis.

3.2. Prenylation

Prenyltransferase is the first true step in adociasulfate biosynthesis. A variety of aromatic prenyltransferases (PTases) are known to generate products resembling the linear adociasulfate precursors shown in Figure 2. The earliest to be characterized of these enzymes is 4HB-PTase, which is involved in ubiquinone biosynthesis [62,64]. 4HB-PTases are present in all forms of life, as ubiquinone is an essential component of biological redox reactions like the electron transport chain. The mechanism of prenyl transfer by UbiA, the 4HB-PTase of *E. coli*, involves activation of the isoprene diphosphate to form a carbocation, initiating the electrophilic addition to 4HB in a Friedel-Crafts type alkylation [65,66]. UbiA and related PTases are broadly substrate selective in vitro, especially with regard to the length of isoprenes that can be incorporated into their product [62,67,68]. UbiA also exhibits broad substrate specificity for prenyl acceptors, provided that these substrates are *para*-alcohol- or amino-substituted benzoates [69]. In fact, membrane-associated aromatic PTases utilize a wide variety of aromatic prenyl acceptors in the biosynthesis of plastoquinones/tocopherols, menaquinone, and even secondary metabolites; a testament to their vast biosynthetic potential [70]. It is likely, owing in particular to their accommodation of variable isoprene chain lengths, that membrane aromatic PTases are involved in sponge meroterpenoid biosynthesis.

Prenylation is not unique to the UbiA-like PTases, however, and could be accomplished by other enzyme families. The ABBA-family of aromatic PTases, so named for their alternating, antiparallel α -β-β-α folds (dubbed the PT-fold or PT-barrel), are soluble aromatic prenyltransferases involved in secondary metabolism of bacterial and fungal natural products [71,72]. Though ABBA PTases are broadly selective with regard to the aromatic prenyl acceptor, they are restricted in the length of the prenyl donor to two or fewer isoprene units. Only one ABBA PTase is known to accept FPP as a prenyl donor [30]. Despite the significant role of ABBA PTases in secondary metabolism, the comparison between PTase families better supports the idea that a membrane-associated PTase is involved in sponge meroterpenoid biosynthesis.

3.3. Cyclization

Cyclization of triterpenes is an electrophilic reaction catalyzed by class II terpene cyclases. Class II triterpene cyclases of the bacterial squalene-hopane cyclase (SHC) and eukaryotic oxidosqualene-lanosterol cyclase (OSC) families are known for both their broad substrate selectivity and their extreme product diversity in vitro [73–76]. This product diversity is related to the proton-initiated mechanism of cyclization. Carbocation-mediated rearrangements occur frequently, and similar substrates may be cyclized differently depending on where they are protonated, which depends on both substrate and enzyme and shape. Despite this, cyclization is a highly stereospecific reaction, resulting in characteristic configurations about the chiral bridgehead and methyl-substituted carbons. The fit of the substrate within the cyclase active site likely plays a large role in determining the arrangement of the rings in the final product. Many adociasulfates display sterol-like stereochemistry within rings A-C, indicative of the "prechair" conformation assumed by group I and II adociasulfates prior to cyclization that is characteristic of both sterol and hopene cyclizations (Figure 4) [73]. Group III and IV adociasulfates exhibit bicyclic skeletons, which are also products of SHC/OSCs in vitro [73,76]. As sponge merotriterpenoids display features characteristic of proton-initiated cyclization, including complex rearrangements and substrate-dependent patterns of carbon skeletons (Figure 2), an SHC- or OSC-like cyclase is likely involved in their synthesis.

Class II terpene cyclases do not utilize linear meroterpenoid substrates in nature, but are nonetheless capable of performing the chemistry required of a putative adociasulfate cyclase. Both SHCs and OSCs can cyclize linear hydroquinone meroterpenoids in vitro [75,77–79]. In these examples, SHCs are able to cyclize the prenyl side chain of the linear meroterpenoid substrate, but their products lack fusion of the aromatic moiety to the terpene ring system [77,79]. The OSC lupeol synthase (LUP1) from *Arabidopsis thaliana*, however, is capable of fusing the aromatic indole ring of its epoxide substrate to the prenyl side chain [78]. This is similar to the epoxide-dependent aromatic ring fusion observed for group I and II adociasulfates (Figures 4 and 5). This would suggest that the presence and position of epoxides determine which type of skeleton will be formed. In this way, a single class II terpene cyclase could be responsible for the production of all sponge merotriterpenoids. In an example of substrate-dependence on cyclization, tetraprenyl-β-curcumene cyclase of *Bacillus subtilis* is capable of utilizing both a linear, monocyclic C_{35} terpenoid substrate to generate a fused four-ring skeleton strongly resembling group I and II adociasulfates, and squalene to produce a fused bicyclic drimane-like skeleton similar to group III and most group IV adociasulfates [80]. In this case, the structural differences between the linear, head-to-tail tetraprenyl-β-curcumene and the tail-to-tail squalene direct the outcome of the cyclization event. Adociasulfate cyclizations sometimes involve heterocycle formation, presumably involving hydroxyls produced by epoxide ring openings. SHCs are capable of heterocycle formation in this way [81]. In general, SHCs exhibit greater substrate flexibility than OSCs and can accept a variety of terpene substrates in vitro, including 2,3-oxidosqualene [73,82]. Thus, it is likely that a bacterial SHC-like enzyme is responsible for adociasulfate cyclization.

Though one could envision the adociasulfate biosynthetic pathway containing an SHC-like terpene cyclase, natural product pathways often include atypical enzymes capable of performing similar chemistry rather than the more recognizable class II terpene cyclases. For example, several fungal indole meroterpenoid pathways utilize a novel family of small, membrane-bound meroterpenoid cyclases (MTCs) capable of proton-initiated cyclization [83]. One of these enzymes, PaxB, has been shown to cyclize doubly epoxidized substrates similar to those predicted for group II sponge merotriterpenoids [84]. The resulting compound, paspaline, is remarkably similar to adociasulfates in that it, too, contains a heterocycle formed after an initial epoxide opening cyclization event, using the resulting hydroxyl group in the second cyclization reaction. MTCs have thus far only been reported to cyclize merosesquiterpenoid and meroditerpenoids, but it appears plausible that such enzymes could catalyze longer cyclizations, such as that predicted for **8** (Figure 7D).

3.4. Epoxidation

Epoxidation of squalene in eukaryotes is carried out by squalene monooxygenase (SM), a membrane-bound flavin-dependent protein that requires molecular oxygen and reduced NADPH, as well as a P450 reductase partner [85]. The requirement for a P450 reductase is unique to SM amongst FMOs, as there is no structural relationship between SM and P450s, but several groups of FMOs are known to require other flavin reductase partners [55]. There is evidence that a second, non-P450 type flavin reductase may be also be able to supply reduced NADPH to SM [86]. There is also a precedent for SM in secondary metabolism. The diterpene phenalinolactone, produced by a *Streptomyces* strain, includes an SM homolog in its biosynthetic gene cluster [87]. This SM homolog is believed to introduce an epoxide at the terminal olefin of the C20 geranylgeranyl diphosphate substrate. SM produces a single isomer of oxidosqualene, introducing an oxirane ring at the terminal 2,3-alkene in the (S) configuration. However, due to the rigid specificity of SM for terminal olefins it is more likely that an unrelated monooxygenase is involved in sponge merotriterpenoid epoxidation. For example, non-SM FMOs related to oxidative genes of the ubiquinone pathway have been identified in fungal indole terpenoid gene clusters, such as that of xiamycin [2,88]. Additionally, P450 monooxygenases are involved in oxidative tailoring reactions in numerous natural product pathways and are capable of performing a wide variety of chemical modifications on diverse substrates, including epoxidation. All P450s obtain reduced flavin via a P450 reductase partner, similar to SM [89]. Owing to their incredible diversity in both function and substrate specificity, either FMOs or P450s are a more likely candidate for epoxidation than SM in the adociasulfate pathway.

3.5. Sulfation

The final step in the synthesis of adociasulfates is sulfation of the hydroquinone moiety. In eukaryotes, sulfation is carried out by sulfotransferases (SULTs) that utilize 3′-phosphoadenosine 5′-phosphosulfate (PAPS) as a sulfonate (SO_3^-) donor. Though SULTs are less prevalent in bacteria than in eukaryotes, sulfation has been incorporated into secondary metabolism. SULT domains have been identified within polyketide synthases to generate sulfated products, or, in one case, sulfation activates a substrate for decarboxylation [90]. The role of sulfation in adociasulfate activity can only be guessed, as the native biological function of adociasulfates is not known. However, with regard to kinesin, the sulfates only prevent membrane penetration and do not affect inhibition [12,91].

Sulfation could be a mechanism for elimination from the sponge to avoid toxicity associated with kinesin inhibition, or it could enhance secretion to facilitate exposure to predators. Not all sponge merotriterpenoids are sulfated, however, but these compounds have not been tested for kinesin inhibition [5,6]. It has been suggested that an analog of **14** containing an esterified glycolic acid moiety and lacking sulfation might be membrane permeable and still inhibit kinesin, making it a good anticancer lead [12]. **21** and haliclotriol triacetate closely resemble this hypothetical analog and should be screened for kinesin inhibition [6]. Nonetheless, sulfation is not essential for adociasulfate biosynthesis, and the genes involved need not reside in the same gene cluster or even the same genome as the rest of the pathway. While a microbial symbiont may produce adociasulfates, the host could be responsible for their sulfation.

4. Concluding remarks

To date, adociasulfates remain the only known natural product kinesin inhibitors that compete with MTs for binding. Until recently adociasulfates were thought to form MT-mimicking aggregates, bringing into question their potential as drugs or mechanistic probes [92]. It is now understood that adociasulfates bind kinesin in a 1:1 interaction [12]. In light of these findings, it is crucial to point out the unlikeliness of kinesin inhibitors RBL and NSC 622124 to behave as expected in biochemical or cell-based investigations. Adociasulfates are the only experimentally validated inhibitors to compete with MTs for binding kinesin at a single-molecule level. Thus, there exists some urgency to achieve sustainable adociasulfate production.

A general biogenetic hypothesis can be made based on the proposed structural relationships between hydroquinone merotriterpenoids (Figure 9A). Proton-initiated cyclization from variable substrates, including non-epoxides, suggests a squalene-hopene cyclase. The positions of the epoxides in the linear precursors suggest that an FMO or P450 may be responsible. The head-to-tail linear triterpene precursor supports the idea that a polyprenyl synthase supplies the precursor of cyclization. These observations, which encompass the more distinct features of the adociasulfate structure, imply a bacterial origin (Figure 9B). This biogenetic hypothesis is supported by the recent discovery of a meroterpenoid pathway from marine cyanobacteria, which are often involved in symbioses with sponges and other invertebrates [93]. The merosterols are meroditerpenoids that greatly resemble adociasulfates. Their biosynthetic pathway incorporates elements of the MEP pathway for isoprene production, and both carbohydrate and tyrosine metabolism for synthesis of the PHB moiety. An UbiA-like PTase and SHC-like cyclase were shown to generate a cyclized meroterpenoid product, and several genes encoding oxidative proteins are present, including two P450s, presumably to introduce modifications to the aromatic ring. Despite these similarities, no biosynthetic pathways for sponge meroterpenoids have ever been identified for comparison. In only one case has a producing organism been claimed to have been identified—for the production of avarol, a merosesquiterpenoid, by the sponge *Dysidea avara*. In these studies, avarol was traced to a specific sponge cell type and production was later observed from an axenic primary sponge culture [94,95]. However, no publications have followed these studies in nearly 18 years. Thus, while the possibility exists that adociasulfates and related meroterpenoids are sponge-derived, or that merosesquiterpenoid biosynthesis may differ substantially with that of merotriterpenoids, the biosynthetic origin of sponge merotriterpenoids that is most consistent with their structure is bacterial.

Though the structure of adociasulfates favors symbiont- over host-derived production, no clear verdict can be reached without experimental investigation. Clues as to what types of enzymes are responsible have been described here. Targeted searches of genes with these functions could help to identify the adociasulfate pathway. Metagenomic approaches may complicate data interpretation in that several to hundreds of gene homologs may be identified within a single metagenome, especially for those genes related to primary pathways, such as *ubiA*. A comparative metagenomics approach may resolve these issues, in which the metagenomes of nonproducing *Chalinidae* sponges are sequenced alongside adociasulfate-producing specimens. Care must be taken to collect and prepare separate samples for analysis by chemical and DNA sequencing approaches. Following the guidelines for pathway identification laid out in this review may result in successful recognition of a meroterpenoid pathway, paving the way for biosynthetic approaches to solve the supply problem that surrounds these valuable compounds. More importantly, a thorough analysis of compound structure can reveal valuable information regarding the compound's origin. This strategy can be used as a general approach in the discovery of natural product biosynthetic pathways.

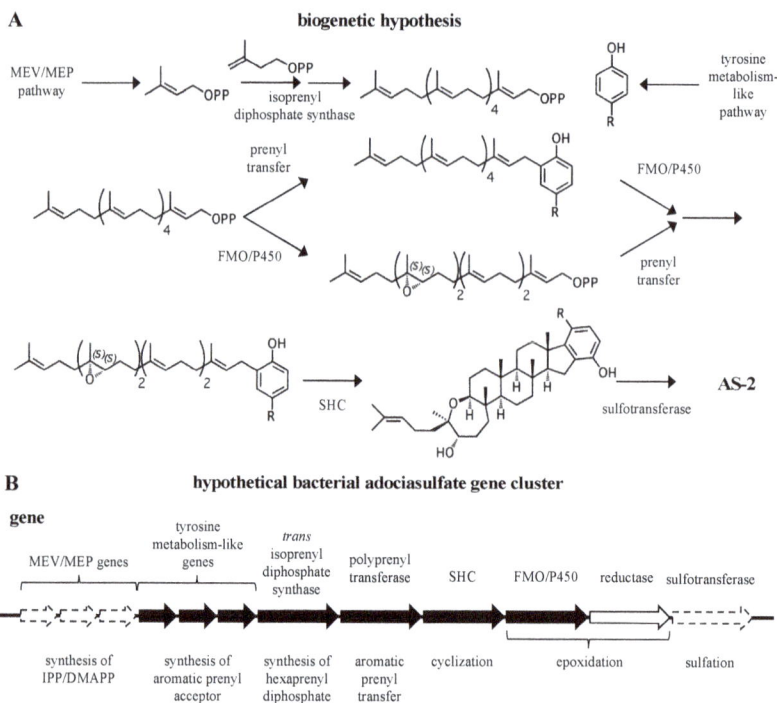

Figure 9. A biogenetic hypothesis for the adociasulfates. (**A**) A biosynthetic scheme summarizing the proposed biogenetic hypothesis for the origin of adociasulfates. (**B**) A hypothetical adociasulfate gene cluster was constructed based on the most probable biosynthetic origin, as addressed in this review. In this scenario, the pathway is assumed to be part of a bacterial genome. Black genes represent those directly involved in biosynthesis, white genes are those indirectly involved in biosynthesis, and those bordered with a dashed line have the potential to be entirely absent from the cluster.

Acknowledgments: I thank Drs. Eric Schmidt, Chris Ireland, and Glenn Prestwich for their advice and critiques regarding this manuscript. This work was supported by the American Foundation for Pharmaceutical Education (AFPE) Predoctoral Fellowship.

Conflicts of Interest: The author declares no conflict of interest.

References

1. Menna, M.; Imperatore, C.; D'Aniello, F.; Aiello, A. Meroterpenes from marine invertebrates: Structures, occurrence, and ecological implications. *Mar. Drugs* **2013**, *11*, 1602–1643. [CrossRef] [PubMed]
2. Matsuda, Y.; Abe, I. Biosynthesis of fungal meroterpenoids. *Nat. Prod. Rep.* **2015**, *33*, 26–53. [CrossRef] [PubMed]
3. Loya, S.; Tal, R.; Hizi, A. Hexaprenoid hydroquinones, novel inhibitors of the reverse transcriptase of human immunodeficiency virus type 1. *J. Nat. Prod.* **1993**, *56*, 2120–2125. [CrossRef] [PubMed]

4. Fukami, A.; Ikeda, Y.; Kondo, S.; Naganawa, H.; Takeuchi, T.; Furuya, S.; Hirabayashi, Y.; Shimoike, K.; Hosaka, S.; Watanabe, Y.; et al. Akaterpin, a Novel Bioactive Triterpene from the Marine Sponge *Callyspongia sp. Tetrahedron Lett.* **1997**, *38*, 1201–1202. [CrossRef]

5. Williams, D.E.; Steino, A.; de Voogd, N.J.; Mauk, A.G.; Andersen, R.J. Halicloic acids A and B isolated from the marine sponge *Haliclona sp.* collected in the Philippines inhibit indoleamine 2,3-dioxygenase. *J. Nat. Prod.* **2012**, *75*, 1451–1458. [CrossRef] [PubMed]

6. Crews, P.; Harrison, B. New triterpene-ketides (merotriterpenes), haliclotriol A and B, from an Indo-Pacific *Haliclona* sponge. *Tetrahedron* **2000**, *56*, 9039–9046. [CrossRef]

7. Isaacs, S.; Hizi, A.; Kashman, Y. Toxicols A-C and Toxiusol—New Bioactive Hexaprenoid Hydroquinones from *Toxiclona toxius. Tetrahedron* **1993**, *49*, 4275–4282. [CrossRef]

8. Blackburn, C.L.; Faulkner, D.J. Adociasulfate 10, a new merohexaprenoid sulfate from the sponge *Haliclona* (aka *Adocia*) *sp. Tetrahedron* **2000**, *56*, 8429–8432. [CrossRef]

9. Blackburn, C.L.; Hopmann, C.; Sakowicz, R.; Berdelis, M.S.; Goldstein, L.S.B.; Faulkner, D.J. Adociasulfates 1–6, inhibitors of kinesin motor proteins from the sponge *Haliclona* (aka *Adocia*) *sp. J. Org. Chem.* **1999**, *64*, 5565–5570. [CrossRef] [PubMed]

10. Kalaitzis, J.A.; Leone, P.; Harris, L.; Butler, M.S.; Ngo, A.; Hooper, J.N.A.; Quinn, R.J. Adociasulfates 1, 7, and 8: New bioactive hexaprenoid hydroquinones from the marine sponge *Adocia sp. J. Org. Chem.* **1999**, *64*, 5571–5574. [CrossRef] [PubMed]

11. Sakowicz, R.; Berdelis, M.S.; Ray, K.; Blackburn, C.L.; Hopmann, C.; Faulkner, D.J.; Goldstein, L.S.B. A marine natural product inhibitor of kinesin motors. *Science* **1998**, *280*, 292–295. [CrossRef] [PubMed]

12. Smith, T.E.; Hong, W.; Zachariah, M.M.; Harper, M.K.; Matainaho, T.K.; Van Wagoner, R.M.; Ireland, C.M.; Vershinin, M. Single-molecule inhibition of human kinesin by adociasulfate-13 and -14 from the sponge *Cladocroce aculeata. Proc. Natl. Acad. Sci. USA* **2013**, *110*, 18880–18885. [CrossRef] [PubMed]

13. Hopkins, S.C.; Vale, R.D.; Kuntz, I.D. Inhibitors of kinesin activity from structure-based computer screening. *Biochemistry* **2000**, *39*, 2805–2814. [CrossRef] [PubMed]

14. Learman, S.S.; Kim, C.D.; Stevens, N.S.; Kim, S.; Wojcik, E.J.; Walker, R.A. NSC 622124 Inhibits Human Eg5 and Other Kinesins via Interaction with the Conserved Microtubule-Binding Site. *Biochemistry* **2009**, *48*, 1754–1762. [CrossRef] [PubMed]

15. McGovern, S.L.; Caselli, E.; Grigorieff, N.; Shoichet, B.K. A Common Mechanism Underlying Promiscuous Inhibitors from Virtual and High-Throughput Screening. *J. Med. Chem.* **2002**, *45*, 1712–1722. [CrossRef] [PubMed]

16. Narasimhan, K.; Pillay, S.; Bin Ahmad, N.R.; Bikadi, Z.; Hazai, E.; Yan, L.; Kolatkar, P.R.; Pervushin, K.; Jauch, R. Identification of a polyoxometalate inhibitor of the DNA binding activity of Sox2. *ACS Chem. Biol.* **2011**, *6*, 573–581. [CrossRef] [PubMed]

17. Prudent, R.; Moucadel, V.; Laudet, B.; Barette, C.; Lafanechere, L.; Hasenknopf, B.; Li, J.; Bareyt, S.; Lacote, E.; Thorimbert, S.; et al. Identification of polyoxometalates as nanomolar noncompetitive inhibitors of protein kinase CK2. *Chem. Biol.* **2008**, *15*, 683–692. [CrossRef] [PubMed]

18. Shah, J.V.; Flanagan, L.A.; Janmey, P.A.; Leterrier, J.-F. Bidirectional Translocation of Neurofilaments along Microtubules Mediated in Part by Dynein/Dynactin. *Mol. Biol. Cell* **2000**, *11*, 3495–3508. [CrossRef] [PubMed]

19. Dictenberg, J.B.; Swanger, S.A.; Antar, L.N.; Singer, R.H.; Bassell, G.J. A direct role for FMRP in activity-dependent dendritic mRNA transport links filopodial-spine morphogenesis to fragile X syndrome. *Dev. Cell* **2008**, *14*, 926–939. [CrossRef] [PubMed]

20. Ewald, A.; Zünkler, C.; Lourim, D.; Dabauvalle, M.-C. Microtubule-dependent assembly of the nuclear envelope in *Xenopus laevis* egg extract. *Eur. J. Cell Biol.* **2001**, *80*, 678–691. [CrossRef]

21. Qiu, D.; Cheng, S.-M.; Wozniak, L.; McSweeney, M.; Perrone, E.; Levin, M. Localization and loss-of-function implicates ciliary proteins in early, cytoplasmic roles in left-right asymmetry. *Dev. Dynam.* **2005**, *234*, 176–189. [CrossRef] [PubMed]

22. Brier, S.; Carletti, E.; DeBonis, S.; Hewat, E.; Lemaire, D.; Kozielski, F. The marine natural product adociasulfate-2 as a tool to identify the MT-binding region of kinesins. *Biochemistry* **2006**, *45*, 15644–15653. [CrossRef] [PubMed]

23. Bogenstätter, M.; Limberg, A.; Overman, L.E.; Tomasi, A.L. Enantioselective Total Synthesis of the Kinesin Motor Protein Inhibitor Adociasulfate 1. *J. Am. Chem. Soc.* **1999**, *121*, 12206–12207. [CrossRef]

24. Darne, C.P. Kinesin Motor Protein Inhibitors: Toward the Synthesis of Adociasulfate Analogs. Master's Thesis, The University of Georgia, Athens, GA, USA, May 2005.

25. Erben, F.; Specowius, V.; Wölfling, J.; Schneider, G.; Langer, P. Benzo-Annulated Steroids: Synthesis of Octahydro-indeno-phenanthrenes by Formal [3+3] Cyclocondensation Reaction with 1,3-Bis[(trimethylsilyl)oxy]buta-1,3-dienes. *Helv. Chim. Acta* **2013**, *96*, 924–930. [CrossRef]

26. Hosoi, H.; Kawai, N.; Hagiwara, H.; Suzuki, T.; Nakazaki, A.; Takao, K.-I.; Umezawa, K.; Kobashi, S. Determination of the Absolute Structure of (+)-Akaterpin. *Chem. Pharm. Bull.* **2012**, *60*, 137–143. [CrossRef] [PubMed]

27. Kalaitzis, J.A.; Quinn, R.J. Adociasulfate-9, a new hexaprenoid hydroquinone from the Great Barrier Reef sponge *Adocia aculeata*. *J. Nat. Prod.* **1999**, *62*, 1682–1684. [CrossRef]

28. Isaacs, S.; Kashman, Y. Shaagrockol B and C; Two Hexaprenylhydroquinone disulfates from the Red Sea Sponge *Toxiclona toxius*. *Tetrahedron Lett.* **1992**, *33*, 2227–2230. [CrossRef]

29. West, L.M.; Faulkner, D.J. Hexaprenoid Hydroquinones from thr Sponge *Haliclona* (aka *Adocia*) sp. *J. Nat. Prod.* **2006**, *69*, 1001–1004. [CrossRef] [PubMed]

30. Bonitz, T.; Zubeil, F.; Grond, S.; Heide, L. Unusual N-prenylation in diazepinomicin biosynthesis: The farnesylation of a benzodiazepine substrate is catalyzed by a new member of the ABBA prenyltransferase superfamily. *PLoS ONE* **2013**, *8*, e85707. [CrossRef] [PubMed]

31. Haagen, Y.; Gluck, K.; Fay, K.; Kammerer, B.; Gust, B.; Heide, L. A gene cluster for prenylated naphthoquinone and prenylated phenazine biosynthesis in *Streptomyces cinnamonensis* DSM 1042. *ChemBioChem* **2006**, *7*, 2016–2027. [CrossRef] [PubMed]

32. Kawasaki, T.; Hayashi, Y.; Kuzuyama, T.; Furihata, K.; Itoh, N.; Seto, H.; Dairi, T. Biosynthesis of a natural polyketide-isoprenoid hybrid compound, furaquinocin A: Identification and heterologous expression of the gene cluster. *J. Bacteriol.* **2006**, *188*, 1236–1244. [CrossRef] [PubMed]

33. Kawasaki, T.; Kuzuyama, T.; Furihata, K.; Itoh, N.; Seto, H.; Dairi, T. A Relationship between the Mevaolnate Pathway and Isoprenoid Production in Actinomycetes. *J. Antibiot.* **2003**, *56*, 957–966. [CrossRef] [PubMed]

34. Kuzuyama, T.; Noel, J.P.; Richard, S.B. Structural basis for the promiscuous biosynthetic prenylation of aromatic natural products. *Nature* **2005**, *435*, 983–987. [CrossRef] [PubMed]

35. McAlpine, J.B.; Banskota, A.H.; Charan, R.D.; Schlingmann, G.; Zazopoulos, E.; Piraee, M.; Janso, J.; Bernan, V.S.; Aouidate, M.; Farnet, C.M.; et al. Biosynthesis of Diazepinomicin/ECO-4601, a *Micromonospora* Secondary Metabolite with a Novel Ring System. *J. Nat. Prod.* **2008**, *71*, 1585–1590. [CrossRef] [PubMed]

36. Saikia, S.; Nicholson, M.J.; Young, C.; Parker, E.J.; Scott, B. The genetic basis for indole-diterpene chemical diversity in filamentous fungi. *Mycol. Res.* **2008**, *112*, 184–199. [CrossRef] [PubMed]

37. Saleh, O.; Gust, B.; Boll, B.; Fiedler, H.P.; Heide, L. Aromatic prenylation in phenazine biosynthesis: Dihydrophenazine-1-carboxylate dimethylallyltransferase from *Streptomyces anulatus*. *J. Biol. Chem.* **2009**, *284*, 14439–14447. [CrossRef] [PubMed]

38. Winter, J.M.; Moffitt, M.C.; Zazopoulos, E.; McAlpine, J.B.; Dorrestein, P.C.; Moore, B.S. Molecular basis for chloronium-mediated meroterpene cyclization: Cloning, sequencing, and heterologous expression of the napyradiomycin biosynthetic gene cluster. *J. Biol. Chem.* **2007**, *282*, 16362–16368. [CrossRef] [PubMed]

39. Zeyhle, P.; Bauer, J.S.; Kalinowski, J.; Shin-ya, K.; Gross, H.; Heide, L. Genome-based discovery of a novel membrane-bound 1,6-dihydroxyphenazine prenyltransferase from a marine actinomycete. *PLoS ONE* **2014**, *9*, e99122. [CrossRef] [PubMed]

40. Hillwig, M.L.; Zhu, Q.; Liu, X. Biosynthesis of ambiguine indole alkaloids in cyanobacterium *Fischerella ambigua*. *ACS Chem. Biol.* **2014**, *9*, 372–377. [CrossRef] [PubMed]

41. Steffensky, M.; Mühlenweg, A.; Wange, Z.-X.; Li, S.-M.; Heide, L. Identification of the Novobiocin Biosynthetic Gene Cluster of *Streptomyces spheroides* NCIB 11891. *Antimicrob. Agents Chemother.* **2000**, *44*, 1214–1222. [CrossRef] [PubMed]

42. Heide, L. Prenyl transfer to aromatic substrates: Genetics and enzymology. *Curr. Opin. Chem. Biol.* **2009**, *13*, 171–179. [CrossRef] [PubMed]

43. Liang, P.-H.; Ko, T.-P.; Wang, A.H.-J. Structure, mechanism and function of prenyltransferases. *Eur. J. Biochem.* **2002**, *269*, 3339–3354. [CrossRef] [PubMed]

44. Crouch, N.P.; Adlington, R.M.; Baldwin, J.E.; Lee, M.-H.; MacKinnon, C.H. A Mechanistic Rationalism for the Substrate Specificity of Recombinant Mammalian 4-Hydroxyphenylpyruvate Dioxygenase (4-HPPD). *Tetrahedron* **1997**, *53*, 6993–7010. [CrossRef]

45. Selmer, T.; Andrei, P.I. *p*-Hydroxyphenylacetate decarboxylase from *Clostridium difficile. Eur. J. Biochem.* **2001**, *268*, 1363–1372. [CrossRef] [PubMed]
46. Yoshida, A.; Rzhetsky, A.; Hsu, L.C.; Chang, C. Human aldehyde dehydrogenase gene family. *Eur. J. Biochem.* **1998**, *251*, 549–557. [CrossRef] [PubMed]
47. Garrattini, E.; Fratelli, M.; Terao, M. The mammalian aldehyde oxidase gene family. *Hum. Genom.* **2009**, *4*, 119–130. [CrossRef]
48. Prescott, A.G.; Lloyd, M.D. The iron(II) and 2-oxoacid-dependent dioxygenases and their role in metabolism (1967 to 1999). *Nat. Prod. Rep.* **2000**, *17*, 367–383. [CrossRef] [PubMed]
49. Urlacher, V.B.; Girhard, M. Cytochrome P450 monooxygenases: An update on perspectives for synthetic application. *Trends. Biotechnol.* **2012**, *30*, 26–36. [CrossRef] [PubMed]
50. Arias-Barrau, E.; Olivera, E.R.; Luengo, J.M.; Fernandez, C.; Galan, B.; Garcia, J.L.; Diaz, E.; Minambres, B. The homogentisate pathway: A central catabolic pathway involved in the degradation of L-phenylalanine, L-tyrosine, and 3-hydroxyphenylacetate in *Pseudomonas putida. J. Bacteriol.* **2004**, *186*, 5062–5077. [CrossRef] [PubMed]
51. Prieto, M.A.; Díaz, E.; García, J.L. Molecular Characterization of the 4-Hydroxyphenylacetate Catabolic Pathway of *Escherichia coli* W: Engineering a Mobile Aromatic Degradative Cluster. *J. Bacteriol.* **1996**, *178*, 111–120. [CrossRef] [PubMed]
52. Liu, A.; Zhang, H. Transition Metal-Catalyzed Nonoxidative Decarboxylation Reactions. *Biochemistry* **2006**, *45*, 10408–10411. [CrossRef] [PubMed]
53. Peng, X.; Masai, E.; Kitayama, H.; Harada, K.; Katayama, Y.; Fukuda, M. Characterization of the 5-Carboxyvanillate Decarboxylase Gene and Its Role in Lignin-Related Biphenyl Catabolism in *Sphingomonas paucimobilis* SYK-6. *Appl. Environ. Microbiol.* **2002**, *68*, 4407–4415. [CrossRef] [PubMed]
54. Yoshida, M.; Fukuhara, N.; Oikawa, T. Thermophilic, reversible gamma-resorcylate decarboxylase from *Rhizobium* sp. strain MTP-10005: Purification, molecular characterization, and expression. *J. Bacteriol.* **2004**, *186*, 6855–6863. [CrossRef] [PubMed]
55. Huijbers, M.M.; Montersino, S.; Westphal, A.H.; Tischler, D.; van Berkel, W.J. Flavin dependent monooxygenases. *Arch. Biochem. Biophys.* **2014**, *544*, 2–17. [CrossRef] [PubMed]
56. Itoh, T.; Tokunaga, K.; Matsuda, Y.; Fujii, I.; Abe, I.; Ebizuka, Y.; Kushiro, T. Reconstitution of a fungal meroterpenoid biosynthesis reveals the involvement of a novel family of terpene cyclases. *Nat. Chem.* **2010**, *2*, 858–864. [CrossRef] [PubMed]
57. Itoh, T.; Tokunaga, K.; Radhakrishnan, E.K.; Fujii, I.; Abe, I.; Ebizuka, Y.; Kushiro, T. Identification of a key prenyltransferase involved in biosynthesis of the most abundant fungal meroterpenoids derived from 3,5-dimethylorsellinic acid. *Chembiochem* **2012**, *13*, 1132–1135. [CrossRef] [PubMed]
58. Lo, H.C.; Entwistle, R.; Guo, C.J.; Ahuja, M.; Szewczyk, E.; Hung, J.H.; Chiang, Y.M.; Oakley, B.R.; Wang, C.C. Two separate gene clusters encode the biosynthetic pathway for the meroterpenoids austinol and dehydroaustinol in *Aspergillus nidulans. J. Am. Chem. Soc.* **2012**, *134*, 4709–4720. [CrossRef] [PubMed]
59. Matsuda, Y.; Awakawa, T.; Abe, I. Reconstituted biosynthesis of fungal meroterpenoid andrastin A. *Tetrahedron* **2013**, *69*, 8199–8204. [CrossRef]
60. Matsuda, Y.; Wakimoto, T.; Mori, T.; Awakawa, T.; Abe, I. Complete biosynthetic pathway of anditomin: Nature's sophisticated synthetic route to a complex fungal meroterpenoid. *J. Am. Chem. Soc.* **2014**, *136*, 15326–15336. [CrossRef] [PubMed]
61. Pojer, F.; Li, S.-M.; Heide, L. Molecular cloning and sequence analysis of the clorobiocin biosynthetic gene cluster: New insights into the biosynthesis of the aminocoumarin antibiotics. *Microbiology* **2002**, *148*, 3901–3911. [CrossRef] [PubMed]
62. Melzer, M.; Heide, L. Characterization of Polyprenyldiphosphate: 4-Hydroxybenzoate Polyprenyltransferase from *Eschericia coli. Biochim. Biophys. Acta* **1994**, *1212*, 93–102. [CrossRef]
63. Norris, S.R.; Barrette, T.R.; DellaPenna, D. Genetic Dissection of Carotenoid Synthesis in *Arabidopsis* Defines Plastoquinone as an Essential Component of Phytoene Desaturation. *Plant Cell* **1995**, *7*, 2139–2149. [CrossRef] [PubMed]
64. Young, L.G.; Leppik, R.A.; Hamilton, J.A.; Gibson, F. Biosynthesis in *Escherichia coli* K-12: 4-Hydroxybenzoate Octaprenyltransferase. *J. Bacteriol.* **1972**, *110*, 18–25. [PubMed]

65. Bräuer, L.; Brandt, W.; Wessjohann, L.A. Modeling the *E. coli* 4-hydroxybenzoic acid oligoprenyltransferase (*ubiA* transferase) and characterization of potential active sites. *J. Mol. Model.* **2004**, *10*, 317–327. [CrossRef] [PubMed]

66. Huang, H.; Levin, E.J.; Liu, S.; Bai, Y.; Lockless, S.W.; Zhou, M. Structure of a membrane-embedded prenyltransferase homologous to UBIAD1. *PLoS Biol.* **2014**, *12*, e1001911. [CrossRef] [PubMed]

67. El Hachimi, Z.; Samuel, O.; Azerad, R. Biochemical study on ubiquinone biosynthesis in *Escherichia coli*: I. Specificity of para-hydroxybenzoate polyprenyltransferase. *Biochimie* **1974**, *56*, 1239–1247. [CrossRef]

68. Okada, K.; Suzuki, K.; Kamiya, Y.; Zhu, X.; Fujisaki, S.; Nishimura, Y.; Nishino, T.; Nakagawa, T.; Kawamukai, M.; Matsuda, H. Polyprenyl diphosphate synthase essentially defines the length of the side chain of ubiquinone. *Biochim. Biophys. Acta* **1996**, *1302*, 217–223. [CrossRef]

69. Wessjohann, L.; Sontag, B. Prenylation of Benzoic Acid Derivatives Catalyzed by a Transferase from Escherichia coli Overproduction: Method Developments and Substrate Specificity. *Angew. Chem. Int. Ed.* **1996**, *35*, 1697–1699. [CrossRef]

70. Li, W. Bringing Bioactive Compounds into Membranes: The UbiA Superfamily of Intramembrane Aromatic Prenyltransferases. *Trends. Biochem. Sci.* **2016**, *41*, 356–370. [CrossRef] [PubMed]

71. Saleh, O.; Haagen, Y.; Seeger, K.; Heide, L. Prenyl transfer to aromatic substrates in the biosynthesis of aminocoumarins, meroterpenoids and phenazines: The ABBA prenyltransferase family. *Phytochemistry* **2009**, *70*, 1728–1738. [CrossRef] [PubMed]

72. Tello, M.; Kuzuyama, T.; Heide, L.; Noel, J.P.; Richard, S.B. The ABBA family of aromatic prenyltransferases: Broadening natural product diversity. *Cell. Mol. Life Sci.* **2008**, *65*, 1459–1463. [CrossRef] [PubMed]

73. Abe, I.; Rohmer, M.; Prestwich, G.D. Enzymatic Cyclization of Squalene and Oxidosqualene to Sterols and Triterpenes. *Chem. Rev.* **1993**, *93*, 2189–2208. [CrossRef]

74. Abe, I.; Tanaka, H.; Noguchi, H. Enzymatic Formation of an Unnatural Hexacyclic C35 Polyprenoid by Bacterial Squalene Cyclase. *J. Am. Chem. Soc.* **2002**, *124*, 14514–14515. [CrossRef] [PubMed]

75. Hammer, S.C.; Dominicus, J.M.; Syrén, P.-O.; Nestl, B.M.; Hauer, B. Stereoselective Friedel–Crafts alkylation catalyzed by squalene hopene cyclases. *Tetrahedron* **2012**, *68*, 7624–7629. [CrossRef]

76. Lodeiro, S.; Xiong, Q.; Wilson, W.K.; Kolesnikova, M.D.; Onak, C.S.; Matsuda, S.P. An Oxidosqualene Cyclase Makes Numerous Products by Diverse Mechanisms: A Challenge to Prevailing Concepts of Triterpene Biosynthesis. *J. Am. Chem. Soc.* **2007**, *129*, 11213–11222. [CrossRef] [PubMed]

77. Tanaka, H.; Noguchi, H.; Abe, I. Enzymatic Formation of Indole-Containing Unnatural Cyclic Polyprenoids by Bacterial Squalene: Hopene Cyclase. *Org. Lett.* **2005**, *7*, 5873–5876. [CrossRef] [PubMed]

78. Xiong, Q.; Zhu, X.; Wilson, W.K.; Ganesan, A.; Matsuda, S.P. Enzymatic Synthesis of an Indole Diterpene by an Oxidosqualene Cyclase: Mechanistic, Biosynthetic, and Phylogenetic Implications. *J. Am. Chem. Soc.* **2003**, *125*, 9002–9003. [CrossRef] [PubMed]

79. Yonemura, Y.; Ohyama, T.; Hoshino, T. Chemo-enzymatic syntheses of drimane-type sesquiterpenes and the fundamental core of hongoquercin meroterpenoid by recombinant squalene-hopene cyclase. *Org. Biomol. Chem.* **2012**, *10*, 440–446. [CrossRef] [PubMed]

80. Sato, T.; Hoshino, H.; Yoshida, S.; Nakajima, M.; Hoshino, T. Bifunctional triterpene/sesquarterpene cyclase: Tetraprenyl-beta-curcumene cyclase is also squalene cyclase in *Bacillus megaterium*. *J. Am. Chem. Soc.* **2011**, *133*, 17540–17543. [CrossRef] [PubMed]

81. Abe, T.; Hoshino, T. Enzymatic cyclizations of squalene analogs with threo- and erythro-diols at the 6,7- or 10,11-positions by recombinant squalene cyclase. Trapping of carbocation intermediates and mechanistic insights into the product and substrate specificities. *Org. Biomol. Chem.* **2005**, *3*, 3127–3139. [CrossRef] [PubMed]

82. Abe, I.; Rohmer, M. Enzymic Cyclization of 2.3-Dihydrosqualene and Squalene 2.3-Epoxide by Squalene Cyclases: From Pentacyclic to Tetracyclic Triterpenes. *J. Chem. Soc. Perkin Trans. 1* **1994**, *7*, 783–791. [CrossRef]

83. Baunach, M.; Franke, J.; Hertweck, C. Terpenoid Biosynthesis Off the Beaten Track: Unconventional Cyclases and Their Impact on Biomimetic Synthesis. *Angew. Chem. Int. Ed.* **2015**, *54*, 2604–2626. [CrossRef] [PubMed]

84. Tagami, K.; Liu, C.; Minami, A.; Noike, M.; Isaka, T.; Fueki, S.; Shichijo, Y.; Toshima, H.; Gomi, K.; Dairi, T.; et al. Reconstitution of Biosynthetic Machinery from Indole-Diterpene Paxilline in *Aspergillus oryzae*. *J. Am. Chem. Soc.* **2013**, *135*, 1260–1263. [CrossRef] [PubMed]

85. Laden, B.P.; Tang, Y.; Porter, T.D. Cloning, heterologous expression, and enzymological characterization of human squalene monooxygenase. *Arch. Biochem. Biophys.* **2000**, *374*, 381–388. [CrossRef] [PubMed]

86. Li, L.; Porter, T.D. Hepatic cytochrome P450 reductase-null mice reveal a second microsomal reductase for squalene monooxygenase. *Arch. Biochem. Biophys.* **2007**, *461*, 76–84. [CrossRef] [PubMed]
87. Dürr, C.; Schnell, H.J.; Luzhetskyy, A.; Murillo, R.; Weber, M.; Welzel, K.; Vente, A.; Bechthold, A. Biosynthesis of the terpene phenalinolactone in *Streptomyces* sp. Tu6071: Analysis of the gene cluster and generation of derivatives. *Chem. Biol.* **2006**, *13*, 365–377. [CrossRef] [PubMed]
88. Xu, Z.; Baunach, M.; Ding, L.; Hertweck, C. Bacterial Synthesis of Diverse Indole Terpene Alkaloids by an Unparalleled Cyclization Sequence. *Angew. Chem. Int. Ed.* **2012**, *51*, 10293–10297. [CrossRef] [PubMed]
89. Hannemann, F.; Bichet, A.; Ewen, K.M.; Bernhardt, R. Cytochrome P450 systems-biological variations of electron transport chains. *Biochim. Biophys. Acta* **2007**, *1770*, 330–344. [CrossRef] [PubMed]
90. McCarthy, J.G.; Eisman, E.B.; Kulkarni, S.; Gerwick, L.; Gerwick, W.H.; Wipf, P.; Sherman, D.H.; Smith, J.L. Structural basis of functional group activation by sulfotransferases in complex metabolic pathways. *ACS Chem. Biol.* **2012**, *7*, 1994–2003. [CrossRef] [PubMed]
91. Sakowicz, R.; Finer, J.T.; Beraud, C.; Crompton, A.; Lewis, E.; Fritsch, A.; Lee, Y.; Mak, J.; Moody, R.; Turincio, R.; Chabala, J.C.; Gonzales, P.; Roth, S.; Weitman, S.; Wood, K.W. Antitumor activity of a kinesin inhibitor. *Cancer Res.* **2004**, *64*, 3276–3280. [CrossRef] [PubMed]
92. Reddie, K.G.; Roberts, D.R.; Dore, T.M. Inhibition of kinesin motor proteins by adociasulfate-2. *J. Med. Chem.* **2006**, *49*, 4857–4860. [CrossRef] [PubMed]
93. Moosmann, P.; Ueoka, R.; Grauso, L.; Mangoni, A.; Morinaka, B.I.; Gugger, M.; Piel, J. Cyanobacterial ent-Sterol-Like Natural Products from a Deviated Ubiquinone Pathway. *Angew. Chem. Int. Ed.* **2017**, *129*, 1–5. [CrossRef]
94. Müller, W.E.G.; Böhm, M.; Batel, R.; De Rosa, S.; Tommonaro, G.; Müller, I.M.; Schröder, H.C. Application of Cell Culture for the Production of Bioactive Compounds from Sponges: Synthesis of Avarol by Primmorphs from *Dysidea avara*. *J. Nat. Prod.* **2000**, *63*, 1077–1081. [CrossRef] [PubMed]
95. Uriz, M.J.; Turon, X.; Galera, J.; Tur, J.M. New light on the cell location of avarol within the sponge *Dysidea avara* (*Dendroceratida*). *Cell. Tissue Res.* **1996**, *285*, 519–527. [CrossRef]

![marine drugs logo] *marine drugs*

MDPI

Article

Stress-Driven Discovery of New Angucycline-Type Antibiotics from a Marine *Streptomyces pratensis* NA-ZhouS1

Najeeb Akhter [1], **Yaqin Liu** [2], **Bibi Nazia Auckloo** [1], **Yutong Shi** [1], **Kuiwu Wang** [3], **Juanjuan Chen** [4], **Xiaodan Wu** [5] **and Bin Wu** [1,*]

[1] Ocean College, Zhejiang University, Hangzhou 310058, China; na_memon@yahoo.com (N.A.); naz22ia@hotmail.com (B.N.A.); 11434028@zju.edu.cn (Y.S.)
[2] Department of Chemistry, Zhejiang University, Hangzhou 301000, China; yaqin86@zju.edu.cn
[3] Department of Chemistry, Zhejiang Gongshang University, Hangzhou 310012, China; wkwnpc@zjgsu.edu.cn
[4] Key Laboratory of Applied Marine Biotechnology, Ningbo University, Chinese Ministry of Education, Ningbo 315211, China; chenjuanjuan@nbu.edu.cn
[5] Centre of Analysis and Measurement, Zhejiang University, Hangzhou 310058, China; wxd_zju@163.com
*   Correspondence: wubin@zju.edu.cn; Tel./Fax: +86-058-0209-2258

Received: 20 July 2018; Accepted: 7 September 2018; Published: 12 September 2018

Abstract: Natural products from marine actinomycetes remain an important resource for drug discovery, many of which are produced by the genus, Streptomyces. However, in standard laboratory conditions, specific gene clusters in microbes have long been considered silent or covert. Thus, various stress techniques activated latent gene clusters leading to isolation of potential metabolites. This study focused on the analysis of two new angucycline antibiotics isolated from the culture filtrate of a marine *Streptomyces pratensis* strain NA-ZhouS1, named, stremycin A (**1**) and B (**2**) which were further determined based on spectroscopic techniques such as high resolution time of flight mass spectrometry (HR-TOF-MS), 1D, and 2D nuclear magnetic resonance (NMR) experiments. In addition, four other known compounds, namely, 2-[2-(3,5-dimethyl-2-oxo-cyclohexyl)-6-oxo-tetrahydro-pyran-4yl]-acetamide (**3**), cyclo[L-(4-hydroxyprolinyl)-L-leucine] (**4**), 2-methyl-3*H*-quinazoline-4-one (**5**), and menthane derivative, 3-(hydroxymethyl)-6-isopropyl-10,12-dioxatricyclo[7.2.1.0]dodec-4-en-8-one (**6**) were obtained and elucidated by means of 1D NMR spectrometry. Herein, we describe the "Metal Stress Technique" applied in the discovery of angucyclines, a distinctive class of antibiotics that are commonly encoded in microbiomes but have never been reported in "Metal Stress" based discovery efforts. Novel antibiotics **1** and **2** exhibited antimicrobial activities against *Pseudomonas aeruginosa*, methicillin resistant *Staphylococcus aureus* (MRSA), *Klebsiella pneumonia*, and *Escherichia coli* with equal minimum inhibitory concentration (MIC) values of 16 µg/mL, while these antibiotics showed inhibition against *Bacillus subtilis* at MIC value of approximately 8–16 µg/mL, respectively. As a result, the outcome of this investigation revealed that metal stress is an effective technique in unlocking the biosynthetic potential and resulting production of novel antibiotics.

Keywords: marine microorganisms; *Streptomyces pratensis*; polyketide antibiotics; metal stress technique; antimicrobial activity

---

1. Introduction

In the microbial world, secondary metabolites may act as natural antibiotics, enzyme inhibitors, pigments, and toxins for microbial protection or behave as signaling agents depending on their concentrations [1–4]. In spite of the fact that above half of all medications are based on terrestrial natural products platforms, the marine habitat that comprises 71% of the Earth's surface may provide an exceptional possibility to explore novel therapeutics because of its unusual chemical diversity and growth conditions [5]. The exploration of new natural products from marine resources led to the isolation of about 15,000 novel secondary metabolites during the period of 2001–2015 [6]. Marine microorganisms usually thrive under distinctive conditions like temperature, pressure, dissolved oxygen, and nutrient availability, leading to the production of structurally and biologically interesting compounds. As such, marine actinomycetes have been revealed as an incredible source of novel secondary metabolites with various biological activities [7–15]. More specifically, marine *Streptomyces* derived compounds have demonstrated their potency to exhibit cytotoxic, anticancer, antifungal, and antimicrobial effects such as warkmycin, 12-deoxo-12-hydroxy-8-*O*-methyltetrangomycin, marizomib, and salinosporamide A [16–19]. Abiotic strategies such as chemical stress (heavy metal), biotic stress (co-cultivation), and changes in fermentation conditions (light, pH, temperature, and various media) are long known to induce notable changes or function to unlock cryptic biosynthetic gene clusters in the microbial metabolome [4,20]. Standard laboratory culture conditions have proven to hinder activation of specific gene clusters which, in turn, hamper the generation of secondary metabolites. Previous work which we have conducted demonstrates the successful utility of the "metal stress" strategy for activating silent gene clusters and subsequent isolation of unique natural products which exhibit potent antimicrobial properties [5,21–23].

The angucycline group of antibiotics belongs to a specific group of polycyclic aromatic polyketides derived from naturally occurring quinone saccharide antibiotics, which exhibit mainly anticancer and antimicrobial activities [24,25]. This type of antibiotic was first discovered as a tetrangomycin isolated from *Streptomyces rimosus* in 1965 and was shown to have a C–C bond connectivity with C-9 linked sugar moieties [17,26]. A large number of angucyclines are produced as C-glycoside antibiotics, and displaythis element as one of the most distinctive and typical structural characteristics. It is also known that theseantibiotics are produced by actinomycetes with Streptomyces as the major producer [27].

In order to discover new secondary metabolites and extend the use of the "Metal Stress" strategy that stimulates the cryptic gene cluster of marine microorganisms, different metal ions were applied to the marine Streptomyces strain NA-ZhouS1. Under one of these implemented conditions, referred to herein as heavy metal nickel (100 µM) followed by their antibacterial capacities together with a comparison of extract in high performance liquid chromatography (HPLC) profile, grasped our consideration, and facilitated interaction with new compounds stremycin A (**1**) and B (**2**). The results in the comparison revealed that the addition of metal induction would streamline natural product development efforts. Further, this study deals with the isolation, structure elucidation and bioactivities of two new aromatic polyketides **1** and **2**, in addition to known compounds **3–6**. The structures are shown in Figure 1.

1 = R = H
2 = R = OH

3

4

5

6

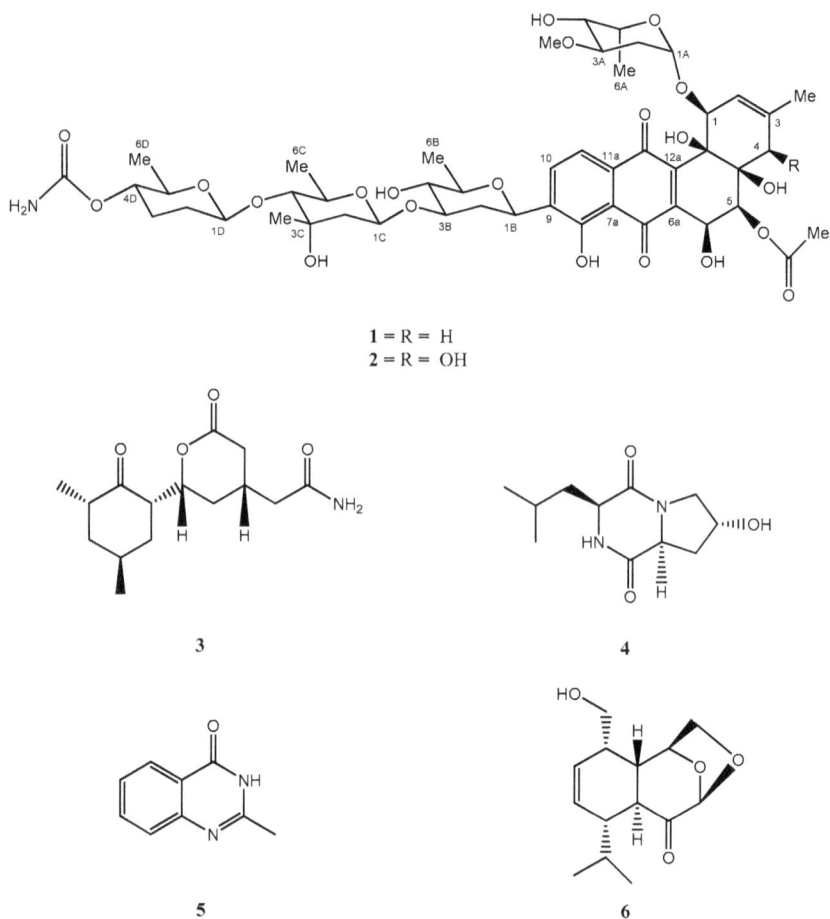

Figure 1. Chemical structures of stress metabolites **1–6**.

2. Results

Streptomyces pratensis strain NA-ZhouS1 was isolated from marine sediment in the waters along the Zhoushan Coast in East China. Throughout this study, the strain was treated with the abiotic stress reagents, such as $NiCl_2 \cdot 6H_2O$; $CoCl_2 \cdot 6H_2O$; $ZnSO_4 \cdot 7H_2O$; $CrCl_3 \cdot 6H_2O$; $MnCl_2 \cdot 6H_2O$ at concentrations of 100, 200, 400, 800 μM, respectively. As a consequence, based on HPLC guided profile, 100 μM nickel ion ($NiCl_2 \cdot 6H_2O$) was chosen as the best elicitor of stress in the *S. pratensis* strain toward the production of antibiotics in comparison to conditions used for normal growth of the strain (Figure 2). The 30 L of nickel treated culture broth was extracted with ethyl acetate (EtOAc) and subjected to reverse phase column using C18 silica gel, Sephadex LH-20, followed by further purification with preparative HPLC (flow rate 10 mL/min, ultraviolet (UV) detector 210 nm), which successfully led to the isolation of two new aromatic polyketides, namely, stremycin A (**1**) and B (**2**) together with a known compound **3**. Moreover, the other culture extract of the same strain, which was induced by zinc ion to a concentration of 100 μM, subjected to analytical HPLC (flow rate 0.8 mL/min, eluted mode 0~30 min 20%~100% (H_2O/MeOH), 30–50 min 100% MeOH, UV detector 210 nm) led to the isolation of three known compounds **4–6**.

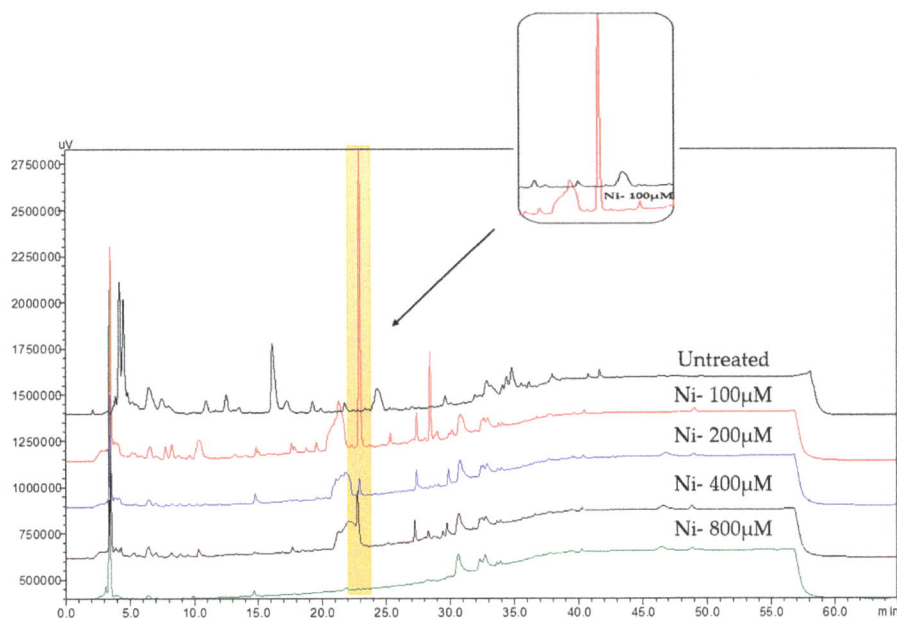

Figure 2. HPLC analysis metabolic profile of NA-ZhouS1 under nickel ion stress condition.

Structural Elucidation of Novel Compounds

Stremycin A (**1**) was detected and isolated as a yellow powder, giving the molecular formula of $C_{48}H_{65}NO_{21}$ according to the HR-TOF-MS analysis in positive ion mode at *m/z* 1014.3937 for $[M + Na]^+$ (Calcd. 1014.3941) and in negative ion mode at *m/z* 990.3964 for $[M - H]^-$ (Calcd. 990.3976), (Supplementary Material (SM), Figures S26 and S27). Carefully analysis of ^1H and ^{13}C NMR spectroscopic data exhibited features characteristic of tetracyclic benz[*a*]anthracene core, with the 1-position *O*-glycosylated and 9-position *C*-glycosylated, which were characterized by resonances corresponding to twenty methines, fourteen quaternary carbons, six methylene, and nine methyl groups (Table 1). These signals were comparable to those of warkmycin and P371A1 [16,28]. Although the structure of **1** was analogous to warkmycin whilea difference was observed in the substituent pattern, i.e., the presence of a carbamoyl group -$CONH_2$ in the region of sugar D instead of sugar A ofwarkmycin [16]. This was further confirmed following observation of heteronuclear multiple bond correlation (HMBC) cross peaks from H-4D to a carbamoyl carbon at (δ_C 159.6). Another difference wasnoted at C-4 (δ_C 36.3) position in the aglycone of **1**, where the methylene protons were seen at δ_H [1.99, 2.36 (d, *J* = 17.6)] instead of an oxygenated methine [δ_H 5.36, δ_C 68.1 (CH-4)] in warkmycin.Hence, the entire assignment of all the ^1H and ^{13}C NMR data of **1** was finally performed by the correlative analysis of its ^1H-^1H correlated spectroscopy (COSY), heteronuclear single quantum correlation (HSQC), HMBC, and nuclear overhauser effect spectroscopy (NOESY) experiments (SM, Figures S4–S6 and S8). The ^1H NMR spectrum displayed a set of *ortho*-coupled aromatic proton signals appeared at δ_H 7.87, 7.64 (d, *J* = 7.9 Hz, H-10/11), two oxygenated methine protons at δ_H 5.84, 4.92 (d, *J* = 6.8 Hz, H-5/6), and an olefinic proton signal at δ_H 5.63 (1H, s, H-2), which were associated with the carbons resonated at δ_C 134.2 (C-10), 120.1 (C-11), 76.2 (C-5), 69.9 (C-6), and 76.2 (C-2), as seen via the HSQC spectrum results. These *ortho*-coupled protons (H-10/H-11 and H-5/H-6), showed diagnostic COSY contacts with typical ^1H-^1H coupling constants, which was extended by HMBC correlations from H-10 to C-7a, C-8, and C-11a; H-11 to C-7a, C-9, C-10, and C-12; H-5 to C-6; H-6 to C-5, C-6a, and C-12a to establish the connectivity of rings. The fusing pattern of another ring was deduced by observing the HMBC correlations from H$_2$-4 to C-4a, C-5, and from H-1 to C-2, C-4a, and C-12b to complete the assignment of aglycone skeleton. In addition, a typical quinone analogs system was identified from the significant carbonyl chemical shifts, which were visible at δ_C 187.24 (C-12) and δ_C 190.52 (C-7) in the ^{13}C NMR spectrum. A substituted singlet methyl resonance noticeable at (δ_H 1.68, δ_C 23.58) was confirmed at C-3 δ_C 136.61 by HMBC correlations of δ_H 1.68 to C-3, C-4, while an acetyl group resonated at δ_C 172.39 (5-COMe) was assigned to an oxygenated methine at [δ_H 5.84, δ_C 76.24 (CH-5)] by cross-peak correlations observed in the HMBC spectrum from H-5 to a quaternary carbon δ_C 172.39. As a consequence, a detailed analysis of two-dimensional (2D) nuclear magnetic resonance spectroscopy data was performed as compared to previously published literature. This revealed the *cis* arrangement with a strong correlation in between H-5 and H-6. Since H-4 showed a diagnostic NOESY cross peak with H-5, it indicated that the acetyl group was in an α-configuration. Since H-5 was α-oriented, no NOESY connection would be present between H-5 and H-4 owing to the bulky OAc group which stayed as equatorial, pushing H-5 away from both H-4. The coupling constant $J_{5–6}$ = 6.8 Hz revealed the hydroxyl group at C-6 to be α-oriented.

Table 1. NMR spectrum data for stremycin A (1), ^{1}H NMR (500 MHz, δ in ppm), ^{13}C NMR (125 MHz, δ in ppm) in MeOD.

Position	δ_C, Type	δ_H (Mult., J in Hz)	HMBC	COSY
1	82.1, CH	4.34, d (4.2)	C-1A, C-2, C-12b, C-4a	H-2
2	120.8, CH	5.63, s	C-3	H-1
3	136.6, C	-	-	-
4	36.3, CH$_2$	1.99, d (17.6); 2.36, d (17.6)	Me-3, C-4a	-
4a	75.6, C	-	—	-
5	76.2, CH	5.84, d, (6.8)	C-6, COMe	H-6
6	69.9, CH	4.92, d, (6.8)	C-5, C-6a, C-12a	H-5
6a	145.2, C	-	-	-
7	190.5, C	-	-	-
7a	115.6, C	-	-	-
8	158.6, C	-	-	-
9	139.3, C	-	-	-
10	134.2, CH	7.87, d (7.9)	C-7a, C-8, C-1B, C11a	H-11
11	120.1, CH	7.64, d (7.9)	C-7a, C-9, C-12	H-10
11a	132.6, C	-	-	-
12	187.2, C	-	-	-
12a	146.0, C	-	-	-
12b	78.5, C	-	-	-
3-Me	23.5, CH$_3$	1.68, s	C-3, C-4	-
5-COMe	172.3, C	-	-	-
	20.9, CH$_3$	2.20, s	-	-
Sugar A				
1A	100.6, CH	4.60, d (4.0)	C-1, C-3A, C-5A	H-2A
2A	30.6, CH$_2$	1.31, overlapped; 1.88, m	-	H-1A, H-3A
3A	78.4, CH	3.29, m	OMe-3A	H-4A
4A	66.9, CH	3.49, m	-	H-5A
5A	63.4, CH	4.29, m	-	H-6A
6A	16.8, CH$_3$	1.18, d (6.6)	C-5A	H-5A
OMe-3A	57.5,C	3.24, s	-	-
Sugar B				
1B	72.4, CH	4.89, s	C-10, C-9, C-8, C7a, C-2B, C-3B, C-4B, C-5B	H-2B
2B	38.6, CH$_2$	1.42, d (12.6);2.50, dd (12.1, 4.2)	-	H-1B, H-3B
3B	82.5, CH	3.84, m	C-1C	H-4B
4B	76.8, CH	3.11, t (8.9)	-	H-5B
5B	77.7, CH	3.46, m	C-6B	H-6B
6B	18.8, CH$_3$	1.38, d (6.1)	-	-

Table 1. *Cont.*

Position	δ_C, Type	δ_H (Mult., *J* in Hz)	HMBC	COSY
Sugar C				
1C	99.3, CH	4.78, dd (9.9, 1.8)	C-3B, C-2C	H-2C
2C	45.6, CH$_2$	1.68, overlapped; 1.95, m	C-3C, C-4C	H-1C
3C	71.5, C	-	-	-
4C	90.5, CH	3.19, d (9.6)	C-5C, C-6C, C-1D	H-5C
5C	71.9, CH	3.55, m	-	H-6C
6C	18.5, CH$_3$	1.31, d (6.1)	C-5C	H-5C
Me-3C	22.6, CH$_3$	1.25, s	C-3C, C-4C	-
Sugar D				
1D	104.5, CH	4.62, s	C-4C, C-2D, C-3D	H-2D
2D	31.4, CH$_2$	1.64, m; 1.99, overlapped	-	H-3D
3D	28.9, CH$_2$	1.57, m; 2.14, m	-	H-4D
4D	73.9, CH	4.24, dd (10.0, 4.4)	C-6D, CONH$_2$	H-5D
5D	75.3, CH	3.65, m	-	H-6D
6D	18.2, CH$_3$	1.22, d (6.2)	C-5D	-
4D-CONH$_2$	159.6, C	-	C-4D	-

In the 1D (^{1}H, ^{13}C) NMR spectrum, three acetal carbon resonances observed at δ_C 99.31 (C-1C), δ_C 100.64 (C-1A) and δ_C 104.57 (C-1D), as well as four doublet methyl proton resonances appeared at δ_H 1.18 (3H, d, *J* = 6.6 Hz, H-6A), 1.38 (3H, d, *J* = 6.1 Hz, 6B), 1.31 (3H, d, *J* = 6.1 Hz, H-6C), and 1.22 (3H, d, *J* = 6.2 Hz, H-6D) revealed the existence of four deoxy sugars, three of which *O*-linked and one needed to be *C*-glycosidically linked to the aglycone of **1**.

A thorough analysis of the 2D NMR experiment was carried out to clarify the connection of four sugar units (A–D) attached to aglycone as shown in Figure 3. As such, in the substituent of sugar A, a small coupling constant (*J* = 4.0 Hz) of an anomeric proton resonated at δ_H 4.60 (H-1A) proved that this unit was α-*O*-glycosidically linked to angucycline core. Further, the observed $^3J_{C-H}$ long-range correlations from H-1A to C-1 (δ_C 82.1) and H-1 to C-1A (δ_C 100.6) in the HMBC spectrum confirmed the connection of C-1-*O*-C-1A between the aglycone and oleandrose. The NOESY cross peaks of H-1 and H-1A revealed an axial orientation of H-1. Similarly the ^{1}H-^{1}H COSY correlations of H-1A/H-2A, H-3A/H-4A, H-5A/H-6A, and the HMBC correlations of H-1A to C-3A, C-5A revealed the presence of a six-membered deoxy sugar. Moreover, the singlet methoxy group resonated at (δ_H 3.24, δ_C 57.50) was confirmed at CH-3A by HMBC $^3J_{C-H}$ long-range cross peaks. Comparison of our conclusions with those found in the literature that the sugar A is a known unit, namely, α-*O*-5-epi-oleandrose [16].

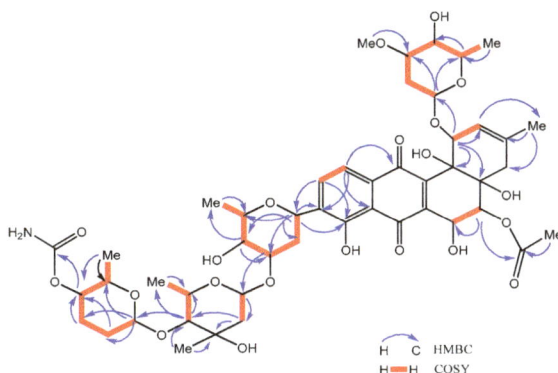

Figure 3. The key ^{1}H-^{1}H COSY, HMBC correlations of stremycin A (**1**).

As such, the significant HMBC long-range correlations from the anomeric methine proton (CH-1B) resonated at (δ_H4.89, δ_C 72.43) to C-8, C-9, and C-10 inferred the presence of *C*-glycosidic bond (C9-C1B) between the aglycone and olivose sugar moiety. The resonance of H-1B showed an overlapped peak in the ^{1}H NMR spectrum, thus it was not possible to determine the exact coupling constant. Correspondingly, the coupling constant (J = 8.9 Hz) of a methine proton resonated at δ_H 3.11 (H-4B) revealed that sugar B assumes the acetal carbon (C-1) conformation in which all protons were axially oriented excluding the H-5B and H-6B. The hydroxyl group at CH-5B (δ_H 3.46, δ_C 77.7) and the methyl group of H-6B considered being equatorial when compared to those of warkamycin [16]. Further analysis was observed by ^{1}H-^{1}H COSY correlations in between H-1B/H-2B, H-2B/H-3B, H-3B/H-4B, H-4B/H-5B, and H-5B/Me-6B, followed by the HMBC correlations of H-1B to C-2, C-3, C4, and C-5 confirmed the presence of the sugar olivose. Hence, the combined results with comparison of published literature led to the identification of sugar B as β-*C*-olivose linked to C-9 on the angucycline core.

Similarly, the substituent of sugar C displayed large coupling constants J_{H-1C} = 9.9, 1.8 Hz and J_{H-4C} = 9.6 Hz resonated at δ_H 4.78 and at δ_H 3.19, signifying this unit as β-glycosidically bonded to sugars by also revealing an axial orientation. In addition, the connectivity of sugar B and C as *O*-glycosidic linkage C-3B-*O*-C-1C was deduced by HMBC long-range correlationsof H-3B to C-1C (δ_C 99.31). Moreover, the ^{1}H-^{1}H COSY correlations of H-1C/H-2C, H-4C/H-5C, H-5C/H-6C, and the HMBC correlations of H-1C to C-2 (δ_C 45.6), H_2-2C to C-3 (δ_C 71.5), and C-4C (δ_C 90.5), and H-4C to C-5C (δ_C 71.9), and C-6C (δ_C 18.5)verified the presence of the sugar unit olivomycose. The NOESY spectrum further confirmed the correlations between H-1C to H-2C, H-1C to methyl proton at H-3C, H-4C to methyl proton at H-6C and established this unit with comparison of previously published literature as β-olivomycose.

Likewise, the unit of sugar D displayed large coupling constant (J_{H-1D} = 9.1 Hz) resonated at δ_H 4.62 revealed an axial orientation of H-1D and confirmed as β-glycosidically boundsugar. The *O*-glycodsidic connectivity of sugar C to D(C-4C-*O*-C-1D) was determined on the basis of HMBC correlations of H-4C to C-1D and H-1D to C-4C. Further, the ^{1}H-^{1}H COSY correlations ofH-1D/H-2D, H-2D/H-3D, H-3D/H-4D, H-4D/H-5D, and H-6D, and the HMBC cross peaks of H-1D to C-2D (δ_C 31.4) and C-3D (δ_C 28.9); H-4D to CONH$_2$ (δ_C 159.6) and C-6D (δ_C 18.2)revealed the presence of sugar amicetose. The structure of **1** exhibited the substituent of a carbamoyl group at δ_H 4.24 (H-4D, dd, J = 10.0, 4.4 Hz) which highlighted the novelty of this compound. Therefore, sugar D was established as 4-*O*-carbamoyl-β-amicetose.

To further confirm the new structure, ESI MS/MS fragmentation experiment of compound **1** was carried out (Figure S28). As such, the positive ion MS^{n}spectrum of the structure gave the major [M + Na]$^+$ ion at *m/z* 1014. As shown in Figure 4, the fragmentation of this precursor ion yielded an interesting product ion at *m/z* 953, which was attributed to the elimination of a neutral molecule CH$_3$NO$_2$ (61 Da) from the precursor ion at *m/z* 1014. The product ions at *m/z* 870 and 709 were generated by the loss of 144 and 161 Da, which were reasonably assigned as the elimination of C$_7$H$_{12}$O$_3$ and C$_7$H$_{15}$NO$_3$, respectively. Further, the fragment ion was observed at *m/z* 852, which indicated the neutral loss of 162 Da (assigned to C$_7$H$_{14}$O$_4$). Similarly, the product ions at *m/z* 792, 774, and 456 were produced by the loss of 60, 18, and 318 Da, which were selected as the elimination of acetic acid (C$_2$H$_4$O$_2$), H$_2$O, and C$_{19}$H$_{10}$O$_5$, respectively. Another major product ion peak with high intensity was observed at *m/z* 713, generated by the loss of 301 Da, specified as the elimination of C$_{14}$H$_{23}$NO$_6$. The fragments at *m/z* 569, 551, 533, 491, and 473, which were in close agreement with the presence of *C*-glycosidic linkage at C-9 position, observed with continuous loss of 144, 18, 18, 60, and 18 Da, assigned to the removal of C$_7$H$_{12}$O$_3$, H$_2$O, and acetic acid (C$_2$H$_4$O$_2$), respectively. The structure of **1**, being a new aromatic polyketide was thus termed as stremycin A.

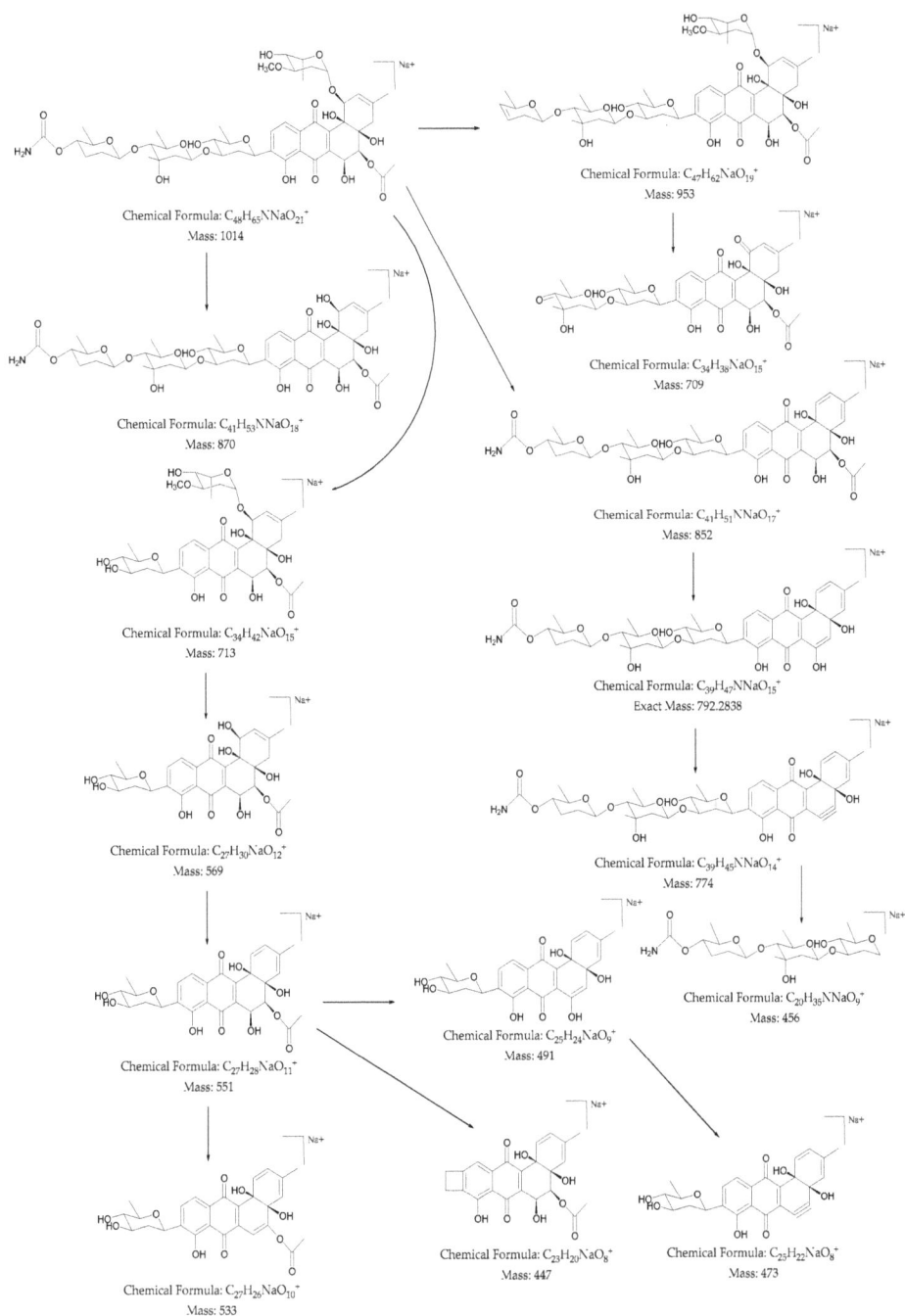

Figure 4. The key plausible MS^n fragmentation pathway of stremycin A (**1**) was confirmed by electrospray ionization mass spectroscopy (ESI-MS/MS) analysis in positive mode.

Stremycin B (**2**) was obtained as a yellow powder. The HR-TOF-MS analysis of **2** yielded a molecular ion peak at *m/z* 1030.3860 [M + Na]$^+$ (Calcd. 1030.3890) in positive mode and at *m/z* 1006.3987 [M − H]$^-$ (Calcd. 1006.3925), giving the molecular formula $C_{48}H_{65}NO_{22}$ (SM, Figures S29 and S30). The general feature of 1D and 2D NMR (^{1}H, ^{13}C, ^{1}H-^{1}H COSY and HMBC) spectrum (Table 2) closely resembled that of **1**, thus strongly suggesting that the structure of **2** was highly similar to the new compound **1** (Figure 1). The main difference between **1** and **2** was 16 Da, suggesting the presence of a hydroxyl group at the C-4 position in **2** instead of a methylene in **1**. According to the 1D NMR spectrum, one proton was seen at δ_H 4.17 (s, H-4), representing CH for compound **2** while two protons were seen at δ_H [1.99, 2.36 (d, *J* = 17.6 Hz, H-4)] representing CH_2 for compound **1**. These deductions were further confirmed on the basis of the HSQC spectrum, where the cross peak notedin between the resonances of H-4 to C-4 (δ_C 70.5), and extended by HMBC correlations from H-4 to C-3, 3-Me, C-4a, and C-12b. Hence, these findings confirmed the presence of a hydroxyl group attached at C-4 to the aglycone unit of **2**. Further confirmation and analysis were carried out by ^{1}H-^{1}H COSY, HMBC, and NOESY experiments, where the correlations of a proton and carbon were in very close agreement to that of compound **1** and clearly confirmed the suggested structure of **2**. However, no NOESY correlation was found between H_β-5 and H-4. The hydroxyl group at C-4 was determined as *β*-oriented. It was also found that the structure of **2** was highly similar to that of 4-*O*-deacetyl-warkmycin, previously reported by Helaly et al., 2015 [16], which was the synthesized version obtained under acidic conditions. Nevertheless, both structures were distinctive in sugar moieties. Finally, based on the results, the structure of **2** was elucidated as a new benz[*a*]anthracene glycoside and named as stremycin B.

Table 2. NMR spectrum data for stremycin B (**2**), ^{1}H NMR (500 MHz, δ in ppm), ^{13}C NMR (125 MHz, δ in ppm) in MeOD.

Position	δ_C, Type	δ_H (Mult., *J* in Hz)	HMBC	COSY
1	81.8, CH	4.43, d (3.8)	C-1A, C-2, C-3, C-12b, C-4a	H-2
2	123.1, CH	5.72, d (3.2)	3-Me, C-4	H-1
3	138.7, C	-	-	-
4	70.5, CH	4.17, s	3-Me, C-3, C-4a, C-12b	-
4a	76.2, C	-	-	-
5	75.9, CH	5.76, d (6.8)	C-6, COMe	H-6
6	69.8, CH	4.91, m (6.8)	C-5, C-6a, C-12a	H-5
6a	144.5, C	-	-	-
7	190.8, C	-	-	-
7a	115.6, C	-	-	-
8	158.7, C	-	-	-
9	139.4, C	-	-	-
10	134.2, CH	7.85 d (7.7)	C-8, C-1B, C11a	H-11
11	120.0, CH	7.61, d (7.7)	C-7a, C-9, C-12	H-10
11a	132.5, C	-	-	-
12	187.3, C	-	-	-
12a	146.7, C	-	-	-
12b	79.7, C	-	-	-
3-Me	21.9, CH_3	1.95, s	C-3, C-4	-
5-COMe	173.0, C	-	-	-
	20.6, CH_3	2.02, s	-	-
Sugar A				
1A	100.9, CH	4.57, d (4.4)	C-1, C-3A, C-5A	H-2A
2A	30.3, CH_2	1.32, overlapped; 1.85, m	C-1A	H-1A, H-3A
3A	78.4, CH	3.27, m	C-4A	H-4A
4A	67.5, CH	3.47, m	-	H-5A
5A	63.7, CH	4.10, d (6.5)	-	H-6A
6A	16.8, CH_3	1.17, d (6.6)	C-5A	H-5A
OMe-3A	57.7, C	3.30, s	-	-

Table 2. *Cont.*

Position	δ_C, Type	δ_H (Mult., *J* in Hz)	HMBC	COSY
Sugar B				
1B	72.4, CH	4.87, m (overlapped)	C-9, C-10	H-2B
2B	38.6, CH$_2$	1.42, m; 2.49, dd (12.3, 4.6)	C-3B, C-4B	H-1B, H-3B
3B	82.3, CH	3.84, m	C-4B, C-1C	H-4B
4B	76.9, CH	3.11, dd (11.1, 6.7)	C-3B, C-5B, C-6B	H-5B
5B	77.7, CH	3.44, dd (6.9, 3.5)	C-6B	H-6B
6B	18.8, CH$_3$	1.37, d (6.1)	C-5B	-
Sugar C				
1C	99.3, CH	4.77, d (9.9)	C-3B, C-2C	H-2C
2C	45.6, CH$_2$	1.95, overlapped; 1.67, m	C-3C, C-4C	H-1C
3C	71.5, C	-	-	
4C	90.5, CH	3.18, d (9.6)	Me-3C, C-5C, C-6C, C-1D	H-5C
5C	72.0, CH	4.81, d (6.5)	—	H-6C
6C	18.5, CH$_3$	1.30, d (6.1)	C-5C	H-5C
Me-3C	22.6, CH$_3$	1.25, s	C-3C, C-4C	-
Sugar D				
1D	104.5, CH	4.61, d (9.2)	C-4C, C-2D, C-3D	H-2D
2D	31.4, CH$_2$	1.99, m; 1.63, m	-	H-3D
3D	28.9, CH$_2$	2.14, m; 1.59, m	-	H-4D
4D	73.9, CH	4.24, dd (9.7, 5.5)	C-6D, CONH$_2$	H-5D
5D	75.3, CH	3.62, m	-	H-6D
6D	18.2, CH$_3$	1.22, d (6.1)	C-5D	-
4D-CONH$_2$	159.6, C	-	C-4D	-

Among the isolates, the known antibiotic 2-[2-(3,5-dimethyl-2-oxo-cyclohexyl)-6-oxo-tetrahydro-pyran-4yl]-acetamide (**3**) from nickel-treated extract was determined by detailed analysis of 1D NMR spectroscopy along with the comparison of data in literature [29]. Moreover, three target stress-induced compounds from the zinc treated filtrate of the same strain NA-ZhouS1 were isolated, namely, cyclo[L-(4-hydroxyprolinyl)-L-leucine] (**4**), 2-methyl-3*H*-quinazoline-4-one (**5**), and menthane derivative, 3-(hydroxymethyl)-6-isopropyl-10,12-dioxatricyclo[7.2.1.0]dodec-4-en-8-one (**6**) which were obtained and further elucidated by a detailed analysis of 1D NMR spectroscopy along with the comparison of data in literature [30–32].

The novel structures of **1** and **2** showed moderate antibiotic activities in comparison to the positive control tetracycline with equal MIC values of 16 µg/mL against *Pseudomonas aeruginosa*, methicillin resistant *Staphylococcus aureus* (MRSA), *Klebsiella pneumonia*, and *Escherchia coli*, while against *Bacillus subtilis*, both compounds showed the inhibition at MIC value of around 8–16 µg/mL, respectively. In earlier bioassay-guided approach it was determined that most of the angucycline related antibiotics possess moderate antibacterial activities against Gram-positive pathogens like warkmycin, chattamycin B, tetrangomycin, and vineomycin A$_1$, while found mayamycin and seitomycin selective inhibit the Gram-negative pathogens [14,16,17,25,33,34]. However, all biologically active angucyclines reported previously are observed to be dependent on the length of the sugar moieties [24]. Similarly, the known antibiotic **3** exhibited antibiotic activity around 16–32 µg/mL against MRSA, *P. aeruginosa*, and *K. pneumonia*. According to previous studies, the antibiotic 2-[2-(3,5-dimethyl-2-oxo-cyclohexyl)-6-oxo-tetrahydro-pyran-4yl]-acetamide (**3**) was known to be bone resorption inhibitor and found to be an active herbicidal component [29,35,36].

3. Discussion

Compound **1** and **2** arestructural analogs that possessa similar polyketide aglycone, which is a tetracyclic benz[*a*]anthracene core like in angucycline-type antibioticsalong with hydrolyzable sugar units which is well known to comprise a large number of diverse representatives. As such, the structural diversity of angucyclines mainly comes from hydroxy substitution, epoxidation, and carbonyl substitutions in the region of C-4, C-5, C-6a, C-7, C-12, C-12a/C-5, C-6 or C-6a, C-12a/C-7, C12, respectively [16,28,37,38]. Moreover, other examples include the amino acid incorporations as in urdamycins and jadomycins, ring cleavages as in grincamycins and gilvocarcins, and the glycosylation at various positions, such as *O*-8 or *O*-3 and C-9 in the landomycins and saquayamycins [27,39–42]. Accordingly, the carbohydrate composition of landomycins and saquayamycins were based on the multiple trisaccharide unit's as β-D-olivose-(4→1)-β-D-olivose-(3→1)-α-L-rhodinose to a said regions of the angucycline backbone. Besides, urdamycin exhibited two glycosylation positions at C-9 and 12b, while warkmycin, chattamycin B, and P371A1 displayed the *C*- and *O*-glycosylation with positions at C-9 and C-1. As such, the warkmycin antibiotic isolated from *Streptomyces* sp. Acta 2930 possessed β-olivose-(4→1)-β-olivomycose-(3→1)-β-amicetose at C-9 region and 4-*O*-carbamoyl-α-5-epi-oleandrose at the C-1 region. In the same way, the structures of **1** and **2** werealso found to possess the *C*- and *O*-glycosylated sugar units belonging to the aquayamycin-type of angucyclines with angular oxygen found in stronger relationship to warkmycinand P371A1 antibiotics. However, the compounds **1** and **2** differed from previous known compounds in the substituents of sugar moieties as β-olivose-(4→1)-β-olivomycose-(3→1)-4-*O*-carbamoyl-β-amicetose. Moreover, it was observed that the sites of attachment of these carbohydrates to the aglycone were the same, albeit, the structures of **1** and **2** had a carbamoyl/carbamate group at sugar D, named, 4-*O*-carbamoyl-β-amicetose. On the contrary, it was found that the antibiotic warkmycinhad the same group at sugar A (4-*O*-carbamoyl-α-*O*-5-epi-oleandrose) while antibiotic P371A1 had a ureido group at sugar C (D-β-amicetose). Another difference was noted in the region of aglycone at a -4 position when compared to warkmycin. The warkmycin possessed anoxygenated methine [δ_H 5.36, δ_C 68.1 (CH-4)] whereacetyl group (OAc) was present as a substituent, while **1** displayed a methylene and **2** possessed a hydroxyl group.

Many metabolites previously isolated from Streptomyces are known to be active against pathogens and display antibacterial activities which are desperately needed on the front line in combating microbial infections. Due to the increasing threat of antibiotic resistance pathogens, scientists are urged to focus on the isolation of more antimicrobial compounds along with the investigation of their mechanisms of action and biosynthetic pathways. As illustrated in this study, heavy metals being applied as elicitors, here referring to the heavy metal nickel ion revealed a distinct HPLC guided profile when compared to the normal one, showing an influence on the secondary metabolome of the *Streptomyces pratensis* strain NA-ZhouS1. We hypothesize that these results are indicative of cryptic gene cluster activation consequential to the metal stress imposed on the strain under study, ultimately resulting in the production of two novel compounds with activity against pathogenic bacterial isolates. It was also observed that the normal products displayed in the untreated culture were considerably lowered when stressed by metals (Figure 2), showing that nickel ion not only stimulates a nonactivated biosynthetic pathway but also impacts the normal biosynthetic capabilities of the strain. We observed that normal growth of *S. pratensis* was repressed when the nickel ion concentration reached around 800 µM, likely having a global effect on processes which occur under normal condition. Moreover, research is required to scrutinize more effective elicitors or ways/techniques for elicitation of cryptic genes clusters of marine microbes may lead to the production of unexpected, albeit potentially potent natural products.

4. Materials and Methods

4.1. General Experimental Procedures

Electrospray ionization mass spectrometry (ESIMS) were recorded on an Agilent 6460 Triple Quade Liquid Chromatography with Mass Spectrometry (Agilent, Beijing, China). HPLC analysis used was composed of a Waters 717 plus Autosampler, a Waters 600 Controller, a Waters 996 Photodiode Array Detector and a Waters Millog workstation (Waters, Shingawa, Tokyo, Japan), while preparative HPLC was performed on an Agilent-1100 system (ChuangXintongheng, Beijing, China) equipped with a Venusil MP-C18 column (10 mm × 250 mm, Agila Technologies, Tianjin, China). Reverse phase column chromatography was performed. ^{1}H NMR (recorded on 500 MHz), ^{13}C NMR (recorded on 125 MHz), DEPT-135, ^{1}H-^{1}H COSY, HMQC, HMBC, and NOESY spectra were measured at 25 °C on a Bruker ADVANCE DMX 500 NMR spectrometer with TMS as internal standard (Bruker, Fällanden, Switzerland). Methanol was used as solvent for NMR experiments. The organic solvents used in chromatographic separation were of analytical grade purchased from Sayfo Technology (Tianjin, China) and chromatographic grade for HPLC analysis purchased from Tedia, Fairfield, OH, USA. Deionized water was prepared by Reverse osmosis Milli-Q water (18 MW) (Millipore, Bedford, MA, USA) and used for all solutions and dilutions. Agar powder for plate culture and other heavy metals including nickel (NiCl$_2$·6H$_2$O) were purchased from Sinopharm Chemical Reagent Co., Ltd. (Shanghai, China).

4.2. Isolation and Identification of Streptomyces sp. NA-ZhouS1

The strain NA-ZhouS1 was isolated from a marine sediment sample, collected from the East China Sea, Zhoushan. The plate dilution method was used to isolate actinomycetes from the sample suspension. Approximately 0.5–1 g of each fresh sediment sample was directly inoculated into the presterilized glass tubes and diluted with artificial sea water. Serially diluted samples were plated in the gauze's (GS), starch casein nutrient (SCN) and Aspergillus minimal (AMM) agar medium in triplicate. All the plates were supplemented with nystatin (0.05 g/L) to prevent fungal contamination. The plates were incubated at 28 °C and actinomycete colonies counted from the 7th day onwards up to the 25th days. Single colony of actinomycete was picked up and grown separately for inoculation an agar slant containing the same isolation medium. 16S ribosomal DNA gene was used to identify the strain. The strain NA-ZhouS1 showed 99.93% resemblance to *Streptomyces pratensis*. Sequences were then searched by online database listed in (SM, Table S29). This species was found with off-white to grey aerial spores on gauze's medium and carried smooth-surfaced spores in straight or flexuous spore chains. This actinomycete sp. was previously known as *Streptomyces flavogriseus*, but then was reclassified as *Streptomyces pratensis* [43,44]. A neighbor-joining tree was constructed using software package of Molecular Evolutionary Genetics Analysis (MEGA), version 7.0, Pennsylvania State University, United States for further phylogenetic analysis (SM, Figure S32).

4.3. Analysis of Normal Culture and Metal Stress Cultivation

For screening and initial analysis of normal culture, the spores of NA-ZhouS1 strain were inoculated in 500-mL Erlenmeyer flasks containing 200-mL liquid Gauze's medium (20 g soluble starch, 1 g KNO$_3$, 0.5 g K$_2$HPO$_4$, 0.5 g MgSO$_4$·7H$_2$O, 0.01 g FeSO$_4$·7H$_2$O, 35 g sea salt per liter at pH 7.4) and was grown on a rotatory shaker at 180 rpm for 7 days at 28 °C. Afterwards, the same actinomycete strain NA-ZhouS1 was stressed under different metal conditions like cobalt (CoCl$_2$·6H$_2$O), nickel (NiCl$_2$·6H$_2$O), zinc (ZnSO$_4$·7H$_2$O), chromium (CrCl$_3$·6H$_2$O), and manganese (MnCl$_2$·6H$_2$O), while each metal was applied with four different concentrations of 100, 200, 400, and 800 μM, respectively. The mycelium was removed and the filtrate was extracted twice with an equal volume of ethyl acetate (EtOAc). Finally, extracts were subjected to analytical reversed phase HPLC-UV for further screening by comparing treated and untreated extracts. Consequently, the comparison of the RP-HPLC profiles of the extracts from the strain NA-ZhouS1 revealed the formation of new metabolites following use of 100 μM nickel ion (NiCl$_2$·6H$_2$O), and thus grabbed our attention as a strong contributing factor toward activation of cryptic gene clusters. Additionally, the extract of both normal and stressed cultures, were assayed after overnight incubation at 37 °C for their antibacterial capacities which boost up our judgment to enlarge 100 μM nickel ion (NiCl$_2$·6H$_2$O) culture due to its effective inhibitory abilities.

4.4. Large Scale Fermentation, Extraction and Isolation

The strain NA-ZhouS1 was cultured in the presence of 100 μM nickel treated agent for extraction into 500-mL Erlenmeyer flasks in 200 mL liquid gauze's medium. A total of 25 L fermentation containing 100 µM nickel ions was carried out at 28 °C on a rotary shaker at 180 rpm for 10 days. Thereafter, the fermentation broth was combined and filtered. Subsequently, the filtrate was extracted with (EtOAc) ethyl acetate (2 × 200 mL) twice and dried in vacuo, to provide an organic extract of (3 g).

The crude extract (3 g) was filtered and dissolved in methanol. The extract was then subjected to silica gel column (reverse phase column), using MeOH-H_2O as an eluent at the ratio of (20:80 → 100:00) to yield 8 fractions. As such, the main fractions obtained were dissolved in methanol and centrifuged at 12,000 rpm for 10 min. The first major fraction was further subjected to preparative HPLC (flow rate 10 mL/min, UV detector 210 nm Ruijia company, Hangzhou, China), using MeOH-H_2O as an eluent, to yielded compound **1** (6.1 mg, 60:40, t_R 22 min) and compound **2** (7.3 mg, 60:40, t_R 21 min). The second yielded fraction was further purified by preparative HPLC (flow rate 10 mL/min, UV detector 210 nm), using MeOH-H_2O as an eluent, to give a known antibiotic **3** (4.5 mg), previously isolated from a soil *Streptomyces* sp. SPRI-70014 and SANK 61296. Similarly, three more known compounds **4**, **5**, and **6** were isolated from the zinc treated (100 µM) crude extract of the same strain using analytical HPLC [flow rate 0.8 mL/min, eluted mode 0~30 min 20%~100% (H_2O/MeOH), 30–50 min 100% MeOH, UV detector 210 nm].

Stremycin A (**1**): Yellow powder; ^{1}H NMR and ^{13}C NMR, see Table 1; HR-TOF-MS *m/z* 1014.3937 [M + Na]$^+$ (Calcd. for $C_{48}H_{65}NNaO_{21}$, 1014.3941).

Stremycin B (**2**): Yellow powder; ^{1}H NMR and ^{13}C NMR, see Table 2; HR-TOF-MS *m/z* 1030.3860 [M + Na]$^+$ (Calcd. for $C_{48}H_{65}NNaO_{22}$, 1030.3890).

4.5. Antimicrobial Activity of Stressed Metabolites

Microbial activity was assessed using the conventional broth dilution assay with Gram-positive and Gram-negative clinical pathogens, namely, *K. pneumoniae* [CMCC (B) 46117], methicillin resistant *S. aureus* (MRSA), *B. subtilis* [CMCC(B) 63501], *E. coli* [CMCC(B) 44102], and *P. aeruginosa* [CMCC(B) 10104]. These pathogens were cultured in nutrient agar medium and left overnight incubation at 37 °C for 12–18 h. Each pathogenic culture was then diluted in 0.9% saline to an inoculum density of 5 × 10^5 cfu by comparison with a McFarland standard. Tetracycline was used as positive control while the solvent methanol was used as negative control. Methanolic solution of 3-[4,5-dimethylthiazol-2-yl]-2,5-diphenyltetrazolium bromide (MTT; Lancaster, PA, USA) was used to observe pathogenic growth by a change in color. 125 µL Muller Hinton broth was distributed into the 96-well plates. Similarly, samples were dispensed into well 1 and serially diluted across the well followed bacterial inoculation. Finally, the plates were incubated at 37 °C for 18 h and the results for bacteriostatic abilities of the compounds were noted in triplicate as MICs.

Supplementary Materials: The following are available online at http://www.mdpi.com/1660-3397/16/9/331/s1, Figures S1, S9, S16, S19, S21 and S24: ^{1}H NMR for compounds **1**, **2**, **3**, **4**, **5**, and **6**. Figures S2, S10, S17, S20, S22 and S25: ^{13}C NMR for compounds **1**, **2**, **3**, **4**, **5**, and **6**. Figures S3, S11, S18 and S23: DEPT135 for compounds **1**, **2**, **3** and **5**. Figures S4 and S12: COSY for compounds **1** and **2**. Figures S5 and S13: HMQC for compounds **1** and **2**. Figures S6, S7 and S15: HMBC for compounds **1** and **2**. Figures S8 and S15: NOESY for compounds **1** and **2**. Figures S26, S27, S29 and S30: HR-TOF-MS in positive and negative mode for compounds **1** and **2**. Figure S28: MS^n spectrum of compound **1**. Figure S31: 16S ribosomal DNA gene, full sequence of *Streptomyces* sp. (strain NA-ZhouS1). Figure S32: Neighbor-joining phylogenetic tree based on 16S rDNA sequence for strain NA-ZhouS1. Table S1: Biological activities of compounds **1–5** (MIC values are given in µg/mL).

Author Contributions: The experimental work was designed and performed by N.A. under the supervision of B.W. who is the corresponding author. X.W., K.W. and Y.L. contributed analysis tools. J.C. conducted the MS analysis. N.A., B.N.A., Y.S. and B.W. wrote the paper.

Funding: This research work was financially supported by the National Key R&D Program of China (No. 2018YFC0311002) and NSFC (Nos. 81573306 and 41876148).

Conflicts of Interest: The authors declare no conflict of interest.

References

1. Magdy, W.; Abdel-Motaal, F.F.; El-Zayat, S.A.; El-Sayed, M.A.E.; Helaly, S.E. Nigragillin, Nigerazine B and five Naphtho-γ-pyrones from *Aspergillus japonicus* isolated from hot desert soil. *Nat. Prod. J.* **2017**, *7*, 216–223. [CrossRef]
2. Ahmad, N.; Fatma, T. Production of indole-3-acetic acid by cyanobacterial strains. *Nat. Prod. J.* **2017**, *7*, 112–120. [CrossRef]
3. Gupta, A.; Singh, S. Characterization of NaCl-tolerant mutant strain of the cyanobacterium *Spirulina platensis* overproducing phycocyanin. *Nat. Prod. J.* **2017**, *7*, 153–164. [CrossRef]
4. Shi, Y.; Pan, C.; Auckloo, B.N.; Chen, X.; Chen, C.T.A.; Wang, K.; Wu, X.; Ye, Y.; Wu, B. Stress-driven discovery of a cryptic antibiotic produced by *Streptomyces* sp. WU20 from Kueishantao hydrothermal vent with an integrated metabolomics strategy. *Appl. Microbiol. Biotechnol.* **2017**, *101*, 1395–1408. [CrossRef] [PubMed]
5. Montaser, R.; Luesch, H. Marine natural products: A new wave of drugs? *Future Med. Chem.* **2012**, *3*, 1475–1489. [CrossRef] [PubMed]
6. Blunt, J.W.; Copp, B.R.; Keyzers, R.A.; Munro, M.H.G.; Prinsep, M.R. Marine natural products. *Nat. Prod. Rep.* **2017**, *34*, 235–294. [CrossRef] [PubMed]
7. Abdelfattah, M.S.; Kharel, M.K.; Hitron, J.A.; Baig, I. Moromycins A and B, Isolation and structure elucidation of C-glycosylangucycline-type antibiotics from *Streptomyces* sp. KY002. *Nat. Prod. J.* **2008**, *10*, 1569–1573. [CrossRef] [PubMed]
8. Rao, M.; Wei, W.; Ge, M.; Chen, D.; Sheng, X. A new antibacterial lipopeptide found by UPLC-MS from an actinomycete *Streptomyces* sp. HCCB10043. *Nat. Prod. Res.* **2013**, *27*, 2190–2195. [CrossRef] [PubMed]
9. Yang, Y.; Yang, X.; Zhang, Y.; Zhou, H.; Zhang, J.; Xu, L.; Ding, Z. A new daidzein derivative from endophytic *Streptomyces* sp. YIM 65408. *Nat. Prod. Res.* **2013**, *27*, 1727–1731. [CrossRef] [PubMed]
10. Awad, H.M.; El-Enshasy, H.A.; Hanapi, S.Z.; Hamed, E.R.; Rosidi, B. A new chitinase-producer strain *Streptomyces glauciniger* WICC-A03: Isolation and identification as a biocontrol agent for plants phytopathogenic fungi. *Nat. Prod. Res.* **2014**, *28*, 2273–2277. [CrossRef] [PubMed]
11. Sun, D.; Sun, W.; Yu, Y.; Li, Z.; Deng, Z.; Lin, S. A new glutarimide derivative from marine sponge-derived *Streptomyces anulatus* S71. *Nat. Prod. Res.* **2014**, *28*, 1602–1606. [CrossRef] [PubMed]
12. Zhang, G.; Zhang, Y.; Yin, X.; Wang, S. Nesterenkonia alkaliphila sp. Nov., an alkaliphilic, halotolerant actinobacteria isolated from the western pacific ocean. *Int. J. Syst. Evol. Microbiol.* **2015**, *65*, 516–521. [CrossRef] [PubMed]
13. Balachandran, C.; Duraipandiyan, V.; Emi, N.; Ignacimuthu, S. Antimicrobial and cytotoxic properties of *Streptomyces* sp. (ERINLG-51) isolated from Southern Western Ghats. *South Indian J. Biol. Sci.* **2015**, *1*, 7–14. [CrossRef]
14. Hu, Z.; Qin, L.; Wang, Q.; Ding, W.; Chen, Z.; Ma, Z. Angucycline antibiotics and its derivatives from marine-derived actinomycete *Streptomyces* sp. A6H. *Nat. Prod. Res.* **2016**, *30*, 2551–2558. [CrossRef] [PubMed]
15. Chen, P.; Zhang, L.; Guo, X.; Dai, X.; Liu, L.; Xi, L.; Wang, J.; Song, L.; Wang, Y.; Zhu, Y.; et al. Diversity, biogeography, and biodegradation potential of actinobacteria in the deep-sea sediments along the southwest Indian ridge. *Front. Microbiol.* **2016**, *7*, 1–17. [CrossRef] [PubMed]
16. Helaly, S.E.; Goodfellow, M.; Zinecker, H.; Imhoff, J.F.; Süssmuth, R.D.; Fiedler, H.P. Warkmycin, a novel angucycline antibiotic produced by *Streptomyces* sp. Acta 2930. *J. Antibiot. (Tokyo)* **2013**, *66*, 669–674. [CrossRef] [PubMed]
17. Ma, M.; Rateb, M.E.; Teng, Q.; Yang, D.; Rudolf, J.D.; Zhu, X.; Huang, Y.; Zhao, L.X.; Jiang, Y.; Li, X.; et al. Angucyclines and Angucyclinones from *Streptomyces* sp. CB01913 featuring C-ring cleavage and expansion. *J. Nat. Prod.* **2015**, *78*, 2471–2480. [CrossRef] [PubMed]
18. Zotchev, S.B. Marine actinomycetes as an emerging resource for the drug development pipelines. *J. Biotechnol.* **2012**, *158*, 168–175. [CrossRef] [PubMed]
19. Feling, R.H.; Buchanan, G.O.; Mincer, T.J.; Kauffman, C.A.; Jensen, P.R.; Fenical, W. Salinosporamide A: A highly cytotoxic proteasome inhibitor from a novel microbial source, a marine bacterium of the new genus *Salinospora*. *Angew. Chem. Int. Ed.* **2003**, *42*, 355–357. [CrossRef] [PubMed]
20. Shi, Y.; Jiang, W.; Auckloo, B.N.; Wu, B. Several classes of natural products with metal ion chelating ability. *Curr. Org. Chem.* **2015**, *19*, 1935–1953. [CrossRef]

21. Auckloo, B.N.; Pan, C.; Akhter, N.; Wu, B.; Wu, X.; He, S. Stress-driven discovery of novel cryptic antibiotics from a marine fungus *Penicillium* sp. BB1122. *Front. Microbiol.* **2017**, *8*, 1450. [CrossRef] [PubMed]
22. Ding, C.; Wu, X.; Auckloo, B.N.; Chen, C.T.A.; Ye, Y.; Wang, K.; Wu, B. An unusual stress metabolite from a hydrothermal vent fungus *Aspergillus* sp. WU 243 induced by cobalt. *Molecules* **2016**, *21*, 105. [CrossRef] [PubMed]
23. Jiang, W.; Zhong, Y.; Shen, L.; Wu, X.; Ye, Y.; Chen, C.-T.; Wu, B. Stress-driven discovery of natural products from extreme marine environment- Kueishantao hydrothermal vent, a case study of metal switch valve. *Curr. Org. Chem.* **2014**, *18*, 925–934. [CrossRef]
24. Shaaban, K.A.; Ahmed, T.A.; Leggas, M.; Rohr, J. Saquayamycins G–K, cytotoxic angucyclines from *Streptomyces* sp. including two analogues bearing the Amino-sugar Rednose. *J. Nat. Prod.* **2012**, *75*, 1383–1392. [CrossRef] [PubMed]
25. Zhou, Z.; Xu, Q.; Bu, Q.; Guo, Y.; Liu, S.; Liu, Y.; Du, Y.; Li, Y. Genome mining-directed activation of a silent angucycline biosynthetic gene cluster in *Streptomyces chattanoogensis*. *Chembiochem* **2015**, *16*, 496–502. [CrossRef] [PubMed]
26. Rohr, J.; Thiericke, R. Angucycline group antibiotics. *Nat. Prod. Rep.* **1992**, *9*, 103–137. [CrossRef] [PubMed]
27. Kharel, M.K.; Pahari, P.; Shepherd, M.D.; Tibrewal, N.; Nybo, S.E.; Shaaban, K.A.; Rohr, J. Angucyclines: Biosynthesis, mode-of-action, new natural products, and synthesis. *Nat. Prod. Rep.* **2012**, *29*, 264–325. [CrossRef] [PubMed]
28. Uesato, S.; Tokunaga, T.; Mizuno, Y.; Fujioka, H.; Kada, S.; Kuwajima, H. Absolute stereochemistry of gastric antisecretory compound P371A1 and its congener P371A2 from *Streptomyces* species P371. *J. Nat. Prod.* **2000**, *63*, 787–792. [CrossRef] [PubMed]
29. Xue, Z.R.; Xu, W.P.; Chen, J.; Tao, L.M. Study of *Streptomyces griseolus* CGMCC 1370 and its herbicidal metabolite. *Chin. J. Pestic. Sci.* **2011**, *13*, 155–161.
30. Cronan, J.M.; Davidson, T.R.; Singleton, F.L.; Colwell, R.R.; Cardellina, J.H. Plant growth promoters isolated from a marine bacterium associated with *Palythoa* sp. *Nat. Prod. Lett.* **1998**, *11*, 271–278. [CrossRef]
31. Maskey, R.P.; Shaaban, M.; Gru, I. Quinazolin-4-one derivatives from *Streptomyces* isolates. *Nat. Prod. J.* **2004**, *7*, 1131–1134. [CrossRef] [PubMed]
32. Sharipov, B.T.; Pilipenko, A.N.; Valeev, F.A. Eleuthesides and their analogs: VIII. Preparation of menthane derivatives from levoglucosenone and (2*E*,4*E*)-6-methylhepta-2,4-dienyl acetate by Diels-Alder reaction. *Russ. J. Org. Chem.* **2014**, *50*, 1628–1635. [CrossRef]
33. Schneemann, I.; Kajahn, I.; Ohlendorf, B.; Zinecker, H.; Erhard, A.; Nagel, K.; Wiese, J.; Imhoff, J.F. Mayamycin, a cytotoxic polyketide from a *Streptomyces* strain isolated from the marine sponge *Halichondria panicea*. *J. Nat. Prod.* **2010**, *73*, 1309–1312. [CrossRef] [PubMed]
34. Abdelfattah, M.; Maskey, R.P.; Asolkar, R.N.; Grün-Wollny, I.; Laatsch, H. Seitomycin: Isolation, structure elucidation and biological activity of a new angucycline antibiotic from a terrestrial *Streptomycete*. *J. Antibiot. (Tokyo)* **2003**, *56*, 539–542. [CrossRef] [PubMed]
35. Gu, X.B.; Xu, W.P.; Ye, G.Y.; Zhang, Y.L.; Ni, W.W.; Tao, L.M. The herbicidal active component derived from soil actinomycete SPRI 70014. *Chin. J. Antibiot.* **2008**, *33*, 461–466.
36. Morishita, T.; Sato, A.; Ando, T.; Oizumi, K.; Miyamoto, M.; Enokita, R.; OKAZAKI, T. A novel bone resorption inhibitor, A-75943 isolated from *Streptomyces* sp. SANK 61296. *J. Antibiot. (Tokyo)* **1998**, *51*, 531–538. [CrossRef] [PubMed]
37. Kalinovskaya, N.I.; Kalinovsky, A.I.; Romanenko, L.A.; Pushilin, M.A.; Dmitrenok, P.S.; Kuznetsova, T.A. New angucyclinones from the marine mollusk associated actinomycete *Saccharothrix espanaensis* An 113. *Nat. Prod. Commun.* **2008**, *3*, 1611–1616.
38. Fotso, S.; Mahmud, T.; Zabriskie, T.M.; Santosa, D.A.; Proteau, P.J. Angucyclinones from an Indonesian *Streptomyces* sp. *J. Nat. Prod.* **2007**, *71*, 61–65. [CrossRef] [PubMed]
39. Fedoryshyn, M.; Nur-e-Alam, M.; Zhu, L.; Luzhetskyy, A.; Rohr, J.; Bechthold, A. Surprising production of a new urdamycin derivative by *S. fradiae*Δ*urdQ/R*. *J. Biotechnol.* **2007**, *130*, 32–38. [CrossRef] [PubMed]
40. Li, L.; Pan, G.; Zhu, X.; Fan, K.; Gao, W.; Ai, G.; Ren, J.; Shi, M.; Olano, C.; Salas, J.A. Engineered jadomycin analogues with altered sugar moieties revealing *JadS* as a substrate flexible *O*-glycosyltransferase. *Appl. Microbiol. Biotechnol.* **2017**, *13*, 5291–5300. [CrossRef] [PubMed]

41. Huang, H.; Yang, T.; Ren, X.; Liu, J.; Song, Y.; Sun, A.; Ma, J.; Wang, B.; Zhang, Y.; Huang, C. Cytotoxic angucycline class glycosides from the deep sea actinomycete *Streptomyces lusitanus* SCSIO LR32. *J. Nat. Prod.* **2012**, *75*, 202–208. [CrossRef] [PubMed]

42. Henkel, T.; Rohr, J.; Beale, J.M.; Schwenen, L. Landomycins, new angucycline antibiotics from *Streptomyces* sp. *J. Antibiot. (Tokyo)* **1990**, *43*, 492–503. [CrossRef] [PubMed]

43. Rong, X.; Doroghazi, J.R.; Cheng, K.; Zhang, L.; Buckley, D.H.; Huang, Y. Classification of *Streptomyces* phylogroup *pratensis* (Doroghazi and Buckley, 2010) based on genetic and phenotypic evidence, and proposal of *Streptomyces pratensis* sp. nov. *Syst. Appl. Microbiol.* **2013**, *36*, 401–407. [CrossRef] [PubMed]

44. Doroghazi, J.R.; Buckley, D.H. Intraspecies comparison of *Streptomyces pratensis* genomes reveals high levels of recombination and gene conservation between strains of disparate geographic origin. *BMC Genomics* **2014**, *15*, 1–14. [CrossRef] [PubMed]

marine drugs — MDPI

Article

Characterization of the Microbial Population Inhabiting a Solar Saltern Pond of the Odiel Marshlands (SW Spain)

Patricia Gómez-Villegas, Javier Vigara and Rosa León *

Laboratory of Biochemistry and Molecular Biology, Faculty of Experimental Sciences, Marine International Campus of Excellence (CEIMAR), University of Huelva, 21071 Huelva, Spain; patgomvil@gmail.com (P.G.-V.); vigara@uhu.es (J.V.)
* Correspondence: rleon@uhu.es; Tel.: +34-959-219-951

Received: 28 June 2018; Accepted: 8 September 2018; Published: 12 September 2018

Abstract: The solar salterns located in the Odiel marshlands, in southwest Spain, are an excellent example of a hypersaline environment inhabited by microbial populations specialized in thriving under conditions of high salinity, which remains poorly explored. Traditional culture-dependent taxonomic studies have usually under-estimated the biodiversity in saline environments due to the difficulties that many of these species have to grow at laboratory conditions. Here we compare two molecular methods to profile the microbial population present in the Odiel saltern hypersaline water ponds (33% salinity). On the one hand, the construction and characterization of two clone PCR amplified-16S rRNA libraries, and on the other, a high throughput 16S rRNA sequencing approach based on the Illumina MiSeq platform. The results reveal that both methods are comparable for the estimation of major genera, although massive sequencing provides more information about the less abundant ones. The obtained data indicate that *Salinibacter ruber* is the most abundant genus, followed by the archaea genera, *Halorubrum* and *Haloquadratum*. However, more than 100 additional species can be detected by Next Generation Sequencing (NGS). In addition, a preliminary study to test the biotechnological applications of this microbial population, based on its ability to produce and excrete haloenzymes, is shown.

Keywords: halo-extremophyles; archaea; 16S rRNA metagenomics; haloenzymes; Odiel marshlands

1. Introduction

The study of the microbial population inhabiting extreme saline environments has gained increasing interest in the last years due to its usually uncompleted characterization, which is essential to understand the ecology of these ecosystems [1] and also because archaea have revealed themselves as the key to understand the origin of eukaryotic cells [2]. Furthermore, these halo-extremophyles microorganisms can be an excellent source of useful compounds and proteins with special properties and potential industrial applications [3,4] such as antioxidant pigments [5,6], haloestable enzymes [7], antimicrobial compounds [8] or antitumor agents [9]. Haloenzymes have unique characteristics that allow them to be stable and functional at saline concentrations as high as 5 M and tolerate high temperatures without losing their activity [10]. This fact makes halotolerant archaea a potential source of enzymes for food, textile, pharmaceutical or chemical industries [11].

The extreme conditions that prevail in salt brines, which include high light intensity, UV radiation, elevated temperatures and salt concentrations near saturation, support a considerable diversity of halophilic microorganisms belonging mainly to the haloarchaea group [12]. Traditional ecological studies, based on serial dilutions or streaking on agar plates for single-cell isolation, have usually underestimated this biodiversity due to the difficulties of some species to grow at lab conditions.

Examples of this are the unsuccessful attempts to culture some generally abundant archaea genera, such as the square-shaped *Haloquadratum* [13]; or the discrepancy commonly found between the characterization of microbial communities by culture-independent and culture-dependent methods.

The application of molecular techniques, based on the comparison of highly conserved DNA reference sequences, such as the genes encoding for ribosomal RNA (16S rRNA, 5S rRNA), has allowed to overcome this limitation, making possible the comparison of different microbial communities and the discovery of a good number of uncultured new species. Examples of these culture-independent methods include: random fragment length polymorphisms (RFLP) [14], fluorescence *in situ* hybridization with rRNA-targeted probes [15], denaturing gradient gel electrophoresis (DGGE) [16] and more recently metagenomic approaches [17,18], which have facilitated the profiling of complex microbial communities. Culture-independent characterization of the microbial assemblages in different halophilic habitats has shown that diversity within the domain archaea is broader than that previously inferred from culture-dependent surveys [19].

The solar salterns located in the Odiel Marshlands (Huelva), at the southwest of Spain, are an excellent example of a hypersaline environment inhabited by microbial populations specialized in thriving under conditions of high salinity, which remains poorly explored [20]. In this work, we have used two independent methods to study the prokaryotic diversity present in these salterns. We have constructed and characterized two libraries of PCR amplified 16S rRNA genes obtained using genomic DNA extracted from a water sample of the brines as template; and we have used a high throughput 16S rRNA massive sequencing approach based on the Illumina MiSeq platform to profile the same genomic sample. This double approach has allowed us, not only to explore the microbial diversity of this water environment but also to validate the results of the PCR gene library by comparison with the NGS approach. Furthermore, the ability of this microbial population to produce and excrete haloenzymes with applied interest has been studied.

2. Results

2.1. Construction of a 16S rRNA Library and Identification of the Obtained Sequences

Two 16S rRNA libraries clone libraries, one for archaea and another one for bacteria, were constructed from an environmental water sample collected at the end of the summer, in the crystallizer ponds located in the Marshlands of the Odiel river in the southwest coast of Spain. The procedure for the libraries construction is detailed in Section 4.4 and the specific primers used listed in Table S1 (Supplementary Material). The main chemical characteristics of the water at the collection time are summarized in Table 1. The salt composition was similar to that reported for other thalassohaline marine solar salterns at this degree of salinity, which was 33.2% at the time of sample collection.

Table 1. Chemical composition of the water sample collected from the evaporation ponds located in the Natural Reserve of Odiel Marshlands in the southwest Spain.

Density (g·mL^{-1})	Brine Composition (g·L^{-1})						Total Salinity
	CaSO$_4$	MgSO$_4$	MgCl$_2$	NaCl	KCl	NaBr	
1.212	1.40	23.06	34.08	265.38	7.51	0.84	332.30

A selection of 50 clones from both clone libraries, 25 clones per each library, were analysed. The preliminary comparison of the obtained sequences with the National Center for Biotechnology Information database (NCBI) revealed that many of the clones in the 16S rRNA libraries were redundant. In the archaeal library, the 25 clones studied corresponded to 11 different species, belonging to six genera (Figure 1). Most clones corresponded to the archaea genus *Halorubrum*, followed by the peculiar square-shaped haloarchaea *Haloquadratum* [13,21]. These two genera represented respectively 32% and 28% of the total archaeal clones obtained, followed by *Halonotius* (12%) and *Halobellus* (8%). The least abundant genera identified were *Haloarcula* and *Halorientalis*, each one represented 4% of the total clones (Figure 1). Three clones could not be directly affiliated to any currently described genera, since they did not reach the 95% of sequence identity with any of the sequences of the database. The Shannon biodiversity index was calculated as previously reported [22], obtaining a value of 2.07. Despite the small number of clones analysed this score indicates a wide range of diversity.

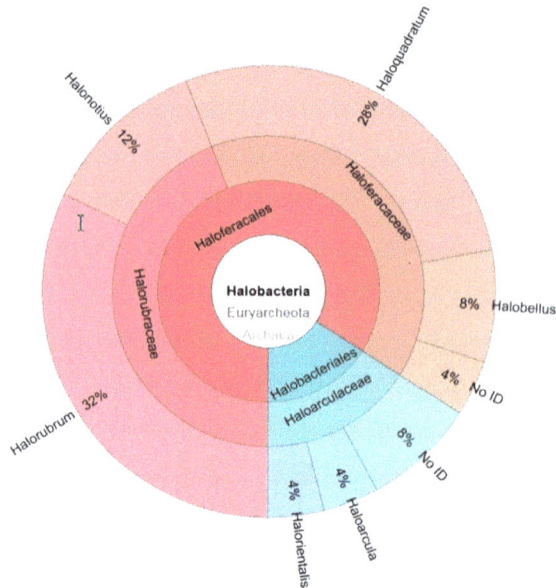

Figure 1. Distribution of the clones from the archaeal 16S rRNA clone library among different genera. 16S rRNA fragments obtained by amplification with archaeal specific primers were cloned into pGEMT vector for the construction of a clone library. The inserted sequence of 25 of the obtained clones were analysed and compared with the NCBI to identify the original genera. Data are expressed as percentages of the total archaeal population. Only sequences that shared over 95% 16S rRNA sequence identity with a known one was assigned to a specific genus.

To obtain additional information, a molecular phylogenetic analysis has been done including all the 16S rRNA encoding sequences obtained from the analysis of the archaeal library and several reference sequences obtained from the NCBI data base. The evolutionary history was inferred by using the Maximum Likelihood method based on the General Time Reversible model [23] (Figure 2).

Some of the amplified 16S rRNA sequences obtained in the archaea library showed 100% of identity or were closely related, showing ≤3% sequence divergence, with species already classified and could be assigned at species level. As it is shown in Figure 2, the sequences Col.10/11/12 and Col. 1/4/14/15/18/19/20 can be assigned to the species *Halonotius pteroides* and *Haloquadratum walsbyi*, respectively. The *Haloquadratum* sequences found (Col.1/14/15/18/19/20) displayed sequence divergences lower than 1% with *H. walsbyi*. By contrast, sequences related to *Halorubrum* genus (Col.21/25, Col.13/22, Col.23/24, Col.3, Col.16), *Halobellus* genus (Col.6/9) and *Haloarcula* genus (Col.17) showed high percentage of identity with several reference species. In addition, sequences clustering within *Halorientalis* genus (Col.7) presented more than 3% divergence with the reference sequences; being impossible their assignation to a particular species in these cases. Finally, two of the sequences which could not be directly assigned to a genus (Col.5, Col.8) clustered within the *Haloarcula* clade, while Col.2 is strongly related to the genus *Hallobelus*.

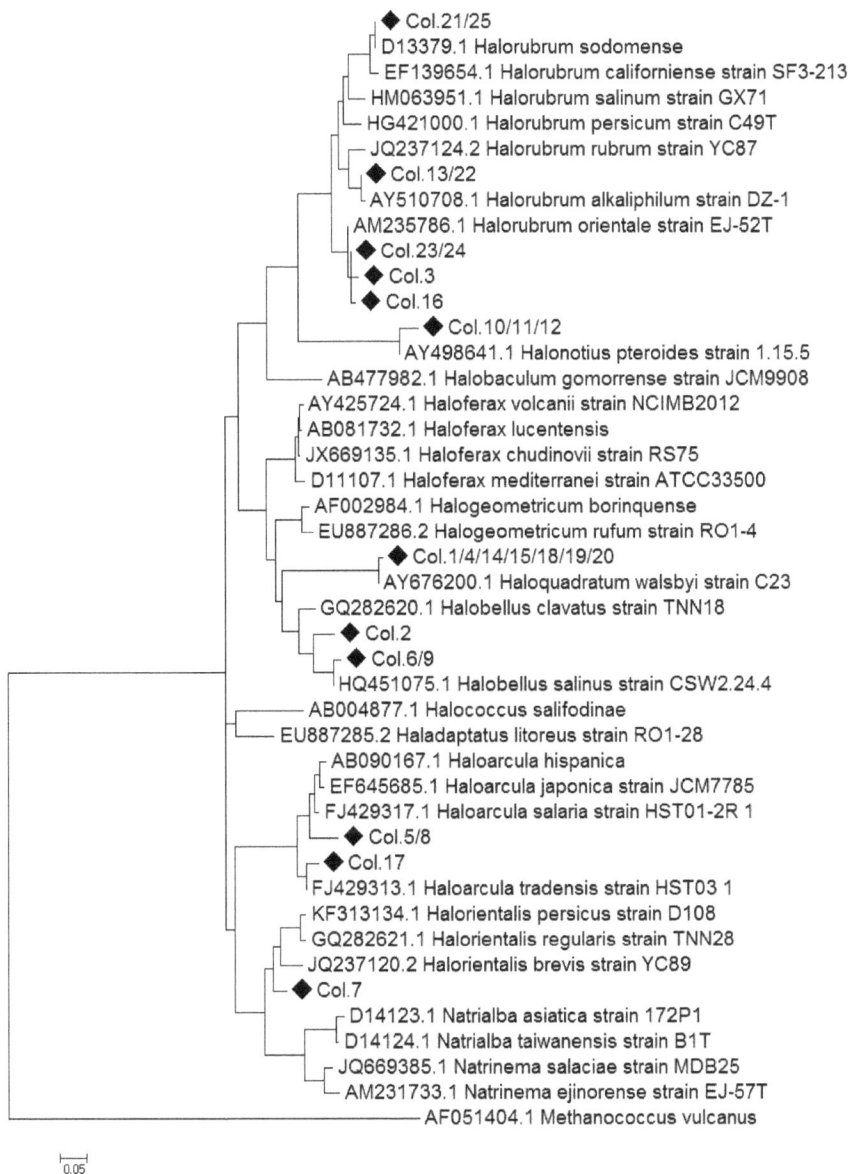

◆ Col.21/25
D13379.1 Halorubrum sodomense
EF139654.1 Halorubrum californiense strain SF3-213
HM063951.1 Halorubrum salinum strain GX71
HG421000.1 Halorubrum persicum strain C49T
JQ237124.2 Halorubrum rubrum strain YC87
◆ Col.13/22
AY510708.1 Halorubrum alkaliphilum strain DZ-1
AM235786.1 Halorubrum orientale strain EJ-52T
◆ Col.23/24
◆ Col.3
◆ Col.16
◆ Col.10/11/12
AY498641.1 Halonotius pteroides strain 1.15.5
AB477982.1 Halobaculum gomorrense strain JCM9908
AY425724.1 Haloferax volcanii strain NCIMB2012
AB081732.1 Haloferax lucentensis
JX669135.1 Haloferax chudinovii strain RS75
D11107.1 Haloferax mediterranei strain ATCC33500
AF002984.1 Halogeometricum borinquense
EU887286.2 Halogeometricum rufum strain RO1-4
◆ Col.1/4/14/15/18/19/20
AY676200.1 Haloquadratum walsbyi strain C23
GQ282620.1 Halobellus clavatus strain TNN18
◆ Col.2
◆ Col.6/9
HQ451075.1 Halobellus salinus strain CSW2.24.4
AB004877.1 Halococcus salifodinae
EU887285.2 Haladaptatus litoreus strain RO1-28
AB090167.1 Haloarcula hispanica
EF645685.1 Haloarcula japonica strain JCM7785
FJ429317.1 Haloarcula salaria strain HST01-2R 1
◆ Col.5/8
◆ Col.17
FJ429313.1 Haloarcula tradensis strain HST03 1
KF313134.1 Halorientalis persicus strain D108
GQ282621.1 Halorientalis regularis strain TNN28
JQ237120.2 Halorientalis brevis strain YC89
◆ Col.7
D14123.1 Natrialba asiatica strain 172P1
D14124.1 Natrialba taiwanensis strain B1T
JQ669385.1 Natrinema salaciae strain MDB25
AM231733.1 Natrinema ejinorense strain EJ-57T
AF051404.1 Methanococcus vulcanus

0.05

Figure 2. Molecular Phylogenetic Analysis by Maximum Likelihood method. The tree represents the relationship among the 16S rRNA sequences from strains isolated from the saltern ponds of the Odiel Marshlands and reference archaeal sequences. Multiple alignments were generated by MUSCLE and the tree was constructed with MEGA 7, using 1000 bootstrap replicates. The name and the NCBI access number are indicated for all the reference sequences. Black diamonds represent 16S rDNA Sequences from the isolates and "Col. N" denotes the colony number. When an identical sequence was obtained from different colonies it was denoted as "Col. N1/N2." The tree is drawn to scale, branch lengths represent the number of substitutions per site. Scale bar indicate 5% sequence divergence. The sequence of *Methanococcus vulcanus* was used as the outgroup.

By contrast, all the clones isolated from the bacterial gene library contained 16S rRNA sequences with 99–100% sequence identity to a unique bacterial species, *Salinibacter ruber*, which is usually present in hypersaline ponds [24]. The chosen primers have been shown to be very specific for each prokaryotic group studied, since archaea sequences have not been obtained in the bacterial library, nor have bacterial sequences been detected in the archaeal library. This specificity makes impossible the use of these primers to create a common bacterial/archaeal library.

2.2. Metagenomic Microbial Profiling by High-throughput 16S rRNA Sequencing

As a second approach, the identification of the microbial population present in the hypersaline water from the Odiel saltern ponds was performed by next-generation sequencing of the 16S rRNA gene, using the Illumina MiSeq platform as detailed in Materials and Methods. The analysis was set up in quadruplicate by two independent sequencing services, Stabvida (SBV) and Life Sequencing (LFS). Bioinformatic processing with the software pipelines described in Materials and Methods allowed us to cluster the obtained reads into a limited number of operational taxonomic units, between 117 and 356, depending on the sequencing reaction and the bioinformatic treatment of the obtained sequences (Table 2). The mean quality of the processed sequences was denoted by the Q scores and the Shannon biodiversity index (H'), calculated including the whole prokaryotic population and following previously described procedures [22,25].

Table 2. Sequence data statistics.

Reaction	Raw Sequence Reads	Mean Read Length (bp)	Sequences after Denoising	Mean Quality (Q Score)	OTUs	Shannon Index
SBV-1	349 726	250	49 100	>28	177	2.75
SBV-2	204 766	250	19 537	>28	117	2.65
LFS-1	57 148	299.8	25 479	37.16	228	2.77
LFS-2	156 520	299.6	71 623	37.25	356	3.03

The number of sequences and Operational Taxonomic Units (OTUs) obtained from Stabvida (SBV) and Life Sequencing (LFS) are shown. The mean quality, expressed as Q scores and the Shannon biodiversity index for each sequencing run have also been included.

Both 16S rRNA NGS analysis indicate that the most abundant reads correspond to the halophilic bacteria *Salinibacter* and the archaeal genera *Halorubrum*, *Haloquadratum* and *Halonotius*, although there are significant discrepancies between their relative abundance. *Salinibacter* represents between 38% and 42% of the total reads. *Halorubrum* (13–19%), *Haloquadratum* (9–18%) and *Halonotius* (8–9%) are the main archaeal genera, followed by *Halobellus* (3–4%), *Natronomonas* (2.5–3%). *Haloplanus* and *Halohasta* each represent around 3% of the total sequences after the analysis of Life Sequencing and only trace amounts (0.1%) in the data from Stabvida. *Halomicroarcula*, *Salinivenus*, *Halovenus*, *Halomicrobium*, *Haloarientalis*, *Haloarcula* and *Halosimplex*, with relative abundances between 0.7% and 1.5%, are also present in both analyses. Genera with relative abundances lower than 0.2% have not been shown in Figure 3, but the complete list of sequenced genera is shown in Supplementary Material (Table S2).

It is interesting to note that the NGS analysis revealed the presence of trace amounts of genus such as *Spiribacter* (0.1%), a moderate halophilic bacteria usually found in medium salinity habitats [26] and other minor bacteria, not revealed in the clone library approach (Table S2).

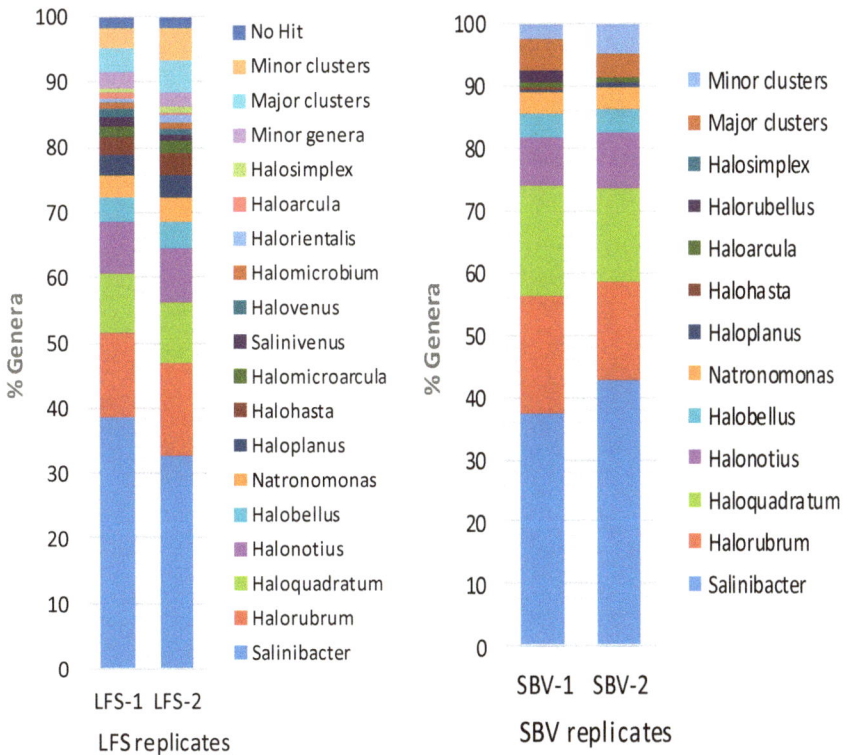

Figure 3. Relative abundance of the genera found by the metagenomic approaches. Operational Taxonomic Units (OTUs) distribution in Odiel saline ponds (33% salinity) obtained from Illumina MiSeq sequencing of the 16S rRNA V3, V4 hypervariable regions. Four data sets were obtained from two different sequencing services. The graphic shows the percentage of the genera with more than 0.2% abundance. Minor genera include all the genera below 0.2%. The sequences that clustered together but could not been affiliated to a genus are named as "clusters" and have been divided in "Major clusters" (>0.2%) and "Minor clusters" (<0.2%). "No hit" represents the sequences which did not cluster with any other obtained sequence.

2.3. Comparison of Clone Library and 16S rRNA Metagenomic Approaches to Identify the Archaeal Microbiota of the Odiel Saltern Ponds Water

The relative abundances of the main genera obtained from the two NGS platforms are compared with those obtained by affiliation of the 25 sequences gathered from the archaeal clone library (Figure 4). Despite the low number of sequences retrieved from the clone library, there was a considerable degree of agreement between the main genera obtained by this method and by the NGS approaches, as it is supported by a correlation coefficient of 0.97 when comparing clone library results versus the mean of NGS results. Similarly, a correlation coefficient of 0.96 was obtained when we compared both NGS methods (Mean LSF vs. Mean SVB).

The estimated percentage of *Halorubrum* and *Haloquadratum* obtained by the clone library approach are almost identical to those obtained by the Stabvida analysis, being the standard deviation (SD) for these values 2 and 0.34 respectively, while the percentages of *Halonotius* and *Halobellus* are of the same order than those obtained by Life Sequencing (SD 2.48 and 0.35, respectively) or Stabvida (SD 1.78 and 1.41, respectively). *Halorientalis* and *Haloarcula* which represented about 4% of the library clones are also present in the massive 16S rRNA analysis but at lower percentage than that estimated by the clone library approach. Standard deviations in these cases were respectively 1.42 and 1.77, comparing the results of the clone library with the Life Sequencing results; and 2.82 and 2.12 respectively, when comparing with Stabvida results. Massive sequencing provides more information about the less abundant species, although the analysis of higher number of archaeal clones could have allowed the identification of more minor genera by the clone library approach.

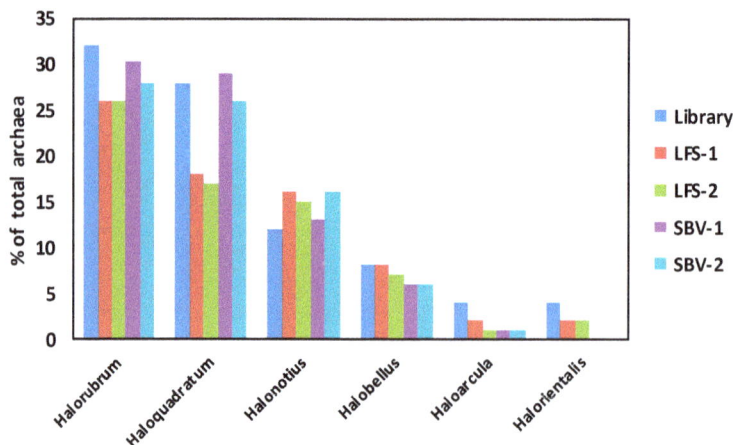

Figure 4. Comparison of the main archaeal genera present in the Odiel saline ponds. Percentage of the different archaea genera over the total archaea population found in the Odiel saline ponds obtained by the two different culture-independent approaches previously described: construction of a clone library (Library) and massive 16S rRNA sequencing, including the two replicates from each Metagenomic Service (LFS and SBV). Only the most abundant genera are shown.

2.4. Evaluation of Halocin Activity

Massive 16S rRNA sequencing has revealed an extremely low representation of the genus *Haloferax* (0.021%) in the Odiel Saltern ponds (Table S2). However, *Haloferax* is a metabolically versatile genus, able to grow on complex substrates and degrade polymeric substances with a wide salt tolerance in laboratory cultures [27]. To investigate the possible reasons for the low presence of *Haloferax* in the Odiel saltern water, we studied the potential ability of the biomass isolated from the Odiel saterns to specifically inhibit the growth of the control *Haloferax* species, *Haloferax lucetense*. The results show the inhibition of *H. lucetense* growth in the presence of the concentrated biomass isolated from the Odiel saltern ponds, indicating the presence of halocin activity (Figure 5A), which could be an important factor to explain the practically absence of species of the *Haloferax* genus in the hypersaline water from the Odiel saltern ponds.

2.5. Haloenzymes Production by the Archaeal Enriched Biomass Isolated from the Odiel Saltern Ponds

Simple and sensitive plate assays were optimized for the detection of archaeal extracellular enzymes produced by the biomass isolated from the hypersaline water (33%) of the Odiel salterns. The biomass was enriched, concentrated and dropped on agar plates supplemented with different carbon sources to detect the excretion of α-amylase, protease, lipase, cellulase and laccase as described in Materials and Methods.

The amylase (Figure 5B), protease (Figure 5C) and lipase (Figure 5D) activities, assayed as described in Material and Methods following the degradation of starch, skimmed milk and Tween 80, respectively, were positive. Cellulase and exo-lacasse activities, assayed in the presence of carboxymethyl cellulose (Figure 5E) and bromophenol blue (Figure 5F), were also detected.

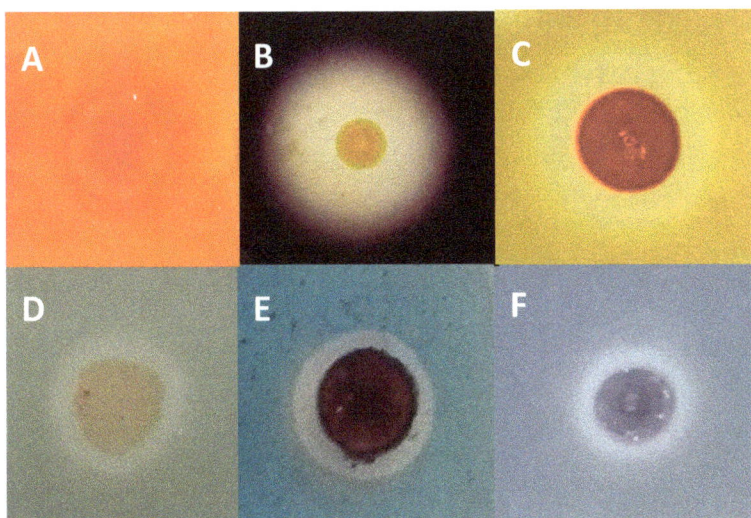

Figure 5. Detection of extracellular halocin and enzymatic activities in the enriched archaeal biomass by plate assay. The biomass obtained from the Odiel salterns water was enriched and concentrated as detailed in Material and Methods and used to test extracellular: halocin (**A**), amylase (**B**), protease (**C**), lipase (**D**), cellulase (**E**) and laccase (**F**) activities by the plate assays described in Materials and Methods.

3. Discussion

3.1. Microbiological Diversity in Hypersaline Solar Saltern Ponds

Despite the existence of many studies about the microbiota inhabiting thalassosaline water ponds all around the word, it is difficult to establish comparisons among them, due to the variety of experimental procedures used to determine microbial composition and the dramatic influence of the water salinity, which can range between the 3.5% of the sea water to the 37% of NaCl saturation.

Numerous studies have described the prokaryotic communities that inhabit saltern crystallizer ponds as distant as the Mediterranean coast of Spain [15,18], Australia [28] or Mexico [29], among others. However, the microbial profile of the Odiel salterns, differently from that of other Spanish salterns, such as Santa Pola in Alicante [24,30] or Isla Cristina in Huelva [18], has been poorly studied. To our knowledge, the only survey about the microbiota of the Odiel Marshlands is a recent study by Vera-Gargallo and Ventosa [20], which focused on studying the assemblage and the metabolic strategies of the microbiota thriving in these hypersaline soils but no information about the aquatic microbial composition of this location has been previously reported.

Most hypersaline ponds, regardless their location, are dominated by the archaeal genera species *Haloquadratum* and *Halorubrum* [31], which coexists and compete for the same hypersaline environments. *Haloquadratum* was the predominant genus followed by *Halorubrum*, in three crystallizer ponds as geographically distant as Australia at 34% salinity [28], Santa Pola (Spain) at 32% salinity [18] and in Bingöl (Turkey) at 25% salinity [32]. Conversely, our results at 33% salinity suggest that in Odiel Salters this relation is inverse, being the most abundant archaeal genus *Halorubrum* followed by *Haloquadratum*, which often dominates the microbial communities in hypersaline waters and was not cultured in a laboratory until 2004 [13,21]. This difference could be attributed to changes in environmental conditions, due to the diverse geographical situations [18]. There are also some interesting exceptions, such as the Maras salterns in the Peruvian Andes, in which around 31% of the detected sequences were related to the usually low abundant *Halobacterium* and no *Halorubrum* was detected [33]; the Adriatic solar saltern crystallizers studied by Pašić et al. [34], where the presence of the usually abundant *Haloquadratum* was rare; or the Pomorie Salterns (Bulgaria) where the predominant archaea genus was *Halanaeroarchaeum*, which reached 28% of the total archaeal community [35]. The aforementioned study carried out by Vera-Gargallo and Ventosa about saline soils from Odiel saltmarshes reveals that *Haloquadratum* was not found in these soils; in contrast to our results from hypersaline water in the same location, which suggest that this is one of the dominant archaeal genus. However, other genera identified in our work, such as *Halorubrum*, *Haloarcula* and *Halobellus*, are also present in the soil samples studied [20]. The predominant microbial genera and their estimated composition in other hypersaline ponds with salinity similar to our study are summarized in Table 3. The most complete studies have been done in Santa Pola, in the Mediterranean coast of Spain and showed that *Haloquadratum* was the most abundant microorganism and that its relative abundance increased with salinity. In hypersaline ponds in Australia (34%) [28] the three most abundant haloarchaeal genera were the same that we observed in the Odiel ponds. However, it is important to notice that in this study and in the study carried out by Kambourova et al. in Bulgaria [35], the bacterial contribution to the prokaryotic community was not considered because the study was based on clone libraries with archaeal specific primers. This comparison confirms that although there are some common genera, which are found in almost all hypersaline ponds, such as *Halorubrum* or *Haloquadratum*, the relative abundance of hypersaline genera is specific of each geographic location. In addition, identifications at species level returned unique lineages which appeared to be specific of the investigated environments.

Salinibacter ruber, was first identified in Santa Pola, Spain (Alicante, Spain) and, despite not being an archaea, it is usually one of the most abundant microorganisms present in hypersaline waters [24]. In Santa Pola ponds with salinities of 19%, 33% and 37% the abundance of *Salinibacter* was reported to be 6.4%, 4.7% and 9.1%, respectively [18]. Our results suggest that the bacteria *Salinibacter ruber* is the dominant microorganism in the Odiel saltern water with a salinity of 33%, reaching around 40% of the total prokaryotic community (Figure 3), which is the highest *Salinibacter* abundance reported in solar salterns to our knowledge. The libraries for the metagenomics studies were built with the universal primers, which have been designed to target V3 and V4 hypervariable regions from both bacterial and archaeal 16S rRNA [36]; however, additional research is needed to determine if the high percentage of reads corresponding to *Salinibacter* obtained in our metagenomic study corresponds to such a percentage of the bacterium abundance or can be influenced by a potential bias in the library construction.

It is interesting to note the low abundance that we have found for the genus *Haloferax* in the Odiel evaporation pond. Although it grows optimally at 2.5 M NaCl (15% salinity) [37], *Haloferax* has been described to grow at salinities of 33.7% with growth rates higher than any other comparable extreme halophile [27]. Despite these characteristics, *Haloferax* is usually found at low percentage, around 1%, in solar saltern ponds with medium and high salinity (10–37%), as described for example in Santa Pola, Spain [30]. Curiously, the relative abundance that we have found in the Odiel evaporation pond for *Haloferax* is even lower, reaching about 0.021% of the total prokaryotic community. There are several studies that have shown the excretion of halocins or archeocins by archaeal species [38,39], demonstrating the importance of such halocins for interspecies competition in hypersaline environments. The results of our growth inhibitory studies support the possible production of halocins against *Haloferax* by the dominant species and give a possible explanation to the limited presence in the Odiel salterns of *Haloferax*, which should be apparently more qualified to dominate hypersaline ecosystems than the usual dominant species [27].

Table 3. Comparison of prokaryotic diversity at genus level in the Odiel hypersaline pond (33%) and other hypersaline ponds with similar salinity.

Sample	Santa Pola (Spain)	Santa Pola (Spain)	Odiel Salterns (Spain) *	Pomorie (Bulgaria) **	Bajool (Australia) **
Salinity	33%	37%	33%	34%	34%
Ref.	[18,30]	[30]	This study	[35]	[28]
%	*Haloquadratum* 29.5 *Haloruburm* 23.1 *Natronomonas* 5.7 *Salinibacter* 4.7 *Haloplanus* 3.4	*Haloquadratum* 58 *Salinibacter* 9.1 *Nanosalina* 4.0 *Haloruburm* 3.2 *Nanosalinarum* 1.7 *Halomicrobium* 1	*Salinibacter* 37.8 *Halorubrum* 15.5 *Haloquadratum* 12.8 *Halonotius* 8.3 *Halobellus* 3.8 *Natromonas* 3.4 *Haploplanus* 2 *Halohasta* 1.7 *Haliorientalis* 0.96 *Haloarcula* 0.67	*Halanaeroarchaeum* 27.8 *Halorubrum* 24 *Halonotius* 15.7 *Halobellus* 6.5 *Halovenus* 6.5 *Natronomonas* 2.8	*Haloquadratum* 47 *Halorubrum* 17.6 *Halonotius* 11.7 *Haloplanus-like* 11.7 *Natronomonas* 2.9

* Mean of four 16S rRNA NGS data sets. ** In these studies, only archaea were considered.

3.2. PCR Library versus 16S rRNA Massive Sequencing

Cloning-based methods have been successfully used for years [40], however the availability of new benchtop NGS technologies has made more popular the use of 16S rRNA high-throughput for sequencing and profiling of microbial communities including that of hypersaline habitats [41].

Here we demonstrate that, although NGS 16S rRNA sequencing offers a more complete view of the microbial community inhabiting the saline ponds of the Odiel Marshlands, providing more information about the less abundant genera, both NGS and 16S rRNA clone library approaches are comparable regarding the estimation of the major genera found in the sample. This is in agreement with the results obtained by other authors, such as González-Pimentel [42], who compared both approaches to study the microbial diversity in lava tubes from Canary Islands.

We have observed that there is a good general agreement in the relative abundance of the main genera distribution obtained in both NGS analysis (SBV and LFS), however there are discrepancies for some genera. Since both massive sequence services have used the same platform (My Seq Illumina), primers and starting genomic DNA, the small discrepancies and the higher sensitivity of LFS analysis can be greatly attributed to differences in the bioinformatic analysis and other factors such as details of PCR libraries preparation. Furthermore, we observed that the use of different tools for the denoising and clustering step (i.e., using DADA2 instead of DEBLUR plugging) caused the removal of potentially valid sequences and yield different results from the same raw data (data not shown).

Shannon index value, which indicates the uniformity of species and its abundance in the obtained OTUs, increases as both the richness and the evenness of the community increase [43]. The Shannon index values calculated for our NGS studies were on average 2.9 for LFS replicates and 2.7 for SVB replicates (Table 2), while the Shannon index calculated for the clone library data was 2.07. This value is in accordance with previous studies carried out by clone library approaches, which reported values between 1.64 and 2.10 in Australia (34% salinity) [28], 1.8 (37% salinity) and 1.6 (38% salinity) in Mexico [29], while Shannon indexes from NGS are substantially higher. This difference is probably due to the small number of sequences analysed by the clone library technique when compared to NGS. It is necessary to remember that both, clone library and 16S rRNA massive sequencing, methods are based on the PCR amplification of a fraction of the highly conserved reference sequence 16S rRNA. Consequently, both methods share the possible bias inherent to PCR amplification of a single gene and the limitations of 16S rRNA for the resolution of closely related species [44]. Some authors have pointed the higher sensitivity and accuracy of whole genome metagenomic sequencing approach [44]. However, the lower cost of the massive sequencing of 16S rRNA and the existence of wide information and databases for the 16S rRNA sequences have converted this approach in the most commonly used method for exploring bacterial communities. It is important to note that to obtain accurate values in the characterization of microbial communities by single gene amplification methods primers must be validated [45], as were the IlluAdp16S primers (Table S1 in Supplementary Material) used in this study [36]; and robust bioinformatic pipelines should be chosen to process the sequencing data, since different bioinformatic treatments usually yield different relative genera abundances.

3.3. Archaeal Halo-Exoenzymes

Haloarchaeal hydrolases are halophilic and usually thermostable exoenzymes able to catalyse clean and ecologically-friendly processes with high specificity. They have interesting features which make them very attractive for many industrial applications, such as paper, textile, food, detergent and pharmaceutical industries [46]. In addition, lipases and esterases could be used in biofuel production [47], while cellulases and laccases could be of interest for the conversion of plant biomass into fuel and renewable products [48] and the detoxification of the treated lignocelluloses substrates [49,50], respectively. Despite all these interesting studies no application of archaeal haloenzymes at industrial scale has been described so far [10,51].

Halophilic termoestable α-amylases have been found in different genera, including *Haloferax* [52], *Haloterrigena* [53] and *Halomonas* [54]. Some haloarchaeal isolates, such as *Haloarcula marismortui*, produce salt-dependent thermoactive lipases and esterases [55]. Extracellular organic-solvent tolerant proteases have been found in *Halobacterium* sp. [56] and *Natrialba magadii* [57]. However, only few cellulases [58–60] and laccases [61,62] producing halophiles have been reported.

In this work, we have demonstrated that the enriched biomass analysed presents different hydrolases activities, including α-amylase, protease, lipase/esterase, cellulose and laccase. Further studies will focus on the isolation and identification of the strains which show the higher activity for each enzyme, followed by the characterization of the parameters that enable the best activity.

4. Materials and Methods

4.1. Sample Collection and Chemical Composition of the Brine

Samples were obtained at the end of the summer from the salt evaporation ponds located in the natural reserve of Odiel Marshlands, at the estuary of the Odiel and Tinto rivers in the Southwest Spain (Latitude: 37.2395, longitude: −6.95287). The salt concentration in the crystallizer pond was 33.2% at the collection time. Climatological features of the location are characteristic of the Mediterranean maritime climate with hot dry summers and rainy autumns and winters. The mean insolation rate exceeds 3000 h per year, the average annual rainfall and air temperature are 506 mm and 18.3 °C, respectively [20]. The ionic composition and the main physicochemical parameters of the seawater brine at the time of sample collection were determined according to following standard methods: ISO 2480-1973 "Determination of sulphate contents-barium sulphate gravimetric method"; ISO 2481-1973 "Determination of halogens expressed as chlorine-mercuric method", ISO 2482-1973 "Determination of calcium and magnesium contents-EDTA complexometric method"; ECSS/SC 2482-1979 "Determination of potassium content by flame atomic absorption spectrophotometric method".

4.2. Genomic DNA Extraction

For genomic DNA extraction, fresh biomass was harvested by centrifugation of a 500 mL water sample at 11,000 rpm. The resulting pellet was washed with ammonium formate 4 M, freeze-dried and used for genomic DNA extraction with the GeneJET Genomic Purification kit (Thermo Fisher Scientific, Waltham, MA, USA), following the instructions of the manufacturer. Quantification of the genomic DNA obtained and assessment of its purity was done on a Nanodrop Spectrophotometer ND-1000 (Thermo Fisher Scientific).

4.3. Amplification of 16S rRNA Encoding Gene and Construction of Clone Libraries

16S rRNA fragments were amplified with the primer sets: Arc340F/Arc1000R [6,63] for archaea and 341F/907R [64,65] for bacteria (Table S1) using 1 μL of genomic DNA, isolated as previously described, as template. Polymerase chain reactions (PCR) were performed in a total volume of 25 μL containing 1 µL of genomic DNA, 10 pM of each primer, 0.2 mM dNTPs, 0.5 U Taq DNA polymerase from Bioline, 2.5 µL of specific 10X buffer and 1.5 µL of 2.5 mM MgCl₂ buffer using an Eppendorff thermo-cycler. The PCR program was 0.5 min at 96 °C, 0.5 min at 50 °C and 1 min at 72 °C for 30 cycles, followed by 10 min of final primer extension.

4.4. Construction and Analysis of Clone Libraries

The PCR products, obtained with both bacterial and archaeal primer sets, were subjected to agarose electrophoretic separation. The bands obtained for each PCR reaction (with around 660 bp for archaea and 560 bp for bacteria) were purified with the GeneJET Gel Extraction Kit (Thermo Fisher Scientific), ligated to pGEM-T vector (Promega, Madison, WI, USA) according to the manufacturer's instructions and cloned into *Escherichia coli* DH5α competent cells to establish two clone libraries, one for archaea and another one for bacteria. A selection of 50 clones, 25 per each library, were analysed by extraction of the plasmidic DNA, Sanger-sequencing of the 16S rRNA DNA encoding fragments (Stabvida, Lisbon, Portugal) and comparison of the obtained sequences with the National Center for Biotechnology Information 16S rRNA database (NCBI, http://www.ncbi.nih.gov) by using the advanced BLASTN search tool. Sequences with more than 98% length coverage and more than 95% of sequence identity were assigned to a described genus. Sequences with high identity (>97%) were assigned to specific species. Sequences which showed percentage of identity lower than 95% can be potential novel species or genera but further evidence is needed to confirm it.

4.5. High-Throughput 16S rRNA Sequencing

For High-throughput 16S rRNA based microbial profiling, the same genomic DNA was analysed on the Illumina MiSeq platform. Analysis was set up in quadruplicate by two independent Sequencing services: Life Sequencing (Valencia, Spain) and Stabvida (Lisbon, Portugal). In both cases the PCR libraries were prepared by targeting the V3-V4 hypervariable regions of the 16S rRNA [36] with the previously validated [66] IlluAdp16S primers (Table S1) and sequenced using the Illumina MiSeq Reagent kits, V2× 250bp or V3× 300 bp, following Illumina recommendations for Library preparation and metagenomic sequencing. R1 and R2 reads were overlapped using PEAR program version 0.9.1 [67]. Raw data were processed for denoising, filtering (minimum quality threshold of Q20) and clustering using different approaches.

Samples sequenced by STABVIDA (SBV) were processed with QIIME2 v2018.02 [68] and Deblur plugin [69]. The resulting sequences were clustered in operational taxonomic Units (OTUs) and taxonomic assignments were done by scikit-learn naïve Bayes machine learning, which was trained using the SILVA database (version 128) with a clustering threshold of 97% similarity. Samples sequenced by Life Sequencing (LFS) were processed with CUTADAPT 1.8.1 [70] and UCHIME [71] programs. The resulting sequences were clustered in operational taxonomic Units (OTUs) with a threshold of 97%. Those clean FASTA files were BLAST [66] against NCBI 16s rRNA database using BLASTN version 2.2.29+. The resulting XML files were processed using a Python script developed by Life sequencing S.L.-ADM (Paterna, Valencia, Spain) in order to annotate each sequence at different phylogenetic levels.

The Q scores, which represent the probability that a base call is erroneous in a logarithmic base and the Shannon biodiversity index (*H'*), which indicates the uniformity of species and its abundance in the obtained OTUs were calculated following previously described procedures [22,25] and included as sequence quality indicators.

4.6. Extracellular Hydrolases Test

Biomass from the saltern water samples was harvested by centrifugation at 11,000 rpm and resuspended in the archaea enrichment medium (ATCC 1176 medium). Typically, 5 L of environmental water were centrifuged to obtain 50 mL of culture, which was incubated at 37 °C and 100 rpm for 7 days. The biomass growth was quantified by measuring the O.D. at 580 nm in a UV–Vis spectrophotometer

Ultrospec 3100 pro. Culture media contained (per litre): 10 g Glucose, 156 g NaCl, 13 g MgCl$_2$·6H$_2$O, 20 g MgSO$_4$·7H$_2$O, 1 g CaCl$_2$·6H$_2$O, 4 g KCl, 0.2 g NaHCO$_3$, 0.5 g NaBr, 5 g yeast extract. The pH of medium was adjusted to 7 before autoclaving. After 7 days of growth, the culture was collected by centrifugation at 11,000 rpm and the pellet was resuspended (1/100) using the aforementioned medium. This enriched-concentrated archaea mixture was used to detect different hydrolase activities by platting 20 µL drops of the concentrated biomass on 1% agar plates with the indicated medium supplemented with starch, skim milk, carboxymethylcellulose, bromophenol blue and Tween 80 as substrates to test for α-amylase, protease, cellulase, laccase and lipase activities, respectively.

To test for amylase activity, starch (1% *w/v*) was added to the glucose-less agar ATCC 1176 medium. After incubation, the plates were flooded with Lugol reagent solution. The presence of a clear zone around the cells indicated starch hydrolysis. The biomass was screened for proteolytic activity by using ATCC 1176 agar medium supplemented with skimmed milk (1% *w/v*). Protease activity detection was based on the presence of a clear zone around the cells growth due to casein hydrolysis. Screening of cellulase production was done on ATCC 1176 agar medium containing carboxymetylcellulose (0.5% *w/v*) instead of glucose as carbon source. Plates were flooded with 0.1% Congo red dye for 20 min followed by treatment with 1 M NaCl for 15 min and finally with 1 M HCl for 5 min in order to increase the halo contrast as described by Sazci et al. [72]. The presence of clearance zones around the cells, as a result of carboxymetylcellulose hydrolysis, indicated production of cellulase. Laccase activity assay was carried out on agar Petri dishes containing the ATCC 1176 medium supplemented with the dye bromophenol blue (0.02% *w/v*), according to Tekere et al. [73]. The formation of discoloration halos around the cells caused by dye degradation showed laccase activity. Lipase activity was screened on nutrient agar plates containing per litre: 10 g peptone, 150 g NaCl, 1 g CaCl$_2$·2H$_2$O Tween 80 (0.1% *v/v*). Opaque halos around the cells resulting from the precipitation of calcium oleate revealed lipase activity [74].

All the plates were incubated at 37 °C and the results were checked periodically from the 3th to 10th day of assay, by measuring the diameters of clearance zones or halos around each archaeal drop. All the tests were done in triplicate.

4.7. Growth Inhibition Test

Halocin activity was determined by observing growth inhibition of a presumably susceptible archaea strain. The enriched biomass, obtained as described for hydrolase activities, was tested against *Haloferax lucetense* (CECT 5871), which was purchased from CECT (Spanish Collection of Culture Type). *H. lucetense* was grown on the medium specified by the CECT (MHE 25 Medium; CECT 188) and 1 mL of the culture was completely spread across the surface of a Petri dish. When the plate was totally dried, a 20 µL drop of the enriched biomass was spotlessly placed in the centre of the Petri dish. The inhibition of *H. lucetense* growth was measured by the formation of a clearance zone around the enriched biomass drop added.

Supplementary Materials: The following are available online at http://www.mdpi.com/1660-3397/16/9/332/s1, Table S1: Sequences of the primers used, Table S2: Complete list of sequenced genera.

Author Contributions: For research articles with several authors, a short paragraph specifying their individual contributions must be provided. The following statements should be used "Conceptualization, R.L. and J.V.; Methodology, P.G.-V.; R.L. and J.V.; Software, P.G.-V.; Validation, P.G.-V.; R.L. and J.V.; Investigation, P.G.-V.; Writing-Original Draft Preparation, R.L.; Writing-Review & Editing, P.G.-V.; R.L. and J.V.; Visualization, P.G.-V.; Supervision, R.L. and J.V.; Funding Acquisition, R.L.", please turn to the CRediT taxonomy for the term explanation. Authorship must be limited to those who have contributed substantially to the work reported.

Funding: This research was funded by INTERREG VA España–Portugal (POCTEP) 2014–2020 Cooperation Program, grant number 0055_ALGARED_PLUS_5_E and SUBV. COOP.ALENTEJO-ALGARVE-ANDALUCIA 2017.

Acknowledgments: The authors would like to thank the company Salinas del Odiel S.L. and J. Ariza from the University of Huelva for kindly providing the water samples.

Conflicts of Interest: "The authors declare no conflict of interest." "The funders had no role in the design of the study; in the collection, analyses, or interpretation of data; in the writing of the manuscript and in the decision to publish the results."

References

1. Oren, A. Halophilic microbial communities and their environments. *Curr. Opin. Biotechnol.* **2015**, *33*, 119–124. [CrossRef] [PubMed]
2. Zaremba-Niedzwiedzka, K.; Caceres, E.F.; Saw, J.H.; Backstrom, D.; Juzokaite, L.; Vancaester, E.; Seitz, K.W.; Anantharaman, K.; Starnawski, P.; Kjeldsen, K.U.; et al. Asgard archaea illuminate the origin of eukaryotic cellular complexity. *Nature* **2017**, *541*, 353–358. [CrossRef] [PubMed]
3. Litchfield, C.D. Potential for industrial products from the halophilic Archaea. *J. Ind. Microbiol. Biotechnol.* **2011**, *38*, 1635–1647. [CrossRef] [PubMed]
4. Coker, J.A. Extremophiles and biotechnology: Current uses and prospects. *F1000Research* **2016**, *5*, 396. [CrossRef] [PubMed]
5. Mandelli, F.; Miranda, V.S.; Rodrigues, E.; Mercadante, A.Z. Identification of carotenoids with high antioxidant capacity produced by extremophile microorganisms. *World J. Microbiol. Biotechnol.* **2012**, *28*, 1781–1790. [CrossRef] [PubMed]
6. de la Vega, M.; Sayago, A.; Ariza, J.; Barneto, A.G.; Leon, R. Characterization of a bacterioruberin-producing Haloarchaea isolated from the marshlands of the Odiel river in the southwest of Spain. *Biotechnol. Prog.* **2016**, *32*, 592–600. [CrossRef] [PubMed]
7. Kumar, S.; Karan, R.; Kapoor, S.; Singh, S.P.; Khare, S.K. Screening and isolation of halophilic bacteria producing industrially important enzymes. *Brazilian J. Microbiol.* **2012**, *43*, 1595–1603. [CrossRef]
8. Price, L.B.; Shand, R.F. Halocin S8: A 36-amino-acid microhalocin from the haloarchaeal strain S8a. *J. Bacteriol.* **2000**, *182*, 4951–4958. [CrossRef] [PubMed]
9. Chen, L.; Wang, G.; Bu, T.; Zhang, Y.; Wang, Y.; Liu, M.; Lin, X. Phylogenetic analysis and screening of antimicrobial and cytotoxic activities of moderately halophilic bacteria isolated from the Weihai Solar Saltern (China). *World J. Microbiol. Biotechnol.* **2010**, *26*, 879–888. [CrossRef]
10. Kumar, S.; Grewal, J.; Sadaf, A.; Hemamalini, R.; Khare, S.K. Halophiles as a source of polyextremophilic α-amylase for industrial applications. *AIMS Microbiol.* **2016**, *2*, 1–26. [CrossRef]
11. Oren, A. Industrial and environmental applications of halophilic microorganisms. *Environ. Technol.* **2010**, *31*, 825–834. [CrossRef] [PubMed]
12. Thombre, R.S.; Shinde, V.D.; Oke, R.S.; Dhar, S.K.; Shouche, Y.S. Biology and survival of extremely halophilic archaeon Haloarcula marismortui RR12 isolated from Mumbai salterns, India in response to salinity stress. *Sci. Rep.* **2016**, *6*, 25642. [CrossRef] [PubMed]
13. Bolhuis, H.; Te Poele, E.M.; Rodriguez-Valera, F. Isolation and cultivation of Walsby's square archaeon. *Environ. Microbiol.* **2004**, *6*, 1287–1291. [CrossRef] [PubMed]
14. Martínez-Murcia, A.J.; Acinas, S.G.; Rodriguez-Valera, F. Evaluation of prokaryotic diversity by restrictase digestion of 16S rDNA directly amplified from hypersaline environments. *FEMS Microbiol. Ecol.* **1995**, *17*, 247–255. [CrossRef]
15. Anton, J.; Llobet-Brossa, E.; Rodriguez-Valera, F.; Amann, R. Fluorescence in situ hybridization analysis of the prokaryotic community inhabiting crystallizer ponds. *Environ. Microbiol.* **1999**, *1*, 517–523. [CrossRef] [PubMed]
16. Benlloch, S.; López-López, A.; Casamayor, E.O.; Øvreås, L.; Goddard, V.; Daae, F.L.; Smerdon, G.; Massana, R.; Joint, I.; Thingstad, F.; et al. Prokaryotic genetic diversity throughout the salinity gradient of a coastal solar saltern. *Environ. Microbiol.* **2002**, *4*, 349–360. [CrossRef] [PubMed]
17. Narasingarao, P.; Podell, S.; Ugalde, J.A.; Brochier-Armanet, C.; Emerson, J.B.; Brocks, J.J.; Heidelberg, K.B.; Banfield, J.F.; Allen, E.E. De novo metagenomic assembly reveals abundant novel major lineage of Archaea in hypersaline microbial communities. *Isme J.* **2011**, *6*, 81. [CrossRef] [PubMed]
18. Fernández, A.B.; Vera-Gargallo, B.; Sánchez-Porro, C.; Ghai, R.; Papke, R.T.; Rodriguez-Valera, F.; Ventosa, A. Comparison of prokaryotic community structure from Mediterranean and Atlantic saltern concentrator ponds by a metagenomic approach. *Front. Microbiol.* **2014**, *5*, 196. [CrossRef] [PubMed]

19. DeLong, E.F.; Pace, N.R. Environmental Diversity of Bacteria and Archaea. *Syst. Biol.* **2001**, *50*, 470–478. [CrossRef] [PubMed]

20. Vera-Gargallo, B.; Ventosa, A. Metagenomic Insights into the Phylogenetic and Metabolic Diversity of the Prokaryotic Community Dwelling in Hypersaline Soils from the Odiel Saltmarshes (SW Spain). *Genes* **2018**, *9*, 152. [CrossRef] [PubMed]

21. Burns, D.G.; Camakaris, H.M.; Janssen, P.H.; Dyall-Smith, M.L. Combined use of cultivation-dependent and cultivation-independent methods indicates that members of most haloarchaeal groups in an Australian crystallizer pond are cultivable. *Appl. Environ. Microbiol.* **2004**, *70*, 5258–5265. [CrossRef] [PubMed]

22. Shannon, C.E. A mathematical theory of communication. *Bell Syst. Tech. J.* **1948**, *27*, 379–423. [CrossRef]

23. Thomas, R.H. Molecular Evolution and Phylogenetics. *Heredity* **2001**, *86*, 385. [CrossRef]

24. Antón, J.; Rosselló-mora, R.; Amann, R.; Anto, J. Extremely Halophilic Bacteria in Crystallizer Ponds from Solar Salterns Extremely Halophilic Bacteria in Crystallizer Ponds from Solar Salterns. *Appl. Environ. Microbiol.* **2000**, *66*, 3052–3057. [CrossRef] [PubMed]

25. Keshri, J.; Mishra, A.; Jha, B. Microbial population index and community structure in saline-alkaline soil using gene targeted metagenomics. *Microbiol. Res.* **2013**, *168*, 165–173. [CrossRef] [PubMed]

26. León, M.J.; Aldeguer-Riquelme, B.; Antón, J.; Sánchez-Porro, C.; Ventosa, A. *Spiribacter aquaticus* sp. nov., a novel member of the genus Spiribacter isolated from a saltern. *Int. J. Syst. Evol. Microbiol.* **2017**, *67*, 2947–2952. [CrossRef] [PubMed]

27. Oren, A.; Hallsworth, J.E. Microbial weeds in hypersaline habitats: The enigma of the weed-like *Haloferax mediterranei*. *FEMS Microbiol. Lett.* **2014**, *359*, 134–142. [CrossRef] [PubMed]

28. Oh, D.; Porter, K.; Russ, B.; Burns, D.; Dyall-Smith, M. Diversity of *Haloquadratum* and other haloarchaea in three, geographically distant, Australian saltern crystallizer ponds. *Extremophiles* **2010**, *14*, 161–169. [CrossRef] [PubMed]

29. Dillon, J.G.; Carlin, M.; Gutierrez, A.; Nguyen, V.; McLain, N. Patterns of microbial diversity along a salinity gradient in the Guerrero Negro solar saltern, Baja CA Sur, Mexico. *Front. Microbiol.* **2013**, *4*, 399. [CrossRef] [PubMed]

30. Ventosa, A.; Fernández, A.B.; León, M.J.; Sánchez-Porro, C.; Rodriguez-Valera, F. The Santa Pola saltern as a model for studying the microbiota of hypersaline environments. *Extremophiles* **2014**, *18*, 811–824. [CrossRef] [PubMed]

31. Cray, J.A.; Bell, A.N.W.; Bhaganna, P.; Mswaka, A.Y.; Timson, D.J.; Hallsworth, J.E. The biology of habitat dominance; can microbes behave as weeds? *Microb. Biotechnol.* **2013**, *6*, 453–492. [CrossRef] [PubMed]

32. Çınar, S.; Mutlu, M.B. Comparative analysis of prokaryotic diversity in solar salterns in eastern Anatolia (Turkey). *Extremophiles* **2016**, *20*, 589–601. [CrossRef] [PubMed]

33. Maturrano, L.; Santos, F.; Rosselló-Mora, R.; Antón, J. Microbial diversity in Maras salterns, a hypersaline environment in the Peruvian Andes. *Appl. Environ. Microbiol.* **2006**, *72*, 3887–3895. [CrossRef] [PubMed]

34. Pašić, L.; Bartual, S.G.; Ulrih, N.P.; Grabnar, M.; Velikonja, B.H. Diversity of halophilic archaea in the crystallizers of an Adriatic solar saltern. *FEMS Microbiol. Ecol.* **2005**, *54*, 491–498. [CrossRef] [PubMed]

35. Kambourova, M.; Tomova, I.; Boyadzhieva, I.; Radchenkova, N.; Vasileva-Tonkova, E. Unusually High Archaeal Diversity in a Crystallizer Pond, Pomorie Salterns, Bulgaria, Revealed by Phylogenetic Analysis. *Archaea* **2016**, *2016*. [CrossRef] [PubMed]

36. Klindworth, A.; Pruesse, E.; Schweer, T.; Peplies, J.; Quast, C.; Horn, M.; Glöckner, F.O. Evaluation of general 16S ribosomal RNA gene PCR primers for classical and next-generation sequencing-based diversity studies. *Nucleic Acids Res.* **2013**, *41*, 1–11. [CrossRef] [PubMed]

37. Payá, G.; Bautista, V.; Camacho, M.; Castejón-Fernández, N.; Alcaraz, L.A.; Bonete, M.J.; Esclapez, J. Small RNAs of Haloferax mediterranei: Identification and potential involvement in nitrogen metabolism. *Genes* **2018**, *9*. [CrossRef] [PubMed]

38. Atanasova, N.S.; Pietilä, M.K.; Oksanen, H.M. Diverse antimicrobial interactions of halophilic archaea and bacteria extend over geographical distances and cross the domain barrier. *Microbiologyopen* **2013**, *2*, 811–825. [CrossRef] [PubMed]

39. Charlesworth, J.C.; Burns, B.P. Untapped resources: Biotechnological potential of peptides and secondary metabolites in archaea. *Archaea* **2015**, *2015*. [CrossRef] [PubMed]

40. Baker, G.C.; Smith, J.J.; Cowan, D.A. Review and re-analysis of domain-specific 16S primers. *J. Microbiol. Methods* **2003**, *55*, 541–555. [CrossRef] [PubMed]

41. Gibtan, A.; Park, K.; Woo, M.; Shin, J.K.; Lee, D.W.; Sohn, J.H.; Song, M.; Roh, S.W.; Lee, S.J.; Lee, H.S. Diversity of extremely halophilic archaeal and bacterial communities from commercial salts. *Front. Microbiol.* **2017**, *8*, 1–11. [CrossRef] [PubMed]

42. Gonzalez-Pimentel, J.L.; Miller, A.Z.; Jurado, V.; Laiz, L.; Pereira, M.F.C.; Saiz-Jimenez, C. Yellow coloured mats from lava tubes of La Palma (Canary Islands, Spain) are dominated by metabolically active Actinobacteria. *Sci. Rep.* **2018**, *8*, 1944. [CrossRef] [PubMed]

43. Magurran, A.E. *Measuring Biological Diversity*; John Wiley & Sons, 2013; ISBN 1118687922.

44. Poretsky, R.; Rodriguez-R, L.M.; Luo, C.; Tsementzi, D.; Konstantinidis, K.T. Strengths and limitations of 16S rRNA gene amplicon sequencing in revealing temporal microbial community dynamics. *PLoS ONE* **2014**, *9*. [CrossRef] [PubMed]

45. Fouhy, F.; Clooney, A.G.; Stanton, C.; Claesson, M.J.; Cotter, P.D. 16S rRNA gene sequencing of mock microbial populations-impact of DNA extraction method, primer choice and sequencing platform. *BMC Microbiol.* **2016**, *16*, 1–13. [CrossRef] [PubMed]

46. Amoozegar, M.A.; Siroosi, M.; Atashgahi, S.; Smidt, H.; Ventosa, A. Systematics of haloarchaea and biotechnological potential of their hydrolytic enzymes. *Microbiol.* **2017**, *163*, 623–645. [CrossRef] [PubMed]

47. Schreck, S.D.; Grunden, A.M. Biotechnological applications of halophilic lipases and thioesterases. *Appl. Microbiol. Biotechnol.* **2014**, *98*, 1011–1021. [CrossRef] [PubMed]

48. Gunny, A.A.N.; Arbain, D.; Edwin Gumba, R.; Jong, B.C.; Jamal, P. Potential halophilic cellulases for in situ enzymatic saccharification of ionic liquids pretreated lignocelluloses. *Bioresour. Technol.* **2014**, *155*, 177–181. [CrossRef] [PubMed]

49. Rezaei, S.; Tahmasbi, H.; Mogharabi, M.; Firuzyar, S.; Ameri, A.; Khoshayand, M.R.; Faramarzi, M.A. Efficient decolorization and detoxification of reactive orange 7 using laccase isolated from paraconiothyrium variabile, kinetics and energetics. *J. Taiwan Inst. Chem. Eng.* **2015**, *56*, 113–121. [CrossRef]

50. Vithanage, L.N.G.; Barbosa, A.M.; Borsato, D.; Dekker, R.F.H. Value adding of poplar hemicellulosic prehydrolyzates: Laccase production by *Botryosphaeria rhodina* MAMB-05 and its application in the detoxification of prehydrolyzates. *BioEnergy Res.* **2015**, *8*, 657–674. [CrossRef]

51. Waditee-Sirisattha, R.; Kageyama, H.; Takabe, T. Halophilic microorganism resources and their applications in industrial and environmental biotechnology. *AIMS Microbiol.* **2016**, *2*, 42–54. [CrossRef]

52. Bajpai, B.; Chaudhary, M.; Saxena, J. Production and Characterization of alpha-Amylase from an Extremely Halophilic Archaeon, *Haloferax* sp. HA10. *Food Technol. Biotechnol.* **2015**, *53*, 11–17. [CrossRef] [PubMed]

53. Santorelli, M.; Maurelli, L.; Pocsfalvi, G.; Fiume, I.; Squillaci, G.; La Cara, F.; Del Monaco, G.; Morana, A. Isolation and characterisation of a novel alpha-amylase from the extreme haloarchaeon *Haloterrigena turkmenica*. *Int. J. Biol. Macromol.* **2016**, *92*, 174–184. [CrossRef] [PubMed]

54. Uzyol, K.S.; Akbulut, B.S.; Denizci, A.A.; Kazan, D. Thermostable a-amylase from moderately halophilic *Halomonas* sp. AAD21. *Turk. J. Biol.* **2012**, *36*, 327–338. [CrossRef]

55. Camacho, R.M.; Mateos, J.C.; Gonzalez-Reynoso, O.; Prado, L.A.; Cordova, J. Production and characterization of esterase and lipase from *Haloarcula marismortui*. *J. Ind. Microbiol. Biotechnol.* **2009**, *36*, 901–909. [CrossRef] [PubMed]

56. Akolkar, A.V.; Deshpande, G.M.; Raval, K.N.; Durai, D.; Nerurkar, A.S.; Desai, A.J. Organic solvent tolerance of *Halobacterium* sp. SP1 (1) and its extracellular protease. *J. Basic Microbiol.* **2008**, *48*, 421–425. [CrossRef] [PubMed]

57. Ruiz, D.M.; De Castro, R.E. Effect of organic solvents on the activity and stability of an extracellular protease secreted by the haloalkaliphilic archaeon *Natrialba magadii*. *J. Ind. Microbiol. Biotechnol.* **2007**, *34*, 111–115. [CrossRef] [PubMed]

58. Wang, C.-Y.; Hsieh, Y.-R.; Ng, C.-C.; Chan, H.; Lin, H.-T.; Tzeng, W.-S.; Shyu, Y.-T. Purification and characterization of a novel halostable cellulase from *Salinivibrio* sp. strain NTU-05. *Enzyme Microb. Technol.* **2009**, *44*, 373–379. [CrossRef]

59. Simankova, M.V.; Chernych, N.A.; Osipov, G.A.; Zavarzin, G.A. *Halocella cellulolytica* gen. nov., sp. nov., a new obligately anaerobic, halophilic, cellulolytic bacterium. *Syst. Appl. Microbiol.* **1993**, *16*, 385–389. [CrossRef]

60. Yu, H.-Y.; Li, X. Alkali-stable cellulase from a halophilic isolate, *Gracilibacillus* sp. SK1 and its application in lignocellulosic saccharification for ethanol production. *Biomass Bioenergy* **2015**, *81*, 19–25. [CrossRef]

61. Uthandi, S.; Saad, B.; Humbard, M.A.; Maupin-Furlow, J.A. LccA, an archaeal laccase secreted as a highly stable glycoprotein into the extracellular medium by *Haloferax volcanii*. *Appl. Environ. Microbiol.* **2010**, *76*, 733–743. [CrossRef] [PubMed]

62. Rezaie, R.; Rezaei, S.; Jafari, N.; Forootanfar, H.; Khoshayand, M.R.; Faramarzi, M.A. Delignification and detoxification of peanut shell bio-waste using an extremely halophilic laccase from an *Aquisalibacillus elongatus* isolate. *Extremophiles* **2017**, *21*, 993–1004. [CrossRef] [PubMed]

63. Gantner, S.; Andersson, A.F.; Alonso-Sáez, L.; Bertilsson, S. Novel primers for 16S rRNA-based archaeal community analyses in environmental samples. *J. Microbiol. Methods* **2011**, *84*, 12–18. [CrossRef] [PubMed]

64. Muyzer, G.; de Waal, E.C.; Uitterlinden, A.G. Profiling of complex microbial populations by denaturing gradient gel electrophoresis analysis of polymerase chain reaction-amplified genes coding for 16S rRNA. *Appl. Environ. Microbiol.* **1993**, *59*, 695–700. [PubMed]

65. Teske, A.; Wawer, C.; Muyzer, G.; Ramsing, N.B. Distribution of sulfate-reducing bacteria in a stratified fjord (Mariager Fjord, Denmark) as evaluated by most-probable-number counts and denaturing gradient gel electrophoresis of PCR-amplified ribosomal DNA fragments. *Appl. Environ. Microbiol.* **1996**, *62*, 1405–1415. [PubMed]

66. Altschul, S.F.; Gish, W.; Miller, W.; Myers, E.W.; Lipman, D.J. Basic local alignment search tool. *J. Mol. Biol.* **1990**, *215*, 403–410. [CrossRef]

67. Zhang, J.; Kobert, K.; Flouri, T.; Stamatakis, A. PEAR: A fast and accurate Illumina Paired-End reAd mergeR. *Bioinformatics* **2014**, *30*, 614–620. [CrossRef] [PubMed]

68. Caporaso, J.G.; Kuczynski, J.; Stombaugh, J.; Bittinger, K.; Bushman, F.D.; Costello, E.K.; Fierer, N.; Pena, A.G.; Goodrich, J.K.; Gordon, J.I.; et al. QIIME allows analysis of high-throughput community sequencing data. *Nat. Methods* **2010**, *7*, 335–336. [CrossRef] [PubMed]

69. Amir, A.; McDonald, D.; Navas-Molina, J.A.; Kopylova, E.; Morton, J.T.; Zech Xu, Z.; Kightley, E.P.; Thompson, L.R.; Hyde, E.R.; Gonzalez, A.; et al. Deblur Rapidly Resolves Single-Nucleotide Community Sequence Patterns. *mSystems* **2017**, *2*. [CrossRef] [PubMed]

70. Martin, M. Cutadapt removes adapter sequences from high-throughput sequencing reads. *EMBnet.journal* **2011**, *17*, 10. [CrossRef]

71. Edgar, R.C.; Haas, B.J.; Clemente, J.C.; Quince, C.; Knight, R. UCHIME improves sensitivity and speed of chimera detection. *Bioinformatics* **2011**, *27*, 2194–2200. [CrossRef] [PubMed]

72. Sazci, A.; Erenler, K.; Radford, A. Detection of cellulolytic fungi by using Congo red as an indicator: A comparative study with the dinitrosalicyclic acid reagent method. *J. Appl. Bacteriol.* **1986**, *61*, 559–562. [CrossRef]

73. Tekere, M.; Mswaka, A.Y.; Zvauya, R.; Read, J.S. Growth, dye degradation and ligninolytic activity studies on Zimbabwean white rot fungi. *Enzym. Microb. Technol.* **2001**, *28*, 420–426. [CrossRef]

74. Lanka, S.; Latha, J.N.L. A short review on various screening methods to isolate potential lipase producers: Lipases-the present and future enzymes of biotech industry. *Int. J. Biol. Chem.* **2015**, *9*, 207–219. [CrossRef]

marine drugs

MDPI

Article

Biosynthetic Potential of a Novel Antarctic Actinobacterium *Marisediminicola antarctica* ZS314^T Revealed by Genomic Data Mining and Pigment Characterization

Li Liao [1,*,†], Shiyuan Su [1,2,†], Bin Zhao [1,3,†], Chengqi Fan [4], Jin Zhang [1], Huirong Li [1] and Bo Chen [1,*]

[1]   SOA Key Laboratory for Polar Science, Polar Research Institute of China, 451 Jinqiao Road, Shanghai 200136, China
[2]   College of Marine Sciences, Shanghai Ocean University, Shanghai 201306, China
[3]   School of Biotechnology, East China University of Science and Technology, Shanghai 200237, China
[4]   Key Laboratory of East China Sea & Oceanic Fishery Resources Exploitation and Utilization, Ministry of Agriculture, East China Sea Fisheries Research Institute, Chinese Academy of Fishery Sciences, Shanghai 200090, China
*   Correspondence: liaoli@pric.org.cn (L.L.); chenbo@pric.org.cn (B.C.); Tel.: +86-21-50385104 (L.L. & B.C.)
†   These authors contributed equally to this work.

Received: 11 June 2019; Accepted: 28 June 2019; Published: 1 July 2019

Abstract: Rare actinobacterial species are considered as potential resources of new natural products. *Marisediminicola antarctica* ZS314^T is the only type strain of the novel actinobacterial genus *Marisediminicola* isolated from intertidal sediments in East Antarctica. The strain ZS314^T was able to produce reddish orange pigments at low temperatures, showing characteristics of carotenoids. To understand the biosynthetic potential of this strain, the genome was completely sequenced for data mining. The complete genome had 3,352,609 base pairs (bp), much smaller than most genomes of actinomycetes. Five biosynthetic gene clusters (BGCs) were predicted in the genome, including a gene cluster responsible for the biosynthesis of C50 carotenoid, and four additional BGCs of unknown oligosaccharide, salinixanthin, alkylresorcinol derivatives, and NRPS (non-ribosomal peptide synthetase) or amino acid-derived compounds. Further experimental characterization indicated that the strain may produce C.p.450-like carotenoids, supporting the genomic data analysis. A new xanthorhodopsin gene was discovered along with the analysis of the salinixanthin biosynthetic gene cluster. Since little is known about this genus, this work improves our understanding of its biosynthetic potential and provides opportunities for further investigation of natural products and strategies for adaptation to the extreme Antarctic environment.

Keywords: *Marisediminicola*; Antarctica; carotenoid; actinobacteria; natural products; gene cluster

1. Introduction

Actinobacteria are considered as one of the most important producers of natural products for drug discovery. To date, more than half of antibiotics were produced by the genus *Streptomyces* [1]. However, the re-discovery of known natural products, especially from the genus *Streptomyces* due to the so-called diminishing marginal effect, has become an issue [1,2]. Meanwhile, the increased prevalence of multi-drug resistance pathogens and the emergence of cross-resistance to antibiotics have driven researchers to seek for novel natural products [3]. Therefore, rare actinobacteria (non-*Streptomyces*), especially those from underexplored extreme environments, draw increasing attention for discovering novel bioactive compounds [4]. Antarctica, a frozen continent surrounded by the Southern Ocean, is characterized by extremely low temperatures, strong ultraviolet radiation and a relatively isolated ecosystem [5]. It represents one of the most under-investigated environments. Hence, it was considered as a huge reservoir of microorganisms with biosynthetic potential of novel natural products.

The genus *Marisediminicola* belonging to rare actinobacteria was first proposed nine years ago [6], with the only type species *Marisediminicola antarctica* and the only type strain *M. antarctica* ZS314^T. The strain ZS314^T was isolated from intertidal sediments of the coast off the Chinese Antarctic Zhongshan Station in East Antarctica (69°22′13″ S, 76°21′41″ E). To date, no other type strains have been reported belonging to this species or this genus. However, many 16S rRNA sequences sharing over 99% identities have been deposited in the NCBI GenBank database, with most of them from the Arctic, Antarctica or glacier-associated environments. It suggests that *M. antarctica* is a phylotype adapted and probably specific to extreme environments. The strain ZS314^T was psychrophilic, with the growth temperature range of 0–26 °C (optimum 18–23°C). It developed a bright reddish orange color, which is usually caused by various carotenoid pigments. Carotenoids are a group of natural isoprenoid pigments mainly isolated from a wide variety of plants, algae, fungi and bacteria [7–12]. The major functions of known carotenoids are to prevent cells from oxidative damage [8], protect against UV radiation and modulate membrane fluidity in bacteria [7]. Therefore, carotenoids are important not only for the survival of the organisms but also for industrial applications.

To investigate the biosynthetic abilities of *M. antarctica* ZS314^T, we performed genome sequencing and bioinformatics data mining. Further experiments including HPLC/TOF-MS (high-performance liquid chromatography/time-of-flight mass spectrometry) and anti-oxidation assays were carried out to characterize the pigments, as supported by the presence of a carotenoid biosynthetic gene cluster and the indicative color. *M. antarctica* was identified and reported nine years ago, but still little is known about this novel genus/species. Here, we report the findings through wet lab experiments and data mining of the genome to increase our understanding of this novel Antarctic actinobacterium. This study represents the first investigation of the novel Antarctic *Marisediminicola* genus for its potential in natural product biosynthesis, and also the first complete genome of the genus for future studies.

2. Results and Discussion

2.1. General Feature of the Genome

The complete genome of *M. antarctica* ZS314^T was circular, containing 3,352,609 bp with an average GC content of 67.17% (Figure S1). Compared with other actinobacteria, *M. antarctica* species tended to have a relatively small genome (3.35 Mb vs. the largest actinobacterial genome of 13.05 Mb from *Nonomuraea* sp. ATCC 55076) and low GC content (67.17% vs. the highest GC content of 74.8 % in *Streptomyces*). To get an idea of the average genome size of the phylum Actinobacteria, a total of 1674 genomes (complete genomes and genomes assembled to the chromosome level) available in the GenBank database were analyzed. The resulting average genome size was 4.52 Mb. Meanwhile, the average GC content was 64.12% and the average number of CDSs was around 3937. Therefore, *M. antarctica* ZS314^T has a genome size below average but a GC content slightly higher than average. A total of 2992 (below average) protein-encoding genes were annotated in the genome, with up to 36% genes annotated as hypothetical proteins or without significant similar matches in the NCBI nr (non-redundant) database. Up to 44%, 41%, 36%, and 23% of genes had no match in the Swiss-Prot, gene ontology (GO), Kyoto encyclopedia of genes and genomes (KEGG) and clusters of orthologous groups (COG) databases, respectively. Two *rrn* operons and 44 tRNA genes were detected in the genome. The complete genome can be further analyzed for genome evolution and genetic adaptation to the extreme environment.

2.2. Prediction of Secondary Metabolite Biosynthetic Gene Clusters

Five secondary metabolite BGCs were predicted by the antiSMASH software (Table 1 and Table S1, and Figure 1). Since most BGCs are large (over 20 kb to around 100 kb), more BGCs are usually found in larger genomes (e.g., about 20 to 40 BGCs are generally predicted in *Streptomyces* genomes over 9 Mb). Although fewer BGCs were found in the genome of *M. antarctica* ZS314^T compared with *Streptomyces*, the difference of relative abundance of BGCs normalized to genome size is small. In the relatively small genome of *M. antarctica* ZS314^T, the presence of the five gene clusters suggests their importance to the strain.

Table 1. Five secondary metabolite biosynthetic gene clusters predicted in *M. antarctica* ZS314^T genome.

Cluster ID	Type	Gene Number	Most Similar Known Gene Cluster	Percentage of Similar Genes
1	T3PKS	7	Alkylresorcinol	100%
2	Terpene	7	Carotenoid	50%
3	Terpene	6	NA *	0
4	NRPS	7	NA *	0
5	Oligosaccharide	30	NA *	0

* NA indicates not available.

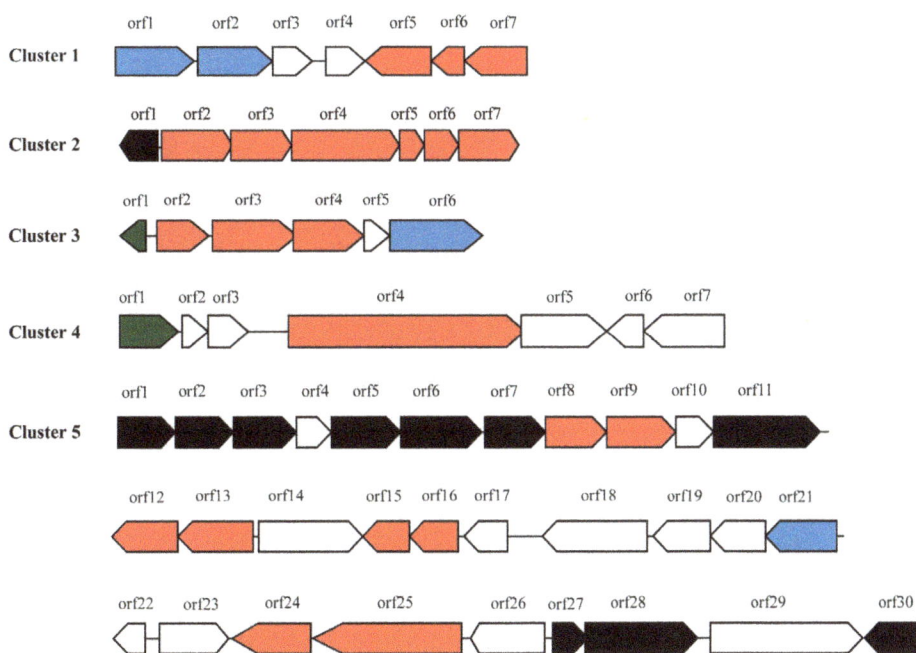

Figure 1. The five gene clusters predicted in the genome of *M. antarctica* ZS314^T. Core biosynthetic genes are indicated in red, additional biosynthetic genes in black, transporter genes in blue, regulator genes in green, and hypothetical protein-encoding genes or unrelated genes in white.

2.2.1. Cluster 1: T3PKS

After manual analysis, cluster 1 (Figure 1) was found to contain three core biosynthetic genes encoding stilbene synthase (*srs*A, orf7), isoprenylcysteine carboxyl methyltransferase (*srs*B, orf6) and monooxygenase (*srs*C, orf5), sharing 51–63% amino acid identities with the *srs* operon of *Streptomyces griseus* NBRC 13350 [13]. Two putative transporter-encoding genes (orf1 and orf2) were detected in the opposite orientation upstream of orf3 and orf4, encoding a hypothetical protein and a haloacid dehalogenase. The two genes (orf3 and orf4) seem not to be related to the *srs* operon since the three core genes (*srs*ABC) alone can biosynthesize alkylresorcinols. SrsA is a specialized type III polyketide synthase (PKS), catalyzing the formation of alkylresorcinols from acyl-CoAs of various chain lengths [13,14]. The alkylresorcinols are further modified by SrsB and SrsC through methylation and hydroxylation [13]. SrsA plays a key role in determining the structure of alkylresorcinols. The most similar record of SrsA from *M. antarctica* ZS314^T in the NCBI nr database was a stilbene synthase from *Agreia* sp. leaf244, sharing 65% amino acid identity. However, SrsA from *M. antarctica* ZS314^T was separated from the known stilbene synthases including the most similar one, as shown in the phylogenetic tree of related stilbene synthases found in the NCBI nr database (Figure S2). It suggests that this specialized type III PKS is distinct from the known enzymes, and hence indicates the production of alkyresorcinols with potential new chemical structure. Alkyresorcinols, a type of non-isoprenoid lipids with an aromatic ring, have important antibiotic activities [13,14]. Alkyresorcinols are relatively rare in nature, with plants as the major sources. Some species of bacteria are known to produce alkyresorcinols, such as *Streptomyces griseus* [13], *Azotobacter* and *Pseudomonas* species [15]. The detection of this putative alkyresorcinol biosynthetic gene cluster extends its distribution to this novel genus.

2.2.2. Cluster 2: Terpene

Cluster 2 was predicted to produce terpene type compounds, a large and diverse group of natural products [2]. Cluster 2 shared some similarity in structure and sequence with four C_{50} carotenoid synthetic gene clusters from *Dietzia* sp. CQ4, *Corynebacterium glutamicum*, *Leifsonia xyli* and *Agromyces mediolanus* (Figure 2 and Table 2). The three genes (orf2 to orf4) in cluster 2 of *M. antarctica* ZS314^T shared 34–61% amino acid identities with geranylgeranyl pyrophosphate synthase gene (*crt*E), phytoene synthase gene (*crt*B) and phytoene desaturase gene (*crt*I) of the four gene clusters, respectively (Table 2). The three genes *crt*EBI were reported to be core genes responsible for biosynthesis of C_{40} carotenoid lycopene. In addition, two genes (orf5 and orf6) encoding C_{50} carotenoid cyclases and orf7 encoding a lycopene elongase were predicted in cluster 2, sharing 35–65% amino acid identities with those in the other four gene clusters (Table 2). These three genes (orf5 to orf7) were expected to convert C_{40} carotenoid lycopene to C_{50} carotenoids, by elongating the carbon chain and forming cyclized end groups finally, as reported in previous studies [9,12]. In certain instances, one C_{50} carotenoid cyclase encoding gene could be fused with the lycopene elongase gene, such as *lbt*BC in *Dietzia* sp. CQ4 (Figure 2). A few additional genes were present in some of the clusters. For example, both cluster 2 of *M. antarctica* ZS314^T and the cluster from *Agromyces mediolanus* contained a gene (orf1/*idi*) encoding isopentenyl-diphosphate Delta-isomerase, which is responsible for isoprenoid biosynthesis. The cluster of *Dietzia* sp. CQ4 contained *crt*X encoding a glycosyl transferase, likely involved in the glucosylation of C_{50} carotenoids [12]. In addition, a transcriptional regulator was encoded with opposite orientation in the gene cluster of *Agromyces mediolanus*. Similarly, a putative transcriptional regulator of MarR family was detected close to the core biosynthetic genes in cluster 2 with opposite orientation.

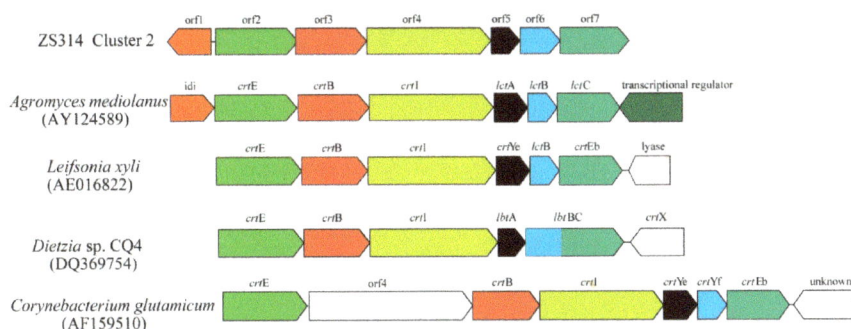

Figure 2. The C_{50} carotenoid synthesis gene clusters of *M. antarctica* ZS314[T], *Dietzia* sp. CQ4, *Corynebacterium glutamicum*, *Leifsonia xyli* and *Agromyces mediolanus*. Homologous genes are presented in the same colors. Genes unrelated to the biosynthesis are denoted in white.

Table 2. A comparison of key genes within the carotenoid biosynthetic gene clusters based on amino acid sequence identities.

Cluster 2 *M. antarctica* ZS314[T]		*Corynebacterium glutamicum*		*Dietzia* sp. CQ4		*Leifsonia xyli*		*Agromyces mediolanus*	
Gene	Annotation	Gene	Identity	Gene	Identity	Gene	Identity	Gene	Identity
orf2	Geranylgeranyl pyrophosphate synthase	crtE	34%	crtE	34%	crtE	44%	crtE	43%
orf3	Phytoene synthase	crtB	46%	crtB	49%	crtB	58%	crtB	61%
orf4	Phytoene desaturase	crtI	52%	crtI	61%	crtI	61%	crtI	58%
orf5	Lycopene cyclase	crtYe	41%	lbtA	40%	crtYe	65%	lctA	50%
orf6	Lycopene cyclase	crtYf	38%	lbtBC[1]	46%	lctB	35%	lctB	43%
orf7	Lycopene elongase	crtEb	56%	lbtBC[2]	59%	crtEb	61%	lctC	64%

lbtBC[1], the N-terminal region of LbtBC (1 to 134 residues) is homologous with lycopene cyclase domain. lbtBC[2], the C-terminal region of LbtBC (135 to 432 residues) is homologous with lycopene elongase domain.

Although the gene clusters responsible for C_{50} carotenoids were quite similar, the final products had different structures. The final product of the gene cluster in *Dietzia* sp. CQ4 was C_{50} β-cyclic carotenoid (C.p.450 monoglucoside) [12], while the final products of the two clusters in *Corynebacterium glutamicum* and *Agromyces mediolanus* were the same, i.e., C_{50} ε-cyclic carotenoid (decaprenoxanthin) [12,16]. The final product of the gene cluster in *Leifsonia xyli* was predicted to be a putative C_{50} carotenoid, without further identification [12]. As previously reported, the C_{50} carotenoid cyclase was the key to determine the final structure and may have multiple catalytic functions [9]. Usually, the C_{50} carotenoid cyclase in Gram-positive bacteria was reported to be encoded by two genes, forming a heterodimeric complex to catalyze the cyclization reaction leading to carotenoids with different end groups [9,17]. Phylogenetic analysis of the C_{50} carotenoid cyclases showed that the two monomers formed two different groups in the tree (Figure 3). The C_{50} carotenoid cyclases (encoded by orf5 an orf6) of *M. antarctica* ZS314[T] clustered with LbtA and LbtBC of *Dietzia* sp. CQ4, respectively (Figure 3). It suggests that cluster 2 of *M. antarctica* ZS314[T] may produce C_{50} β-cyclic carotenoid-like products, as in *Dietzia* sp. CQ4. However, it is hard to predict the final products of the gene cluster 2 of *M. antarctica* ZS314[T] solely based on the gene cluster analysis.

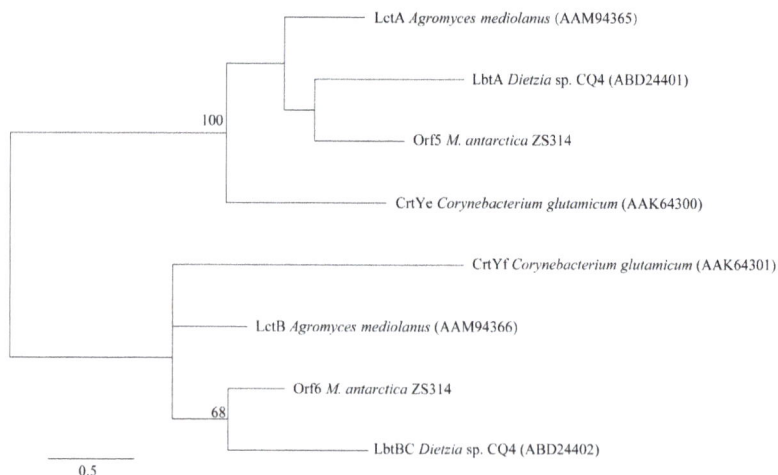

Figure 3. Maximum likelihood phylogenetic tree of C$_{50}$ carotenoid cyclases based on amino acid sequences. Bootstrap values below 50 were not shown.

2.2.3. Cluster 3: Terpene

Cluster 3 was predicted to be a terpene-producing gene cluster due to the presence of a core biosynthetic gene (orf3) encoding lycopene cyclase (Figure 1 and Table S1). Next to the orf3 was a gene encoding xanthorhodopsin (orf2). Xanthorhodopsin is a light-driven proton pump with a dual chromophore, containing one retinal and one salinixanthin [18,19]. Xanthorhodopsin is more effective than bacteriorhodopsin for collecting light due to the presence of salinixanthin. Salinixanthin is a C-40 acyl glycoside carotenoid, a kind of rare carotenoid produced mostly from haloarchaea and halophilic bacteria [20]. However, the biosynthesis of salinixanthin has not been reported yet, according to our knowledge. A gene annotated as β-carotene 15,15'-monooxygenase (orf4) was found downstream of the core biosynthetic gene (orf3), which was reported to catalyze the breakdown of β-carotene into two molecules of retinal. A putative pathway for salinixanthin and retinal biosynthesis was hence proposed (Figure 4), according to reactions in the KEGG database and the genes existed in cluster 3 and the genome. Lycopene, which can be biosynthesized by the three genes *crt*EBI in cluster 2, was further converted to 3,4-dehydrolycopene by CtrI. The later was cyclized at one end to form torulene by most likely the lycopene cyclase in cluster 3. The reactions from lycopene to torulene were the same as the steps in the biosynthesis of neurosporaxanthin [21]. From torulene to salinixanthin, hydroxylase, dehydrogenase or ketolase, glucosyltransferase, and acyltransferase were proposed to catalyze the reactions, according to similar reactions in the biosynthesis of canthaxanthin [22] and myxol [23] (Figure 4). Gene candidates encoding for these enzymes have been annotated in the genome, however, not in cluster 3. We also performed metabolic pathway reconstruction for the biosynthesis of salinixanthin using PathPred, a web-based server integrated into the KEGG database to predict plausible enzyme-catalyzed reaction pathways using the information of biochemical structure transformation patterns and chemical structure alignments of substrate-product pairs [24]. The reconstructed pathway from lycopene to the core structure of salinixanthin without glucoside ester modification was shown in Figure S3 (since pathway reconstruction from lycopene to salinixanthin was frequently disrupted due to some unknown reason). The predicted pathway was quite complex, involving up to 19 reactions different from the pathway we proposed. It is unknown whether both pathways are possible without further experimental verification. Genetic manipulation and chemical identification are encouraged to verify the hypothetical pathways and identify the enzymes responsible for the catalysis of these reactions. Moreover, genome-scale metabolic pathway reconstruction can be used to further discover new metabolic pathways and analyze metabolic networks of *M. antarctica* ZS314^T, as shown in previous studies [25,26].

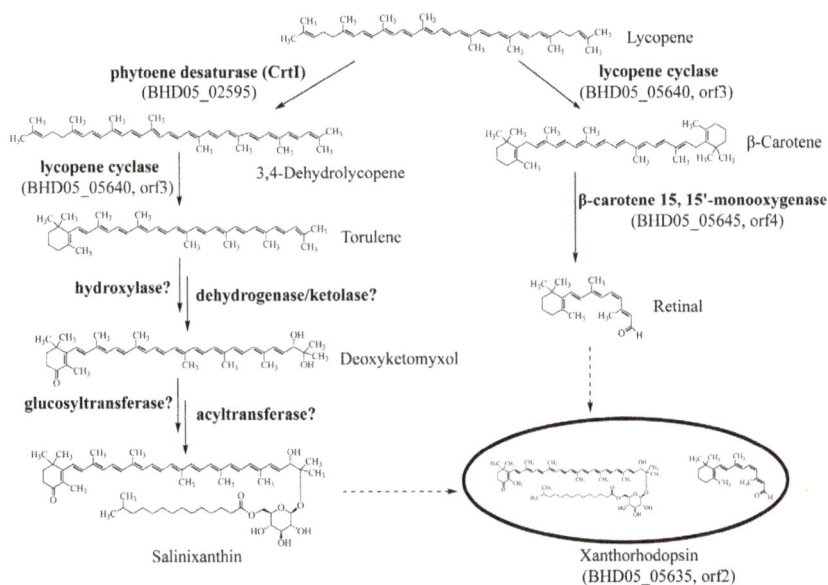

Figure 4. Proposed pathways for the biosynthesis of the two chromophores, salinixanthin and retinal, of xanthorhodopsin from *M. antarctica* ZS314^T.

Xanthorhodopsin was first discovered in *Salinibacter ruber*, a halophilic bacterium [18]. Later, two *Octadecabacter* strains from the Antarctic and the Arctic were reported to have a new subgroup of xanthorhophosins and were predicted to be an adaptation advantage to icy environments [27]. The xanthorhodopsin from *M. antarctica* ZS314^T clustered with those from Actinobacteria within the subgroup I as proposed previously (Figure 5) [27]. However, it was separated from the known xanthorhodopsin genes within the subgroup I with the highest similarity of around 57%, indicating that it was a new xanthorhodopsin gene. Although xanthorhodopsin genes were horizontally transferred, it seems that Actinobacteria shared a common ancestor distinct from non-actinobacterial genes. The residues involved in keto-carotenoid binding identified in *Salinibacter ruber* DSM 13855 were not conserved, except for two residues highlighted in red boxes (Figure 5b). It suggests that xanthorhodopsins may adopt different mechanisms for chromophore binding.

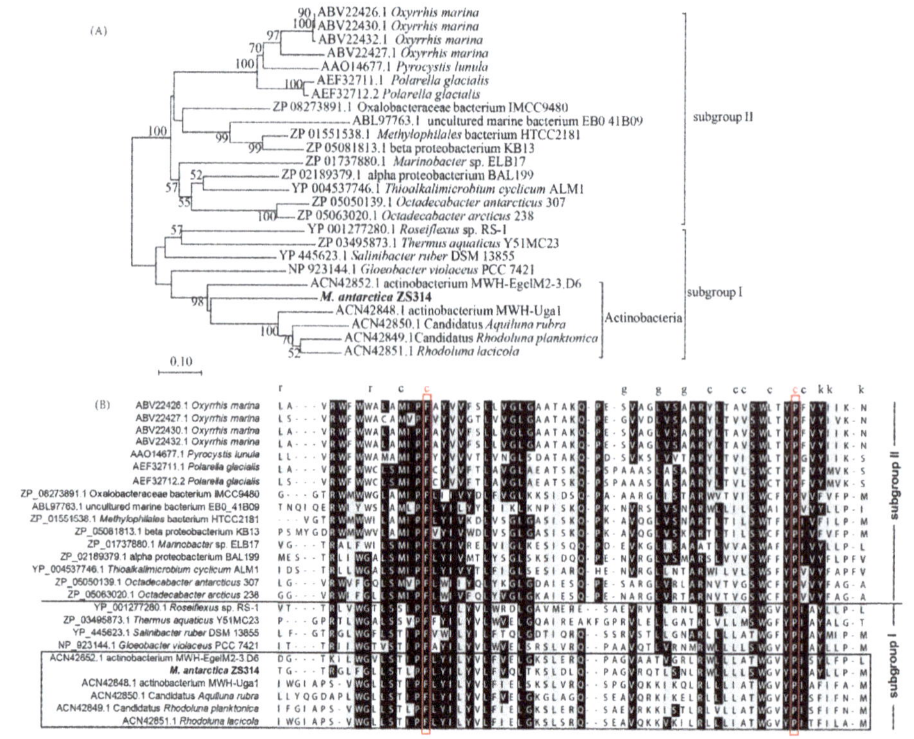

Figure 5. Comparison of xanthorhodopsins from *M. antarctica* ZS314^T and other representative microbial groups used in a previous study [27]. (**A**) Neighbor-joining tree of xanthorhodopsins based on amino acid sequences. Bootstrap values below 50 were not shown. (**B**) Multiple alignments of the putative keto-carotenoid-binding region of the xanthorhodopsins in (**A**), as identified previously [27,28]. Same as in the previous study [27], residues are marked by the letters r, c, g, and k to indicate contact with the ring, chain, glucoside and keto group of the carotenoid. The alignment was made using the ClustalW program online (https://www.genome.jp/tools-bin/clustalw) and shaded by the BoxShade program (https://embnet.vital-it.ch/software/BOX_form.html). Similar and identical residues in the alignment are highlighted with grey and black colors on the background, respectively.

2.2.4. Cluster 4: NRPS

A putative cluster containing an NRPS gene as the core biosynthetic gene (orf4) was predicted right next to cluster 3 (Figure 1 and Table S1). However, the NRPS only contained an adenylation (A) domain, a peptidyl carrier protein (PCP) domain, and a termination (T) domain, lacking the required condensation (C) domain for elongation. In addition, the NRPS gene is much smaller than those multimodular assembly lines for non-ribosomal peptide compound biosynthesis. A peptidase M1 gene (orf5) followed the NRPS gene. No other biosynthetic gene was detected in the cluster (Figure 1). Therefore, it seems that cluster 4 may not produce large non-ribosomal peptide compounds or may produce single amino acid-derived compounds. However, homologs of this NRPS gene were found in other bacteria including mostly *Cryobacterium* species within Microbacteriaceae of Actinobacteria, with 76.86% as the highest amino acid identity. It is unclear whether this NRPS gene is inactivated or not. The function of this NRPS gene requires further investigation.

2.2.5. Cluster 5: Oligosaccharide

Cluster 5 was predicted to biosynthesize oligosaccharides, including genes encoding polysaccharide biosynthesis protein, glycosyltransferases and epimerases etc. (Figure 1 and Table S1). However, no known gene cluster was found to be similar. Currently, only a few gene clusters biosynthesizing oligosaccharides have been identified, such as gene clusters producing orthosomycins, moenomycins, saccharomucins, and acarviostatins [29]. In general, these gene clusters are complex and large, usually including multiple genes for sugar synthesis and modification. For example, there are 39 biosynthetic related genes in the gene cluster producing avilamycin (a member of orthosomycins), including four glycosyltransferases, 22 sugar synthesis and tailoring genes, and other genes involved in further modification, resistance, regulation, and transport [30]. Due to the complexity of oligosaccharide producing gene clusters, the biosynthesis of oligosaccharides is lagged behind. Indeed, oligosaccharide natural products are under-represented in pharmacopeia compared with other classes, although oligosaccharides have a wide range of biological activities and high structural diversity [29]. Possible reasons were proposed including low levels of expression in nature, difficulties in detecting, isolation, and identification, as well as low stability [29]. Since oligosaccharide secondary metabolite BGCs are rarely identified and biochemically characterized, the prediction of these gene clusters in genomes is also less frequent than other classes of natural products. Therefore, further identification of the putative oligosaccharide gene cluster 5 is important but challengeable.

2.3. Analysis of the Pigments from M. antarctica ZS314^T

As revealed by the genomic data mining, a putative carotenoid biosynthetic gene cluster was detected, suggesting the production of carotenoids. The physiological characteristics indicated the production of carotenoid-like pigments as well. Colonies of *M. antarctica* ZS314^T were reddish orange on Marine Agar 2216 (Difco) (Figure 6A). The supernatant of liquid culture was almost colorless, and the cell pellets were reddish orange after centrifugation (Figure 6B,C). It suggested that the pigments were stored in the cells rather than being secreted into the broth. The UV-visible absorption spectrum of the concentrated methanolic extracts showed three major absorption peaks at 450, 472, and 500 nm, with the highest at 472 nm (Figure 7A). The shape of the absorption spectrum resembled that of carotenoids in general, and the wavelengths fell within the range of carotenoids as well, indicating the presence of carotenoids in the extracts. The concentrated methanolic extracts were separated by using HPLC, resulting in three major peaks (peak 1, peak 2 and peak 3) at retention times of 24.786 min, 29.661 min and 33.284 min (Figure 7B). The three peaks showed UV-visible absorption spectrum characteristic of carotenoids, indicating that carotenoids may exist in the peaks. The absorption spectra of the three peaks were slightly different from the overall spectrum of total methanolic extracts. Peak 1 showed maximum absorptions at around 468 nm, 490 nm, and 520 nm (Figure 7C). Peaks 2 and 3 had similar absorption spectra, with three maximum absorptions at around 448 nm, 472 nm, and 505 nm (Figure 7C). In order to isolate putative carotenoids indicated by HPLC, three pooled fractions showing indicative color and relatively good purity were obtained after column chromatography with silica gel and Sephadex LH-20. Fraction 1 had one peak showing characteristic absorption spectrum of carotenoids (λ_{max} 452 and 480) (data not shown), with the major MS signal MH$^+$ at 704.54898 (Figure S4), which is close to the molecular mass of C.p.450, sarcinaxanthin and flavuxanthin [9,12]. In addition, the same absorption spectrum was reported for C.p.450, rather than the other two carotenoids. Therefore, the preliminary data suggested that fraction 1 was likely C.p.450. Fraction 2 had one putative carotenoid-like peak (λ_{max} 450 and 475) (data not shown), showing the major MS signal MH$^+$ at 866.59451 (Figure S4). Fraction 3 had two putative carotenoid-like peaks, showing similar absorption spectra (λ_{max} 450 and 475 for the first peak, and λ_{max} 452 and 477 for the second peak) (data not shown). The major MS signal MH$^+$ for the first peak was at 866.58174, while it was at 865.57678 for the second peak (Figure S4). The major MS signals in fractions 2 and 3 had close *m/z* value and absorption spectra, which generally agreed with that of C.p.450 glucoside (MH$^+$: 867). However, it should be interpreted with caution, since high-resolution mass spectrometry accepts small deviation and the reported MS data of these C_{50} carotenoids were obtained using low resolution mass spectrometry. As suggested by cluster 2, *M. antarctica* ZS314[T] is expected to produce C50 carotenoids probably similar to C.p.450 and its glucoside derivatives, the products of *Dietzia* sp. CQ4. Considering the analysis of the gene cluster and the preliminary characterization of the pigments, the carotenoids produced by *M. antarctica* ZS314[T] are most likely C.p.450 and its glucosylated derivatives.

Figure 6. Reddish orange pigments produced by *M. antarctica* ZS314[T] as observed on Marine Agar 2216 plate (**A**) and in Marine Broth 2216 before (**B**) and after (**C**) centrifugation.

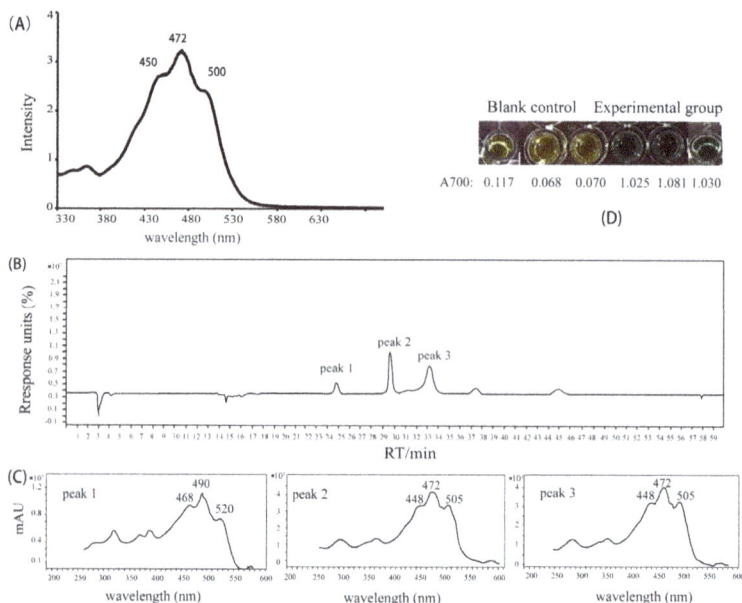

Figure 7. Analysis of the concentrated methanolic extracts of *M. antarctica* ZS314[T]. (**A**) UV-visible absorption spectra of total methanolic extracts. (**B**) HPLC chromatogram of the concentrated methanolic extracts showing three peaks (1, 2, 3) with typical UV absorption spectra of carotenoids. (**C**) The UV-visible absorption spectra of peaks 1, 2, and 3. (**D**) The anti-oxidation assay (FRAP assay) of total methanolic extracts.

Carotenoids have many biological activities and biotechnological applications, including antioxidant activity, absorbing light and energy, transporting oxygen, antitumor activity and enhancing antibody production etc. [31,32]. The most apparent activity of carotenoids is their excellent antioxidant activity, due to their electron rich polyene chains proficient in quenching reactive oxygen species [8]. The formation of iron ferrocyanide (Prussian blue) in the ferric reducing antioxidant power (FRAP) assay due to the presence of the reddish orange pigment extracts suggested antioxidant activity of the pigments, compared with the blank control showing yellow color of potassium ferricyanide solution (Figure 7D). The absorbance of iron ferrocyanide (Prussian blue) at 700 nm (A700) further supported the above observation, with A700 of 1.045 in the pigment group on average vs. A700 of 0.085 in the control group on average. The A700 was significantly elevated in the experimental group compared with the control group, with p value < 0.01 (Student's t-test). Since *M. antarctica* ZS314^T inhabits the intertidal sediments in Antarctica, it has to deal with UV radiation, increased oxidative stress and all other challenges caused by low temperature. Therefore, the production of carotenoids contributes to survival and increases the fitness of *M. antarctica* ZS314^T under the extreme environmental conditions by increasing the antioxidant capability. Moreover, the carotenoids produced by *M. antarctica* ZS314^T may be further explored for potential applications in biotechnology. Since *M. antarctica* ZS314^T has few biosynthetic gene clusters and hence a relatively simple background, carotenoids are the major products in the crude extracts, as indicated by the HPLC analysis as well (Figure 7B). Therefore, it simplifies further purification and isolation of carotenoids from *M. antarctica* ZS314^T for industrial purposes.

3. Materials and Methods

3.1. Strain and Culture Conditions

M. antarctica ZS314^T was isolated on Marine Agar 2216 (Difco) from intertidal sediments of the coast off the Chinese Antarctic Zhongshan Station in East Antarctica. For further morphological observation, the strain ZS314^T was grown on Marine Agar 2216 (Difco) and in Marine Broth 2216 (Difco) at 20 °C for 7 days. For analysis of reddish orange pigments, the strain ZS314^T was cultured in Erlenmeyer flasks containing one liter of Marine Broth 2216 (Difco) at 20 °C for seven days, under rigorous shaking at 200 rpm.

3.2. Genome Sequencing and Data Mining of M. antarctica ZS314^T

Genomic DNA was extracted using QIAamp® DNA Mini kit (QIAGEN, Shanghai, China). The genome was sequenced by a combination of Illumina Hiseq 4000 and PacBio RSII sequencing platforms. Two genomic libraries with inserts of 500 bp and 10 kb were constructed for Illumina and PacBio sequencing, respectively. Clean data from Illumina and PacBio sequencing were assembled using SOAPdenovo software version 2 [33] and RS_HGAP assembly version 3 included in the SMRT Analysis version 2.3.0 (https://github.com/PacificBiosciences/SMRT-Analysis). Circularization of the assemblies was evaluated based on overlaps and using SSPACE-LongRead software [34]. Genes were predicted by using Glimmer version 3.0.2 [35]. Gene annotations were obtained by querying the databases of GenBank non-redundant protein [36], Swiss-Prot [37], KEGG [38], COG [39], GO [40], and carbohydrate-active enzymes database (CAZy) [41]. The presence of rRNA and tRNA was predicted by using rRNAmmer [42] and tRNAscan [43], respectively. The secondary metabolite BGCs were predicted by the antiSMASH program, using the relaxed parameters recommended by the program [44]. Circular map of the genome was drawn using the Circos package [45]. Phylogenetic trees were constructed by using the MEGA 7.0 software with bootstrap replication set at 1000 [46]. Metabolic pathway reconstruction was performed using the web-based server PathPred, following the reference pathway of biosynthesis of secondary metabolites (plants) with default parameters [24].

3.3. Extraction and Analysis of Reddish Orange Pigments

Five liters of liquid Marine Broth 2216 cultures were centrifuged at 7,000 rpm at 4 °C for 10 min. After centrifugation, the supernatant was almost colorless, and the cell pellets were reddish orange. The pigments were then extracted repeatedly with methanol from freeze-dried cell pellets. To avoid any chemical change caused by light, all manipulations during the extraction of the pigments were carried out in the dark and the containers were wrapped in aluminum foil. The extracts were then pooled and concentrated by rotary evaporation at 45 °C.

The UV-visible absorption spectra scan of concentrated methanolic extracts was performed by using Epoch spectrophotometer system with the spectrum range of 330–700 nm. The extracts were analyzed by HPLC (waters e2695) using a C18 reversed-phase column (3.5 μm, 4.6 × 150 mm, waters) with a flow rate of 1 ml·min^{-1} and detection wavelength at 480 nm. The mobile phase was a gradient of 10 to 100% acetonitrile with 0.1% formic acid for 10 min followed by 100% acetonitrile with 0.1% formic acid for 30 min, with a flow rate of 1 mL·min^{-1}. The extracts were isolated and purified by column chromatography with silica gel (200–300 mesh, Merck, Darmstadt, Germany). Thin-layer chromatography (TLC) was used to check the presence and purity of carotenoid-like compounds in the fractions separated by column chromatography before pooling. TLC was performed in TLC Silica gel 60F plates (Merck, Germany), and was developed with a 3:2 (v/v) solvent mixture of acetone and petroleum ether. Carotenoids were indicated by yellow and red colors on TLC plates (Figure S5). Fractions containing putative carotenoids were further purified by column chromatography with Sephadex LH-20 (Pharmacia Biotech, Uppsala, Sweden). Purified fractions were identified using an Agilent 6224 TOF LC/MS spectrometer, with C18 reversed-phase column (3 μm, 4.6 × 75 mm, YMCSep. Technol., Japan). The mobile phase was a gradient of 20 to 100% acetonitrile with 0.1% formic acid for 12 min, followed by 100% acetonitrile with 0.1% formic acid for 48 min, with a flow rate of 0.3 mL·min^{-1}. The eluted compounds were monitored at 480 nm using a diode array detector (DAD) detector.

3.4. Anti-Oxidation Assay of Reddish Orange Pigments

The anti-oxidation ability of the reddish orange pigments was measured via the ferric reducing antioxidant power (FRAP) assay [47]. The mixture of 50 μL of concentrated methanolic extracts and 50 μL of 1% potassium ferricyanide $K_3Fe[CN]_6$ were incubated in the dark for 20 min at 50 °C. Then, 50 μL of 10% trichloroacetic acid was added to the above mixture, and centrifuged for 10 min at 7000 rpm at 4 °C. After centrifugation, 20 μL of sterile deionized-distilled H_2O and 20 μL of 0.1% ferric chloride were added to the supernatant, resulting in the production of iron ferrocyanide, i.e., Prussian blue. The absorbance of Prussian blue at 700 nm was measured by Epoch spectrophotometer system. Methanol was used as a blank control and was manipulated at the same time under the same conditions. Both the experimental group and the blank group were performed in triplicate.

3.5. Data Accessibility

The complete genome of *M. antarctica* ZS314^T has been deposited in the GenBank database under the accession number CP017146.

4. Conclusions

M. antarctica ZS314^T represents a novel type of under-investigated Actinobacteria. The presence of the five putative secondary metabolite gene clusters suggested the genetic potential of *M. antarctica* ZS314^T in producing terpenes, alkylresorcinols, salinixanthin, oligosaccharides or amino acid-derived compounds. Combined with further experimental characterization and the gene cluster analysis provided evidence for the production of C.p.450-like carotenoids. In addition, a new xanthorhodopsin gene was discovered during the analysis of cluster 3. This work contributes to the investigation of natural product biosynthesis of *M. antarctica* ZS314^T. The report of the complete genome also provides opportunities for further investigation of the novel actinobacterium in adaptation to the extreme Antarctic environment.

Supplementary Materials: The following are available online at http://www.mdpi.com/1660-3397/17/7/388/s1, Figure S1: Circular genome map of *M. antarctica* ZS314^T. From outer to inner circle: genome size, protein-encoding genes in forward orientation, protein-encoding genes in reverse orientation, ncRNA in forward orientation, ncRNA in reverse orientation, repeat sequences, GC plot and GC skew., Figure S2: Phylogenetic tree of SrsA from *M. antarctica* ZS314^T and close deposits in the NCBI non-redundant protein sequences (nr) database., Figure S3: Reconstructed biosynthetic pathway of the core structure of salinixanthin without the glucose ester modification from lycopene., Figure S4: Mass spectrometry of the purified fractions from the methanolic extracts of the pigments produced by *M. antarctica* ZS314^T., Figure S5: TLC profile of carotenoid-like pigments indicated by yellow and red colors., Table S1: Gene organization of the five gene clusters predicted in the genome of *M. antarctica* ZS314^T.

Author Contributions: conceptualization, L.L.; methodology, S.S., B.Z. and J.Z.; validation, L.L., and B.C.; formal analysis, C.F. and H.L.; investigation, L.L.; resources, H.L.; data curation, C.F.; writing—original draft preparation, L.L. and B.Z.; writing—review and editing, L.L. and B.C.; funding acquisition, L.L.

Funding: This research was funded by the National Key R&D Program of China (2018YFC1406702), National Natural Science Foundation of China (Grant No. 41406181), and State Key Laboratory of Microbial Technology Open Projects Fund (Project NO. M2017-03).

Acknowledgments: The authors acknowledge the assistance of the Chinese Arctic and Antarctic Administration (CAA) in supporting the CHINARE 23 field work. The authors thank Xiaoqing Tian for technical assistance and the Key Laboratory of East China Sea & Oceanic Fishery Resources Exploitation and Utilization for providing the HPLC/TOF-MS. The authors thank Yuanhua Liu at bioinformatics platform of Institut Pasteur of Shanghai, CAS, for her assistance and ideas in metabolic pathway reconstruction.

Conflicts of Interest: The authors declare no conflict of interest.

References

1. Azman, A.S.; Othman, I.; Velu, S.S.; Chan, K.G.; Lee, L.H. Mangrove rare actinobacteria: Taxonomy, natural compound, and discovery of bioactivity. *Front. Microbiol.* **2015**, *6*, 856. [CrossRef] [PubMed]
2. Zhao, B.; Li, L.; Yu, Y.; Bo, C. Complete genome of *Brachybacterium* sp. P6-10-X1 isolated from deep-sea sediments of the Southern Ocean. *Mar. Genom.* **2017**, *35*, 27–29. [CrossRef]
3. Xiong, Z.Q.; Wang, J.F.; Hao, Y.Y.; Wang, Y. Recent advances in the discovery and development of marine microbial natural products. *Mar. Drugs* **2013**, *11*, 700–717. [CrossRef] [PubMed]
4. Tiwari, K.; Gupta, R.K. Diversity and isolation of rare actinomycetes: an overview. *Crit. Rev. Microbiol.* **2013**, *39*, 256–294. [CrossRef] [PubMed]
5. Rampelotto, P.H. Polar microbiology: Recent advances and future perspectives. *Biology (Base)* **2014**, *3*, 81–84. [CrossRef]
6. Li, H.R.; Yu, Y.; Luo, W.; Zeng, Y.X. *Marisediminicola antarctica* gen. nov., sp. nov., an actinobacterium isolated from the Antarctic. *Int. J. Syst. Evol. Microbiol.* **2010**, *60*, 2535–2539. [CrossRef] [PubMed]
7. Jagannadham, M.V.; Chattopadhyay, M.K.; Subbalakshmi, C.; Vairamani, M.; Narayanan, K.; Rao, C.M.; Shivaji, S. Carotenoids of an Antarctic psychrotolerant bacterium, *Sphingobacterium antarcticus*, and a mesophilic bacterium, *Sphingobacterium multivorum*. *Arch. Microbiol.* **2000**, *173*, 418–424. [CrossRef]
8. Joanna, F.; Květoslava, B. Potential role of carotenoids as antioxidants in human health and disease. *Nutrients* **2014**, *6*, 466–488.
9. Roman, N.; Stafsnes, M.H.; Trygve, A.; Audun, G.Y.; Per, B.; Trygve, B. Biosynthetic pathway for γ-cyclic sarcinaxanthin in *Micrococcus luteus*: heterologous expression and evidence for diverse and multiple catalytic functions of C(50) carotenoid cyclases. *J. Bacteriol.* **2010**, *192*, 5688–5699.
10. Wang, F.; Jiang, J.G.; Chen, Q. Progress on molecular breeding and metabolic engineering of biosynthesis pathways of C-30, C-35, C-40, C-45, C-50 carotenoids. *Biotechnol. Adv.* **2007**, *25*, 211–222. [CrossRef]
11. Takaichi, S.; Inoue, K.; Akaike, M.; Kobayashi, M.; Ohoka, H.; Madigan, M.T. The major carotenoid in all known species of heliobacteria is the C_{30} carotenoid 4,4'-diaponeurosporene, not neurosporene. *Arch. Microbiol.* **1997**, *168*, 277–281. [CrossRef] [PubMed]
12. Tao, L.; Yao, H.; Cheng, Q. Genes from a *Dietzia* sp. for synthesis of C_{40} and C_{50} β-cyclic carotenoids. *Gene* **2007**, *386*, 90–97. [CrossRef] [PubMed]

13. Funabashi, M.; Funa, N.; Horinouchi, S. Phenolic lipids synthesized by type III polyketide synthase confer penicillin resistance on *Streptomyces griseus*. *J. Biol. Chem.* **2008**, *283*, 13983–13991. [CrossRef] [PubMed]

14. Yu, D.; Xu, F.; Zeng, J.; Zhan, J. Type III polyketide synthases in natural product biosynthesis. *IUBMB Life* **2012**, *64*, 285–295. [CrossRef] [PubMed]

15. Kozubek, A.; Pietr, S.; Czerwonka, A. Alkylresorcinols are abundant lipid components in different strains of *Azotobacter chroococcum* and *Pseudomonas* spp. *J. Bacteriol.* **1996**, *178*, 4027–4030. [CrossRef]

16. Krubasik, P.; Kobayashi, M.; Sandmann, G. Expression and functional analysis of a gene cluster involved in the synthesis of decaprenoxanthin reveals the mechanisms for C_{50} carotenoid formation. *FEBS J.* **2001**, *268*, 3702–3708. [CrossRef]

17. Sandmann, G. Molecular evolution of carotenoid biosynthesis from bacteria to plants. *Physiol. Plant* **2002**, *116*, 431–440. [CrossRef]

18. Balashov, S.P.; Imasheva, E.S.; Boichenko, V.A.; Antón, J.; Wang, J.M.; Lanyi, J.K. Xanthorhodopsin: A proton pump with a light-harvesting carotenoid antenna. *Science* **2005**, *309*, 2061–2064. [CrossRef]

19. Lanyi, J.K.; Balashov, S.P. Xanthorhodopsin: A bacteriorhodopsin-like proton pump with a carotenoid antenna. *Biochim. Biophys. Acta-Bioenerg.* **2008**, *1777*, 684–688. [CrossRef]

20. Rodrigobaños, M.; Garbayo, I.; Vílchez, C.; Bonete, M.J.; Martínezespinosa, R.M. Carotenoids from haloarchaea and their potential in biotechnology. *Mar. Drugs* **2015**, *13*, 5508–5532. [CrossRef]

21. Estrada, A.F.; Maier, D.; Scherzinger, D.; Avalos, J.; Al-Babili, S. Novel apocarotenoid intermediates in *Neurospora crassa* mutants imply a new biosynthetic reaction sequence leading to neurosporaxanthin formation. *Fungal Genet Biol.* **2008**, *45*, 1497–1505. [CrossRef] [PubMed]

22. Hannibal, L.; Lorquin, J.; Dortoli, N.A.; Garcia, N.; Chaintreuil, C.; Massonboivin, C.; Dreyfus, B.; Giraud, E. Isolation and characterization of canthaxanthin biosynthesis genes from the photosynthetic bacterium *Bradyrhizobium* sp. strain ORS278. *J. Bacteriol.* **2000**, *182*, 3850–3853. [CrossRef] [PubMed]

23. Graham, J.E.; Bryant, D.A. The biosynthetic pathway for Myxol-2′fucoside (myxoxanthophyll) in the cyanobacterium *Synechococcus* sp. strain PCC 7002. *J. Bacteriol.* **2009**, *191*, 3292–3300. [CrossRef] [PubMed]

24. Moriya, Y.; Shigemizu, D.; Hattori, M.; Tokimatsu, T.; Kotera, M.; Goto, S.; Kanehisa, M. PathPred: an enzyme-catalyzed metabolic pathway prediction server. *Nucleic Acids Res.* **2010**, *38*, W138–W143. [CrossRef] [PubMed]

25. Ullrich, A.; Rohrschneider, M.; Scheuermann, G.; Stadler, P.F.; Flamm, C. In silico evolution of early metabolism. *Artif Life* **2011**, *17*, 87–108. [CrossRef] [PubMed]

26. Thiele, I.; Palsson, B.O. A protocol for generating a high-quality genome-scale metabolic reconstruction. *Nat. Protoc.* **2010**, *5*, 93–121. [CrossRef]

27. Vollmers, J.; Voget, S.; Dietrich, S.; Gollnow, K.; Smits, M.; Meyer, K.; Brinkhoff, T.; Simon, M.; Daniel, R. Poles Apart: Arctic and Antarctic *Octadecabacter* strains share high genome plasticity and a new type of xanthorhodopsin. *PLOS ONE* **2013**, *8*, e63422. [CrossRef]

28. Imasheva, E.S.; Balashov, S.P.; Choi, A.R.; Jung, K.H.; Lanyi, J.K. Reconstitution of *Gloeobacter violaceus* rhodopsin with a light-harvesting carotenoid antenna. *Biochemistry* **2009**, *48*, 10948–10955. [CrossRef]

29. Mccranie, E.K.; Bachmann, B.O. Bioactive oligosaccharide natural products. *Nat. Prod. Rep.* **2014**, *31*, 1026–1042. [CrossRef]

30. Weitnauer, G.; Mühlenweg, A.; Trefzer, A.; Hoffmeister, D.; Süßmuth, R.D.; Jung, G.; Welzel, K.; Vente, A.; Girreser, U.; Bechthold, A. Biosynthesis of the orthosomycin antibiotic avilamycin A: deductions from the molecular analysis of the avi biosynthetic gene cluster of *Streptomyces viridochromogenes* Tü57 and production of new antibiotics. *Chem. Biol.* **2001**, *8*, 569–581. [CrossRef]

31. Sarvari, S.; Moakedi, F. High carotenoid production by a halotolerant bacterium, *Kocuria* sp. strain QWT-12 and anticancer activity of its carotenoid. *EXCLI J.* **2017**, *16*, 840–851.

32. Britton, G.; Liaaenjensen, S.; Hanspeter, P. Carotenoids, Volume 5: Nutrition and health. *Carotenoids* **2009**, *24*, 497–522.

33. Luo, R.; Liu, B.; Xie, Y.; Li, Z.; Huang, W.; Yuan, J.; He, G.; Chen, Y.; Pan, Q.; Liu, Y.; et al. SOAPdenovo2: an empirically improved memory-efficient short-read de novo assembler. *Gigascience* **2012**, *1*, 18. [CrossRef] [PubMed]

34. Boetzer, M.; Pirovano, W. SSPACE-LongRead: scaffolding bacterial draft genomes using long read sequence information. *BMC Bioinform.* **2014**, *15*, 211. [CrossRef] [PubMed]

35. Delcher, A.L.; Bratke, K.A.; Powers, E.C.; Salzberg, S.L. Identifying bacterial genes and endosymbiont DNA with Glimmer. *Bioinformatics* **2007**, *23*, 673–679. [CrossRef] [PubMed]
36. Pruitt, K.D.; Tatusova, T.; Maglott, D.R. NCBI reference sequences (RefSeq): A curated non-redundant sequence database of genomes, transcripts and proteins. *Nucleic Acids Res.* **2007**, *35*, D61–D65. [CrossRef] [PubMed]
37. Boeckmann, B.; Bairoch, A.; Apweiler, R.; Blatter, M.C.; Estreicher, A.; Gasteiger, E.; Martin, M.J.; Michoud, K.; O'Donovan, C.; Phan, I.; et al. The SWISS-PROT protein knowledgebase and its supplement TrEMBL in 2003. *Nucleic Acids Res.* **2003**, *31*, 365–370. [CrossRef] [PubMed]
38. Kanehisa, M.; Goto, S. KEGG: Kyoto encyclopedia of genes and genomes. *Nucleic Acids Res.* **1999**, *28*, 27–30. [CrossRef] [PubMed]
39. Tatusov, R.L.; Fedorova, N.D.; Jackson, J.D.; Jacobs, A.R.; Kiryutin, B.; Koonin, E.V.; Krylov, D.M.; Mazumder, R.; Mekhedov, S.L.; Nikolskaya, A.N. The COG database: An updated version includes eukaryotes. *BMC Bioinform.* **2003**, *4*, 41. [CrossRef]
40. Harris, M.A.; Clark, J.; Ireland, A.; Lomax, J.; Ashburner, M.; Foulger, R.; Eilbeck, K.; Lewis, S.; Marshall, B.; Mungall, C.; et al. The Gene Ontology (GO) database and informatics resource. *Nucleic Acids Res.* **2004**, *32*, D258–D261. [PubMed]
41. Lombard, V.; Golaconda Ramulu, H.; Drula, E.; Coutinho, P.M.; Henrissat, B. The carbohydrate-active enzymes database (CAZy) in 2013. *Nucleic Acids Res.* **2014**, *42*, D490–D495. [CrossRef] [PubMed]
42. Lagesen, K.; Hallin, P.; Rodland, E.; Staerfeldt, H.; Rognes, T.; Ussery, D. RNAmmer: consistent and rapid annotation of ribosomal RNA genes. *Nucleic Acids Res.* **2007**, *35*, 3100–3108. [CrossRef] [PubMed]
43. Lowe, T.M.; Eddy, S.R. tRNAscan-SE: A program for improved detection of transfer RNA genes in genomic sequence. *Nucleic Acids Res.* **1997**, *25*, 955–964. [CrossRef] [PubMed]
44. Blin, K.; Wolf, T.; Chevrette, M.G.; Lu, X.; Schwalen, C.J.; Kautsar, S.A.; Suarez Duran, H.G.; de Los Santos, E.L.C.; Kim, H.U.; Nave, M.; et al. antiSMASH 4.0-improvements in chemistry prediction and gene cluster boundary identification. *Nucleic Acids Res.* **2017**, *45*, 36–41. [CrossRef] [PubMed]
45. Krzywinski, M.; Schein, J.I. Circos: an information aesthetic for comparative genomics. *Genome Res.* **2009**, *19*, 1639–1645. [CrossRef] [PubMed]
46. Kumar, S.; Stecher, G.; Tamura, K. MEGA7: Molecular evolutionary genetics analysis version 7.0 for bigger datasets. *Mol. Biol. Evol.* **2016**, *33*, 1870–1874. [CrossRef] [PubMed]
47. Bolanos de la Torre, A.A.S.; Henderson, T.; Nigam, S.P.; Owusu-Apenten, K.R. A universally calibrated microplate ferric reducing antioxidant power (FRAP) assay for foods and applications to Manuka honey. *Food Chem.* **2015**, *174*, 119–123. [CrossRef] [PubMed]

marine drugs

MDPI

Article

Genome Mining of Marine-Derived *Streptomyces* sp. SCSIO 40010 Leads to Cytotoxic New Polycyclic Tetramate Macrolactams

Wei Liu [1,2], Wenjun Zhang [2], Hongbo Jin [2], Qingbo Zhang [2], Yuchan Chen [3], Xiaodong Jiang [2], Guangtao Zhang [2], Liping Zhang [2], Weimin Zhang [3], Zhigang She [1] and Changsheng Zhang [2,*]

[1]   South China Sea Resource Exploitation and Protection Collaborative Innovation Center (SCS-REPIC)/School of Marine Sciences, Sun Yat-sen University, Guangzhou 510006, China; liuw69@mail2.sysu.edu.cn (W.L.); cesshzhg@mail.sysu.edu.cn (Z.S.)
[2]   Key Laboratory of Tropical Marine Bio-resources and Ecology, Guangdong Key Laboratory of Marine Materia Medica, South China Sea Institute of Oceanology, Innovation Academy for South China Sea Ecology and Environmental Engineering, Chinese Academy of Sciences, 164 West Xingang Road, Guangzhou 510301, China; wzhang@scsio.ac.cn (W.Z.); khbmail@126.com (H.J.); gudaobo@163.com (Q.Z.); jxd374487986@163.com (X.J.); gtzhang@scsio.ac.cn (G.Z.); zhanglp@scsio.ac.cn (L.Z.)
[3]   State Key Laboratory of Applied Microbiology Southern China, Guangdong Institute of Microbiology, 100 Central Xianlie Road, Guangzhou 510070, China; chenyc@gdim.cn (Y.C.); wmzhang@gdim.cn (W.Z.)
*   Correspondence: czhang2006@gmail.com; Tel.: +86-20-8902-3038

Received: 29 August 2019; Accepted: 16 November 2019; Published: 25 November 2019

Abstract: Polycyclic tetramate macrolactams (PTMs) biosynthetic gene cluster are widely distributed in different bacterial types, especially in *Streptomyces* species. The mining of the genomic data of marine-derived *Streptomyces* sp. SCSIO 40010 reveals the presence of a putative PTM-encoding biosynthetic gene cluster (*ptm′* BGC) that features a genetic organization for potentially producing 5/5/6 type of carbocyclic ring-containing PTMs. A fermentation of *Streptomyces* sp. SCSIO 40010 led to the isolation and characterization of six new PTMs **1–6**. Comprehensive spectroscopic analysis assigned their planar structures and relative configurations, and their absolute configurations were deduced by comparing the experimental electronic circular dichroism (ECD) spectra with the reported spectra of the known PTMs. Intriguingly, compounds **1–6** were determined to have a *trans*-orientation of H-10/H-11 at the first 5-membered ring, being distinct from the *cis*-orientation in their known PTM congeners. PTMs **1–5** displayed cytotoxicity against several cancer cell lines, with IC_{50} values that ranged from 2.47 to 17.68 μM.

Keywords: *Streptomyces* sp. SCSIO 40010; marine; genome mining; polycyclic tetramate macrolactams; cytotoxicity

1. Introduction

Polycyclic tetramate macrolactams (PTMs) are a unique class of natural products that consist of a tetramate-embedding macrocyclic lactam core and a varying carbocycle with 5/6, 5/5, 5/6/5, or 5/5/6 ring system [1]. PTMs display a wide range of antifungal, antibiotic, antiprotozoal, and antitumor properties [2–5], and they have significant potential for applications in agricultures and medicines [1,6]. HSAF (also known as dihydromaltophilin) [7], a typical representative of 5/5/6 type of PTMs, exhibits a broad spectrum of antifungal activities and it has been used as an antifungal agent to control plant diseases [8]. The anticancer agent ikarugamycin [9], a typical 5/6/5 type of PTMs, shows activity as an inhibitor of clathrin-mediated endocytosis [10]. Therefore, PTMs draw the attention of synthetic chemists; however, multiple chiral centers in PTMs greatly enhance the structure diversity and increase the difficulty for the total synthesis [11–14]. In a sharp contrast, in nature, a conserved and compact biosynthetic pathway has been evolved to simply assemble such kinds of complex structures [1]. Recent studies reveal that PTMs are derived from a conserved hybrid polyketide synthase (PKS)/non-ribosomal peptide synthethase (NRPS) pathway [1,15]. The PKS portion of the hybrid PKS/NRPS enzyme is iteratively used to produce two separate polyketide chains, which are respectively condensed with the α- and δ-amino groups of an L-orinithine that is tethered in the NRPS portion to generate a common polyene tetramate precursor [7,15–21]. Afterwards, a set of oxidoreductases catalyzes divergent cyclization reactions to control the formation of diverse carbocyclic ring systems in PTMs [18,20,22–24]. Particularly, the iterative assembly of two separate polyketide chains by the same single-module bacterial polyketide synthase has been demonstrated in vitro in the biosynthesis of HSAF [16], and the biocatalytic total synthesis of ikarugamycin has been recently achieved [22].

Genome Mining has been successfully utilized to discover new PTMs [9,15,17,20,25,26]. For example, we have reported the activation of a silent PTM biosynthetic gene cluster (BGC) by promoter engineering in marine-derived *Streptomyces pactum* SCSIO 02999 to produce a series of new PTMs pactamides with a 5/5/6 ring system [20]. In addition, we have characterized three new PTMs containing a 5/6/5 ring system from a South China Sea-derived *Streptomyces* sp. SCSIO 40060 while using a genomics-guided approach [26]. We identified a mangrove-derived *Streptomyces* sp. SCSIO 40010 harboring a putative PTM BGC during our continuous search for PTM-producing strains. Herein, we reported the isolation, structural elucidation and biological evaluation of six new PTMs **1–6** (Figure 1).

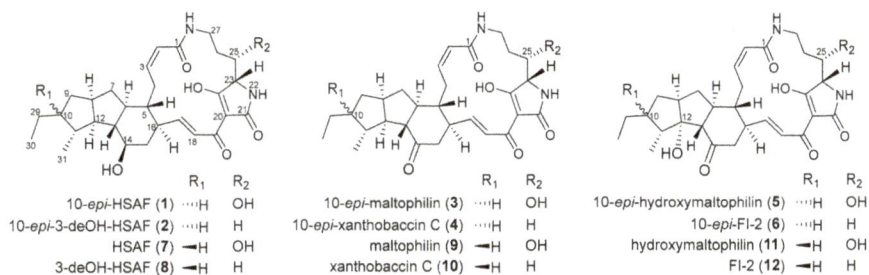

Figure 1. Chemical structures of polycyclic tetramate macrolactams (PTMs). Compounds **1–6** were isolated from *Streptomyces* sp. SCSIO 40010. The known compounds **7–12** with the same planar structures as those of **1–6**, respectively, are shown here for comparison.

2. Results and Discussion

2.1. Genome Mining of a PTM Biosynthetic Gene Cluster

The strain SCSIO 40010 was isolated from the mangrove sediment in Penang, Malaysia, and it was identified to be a *Streptomyces* species on the basis of its 16S rDNA sequence (GenBank accession number MN224032). The mining of the sequenced genome of *Streptomyces* sp. SCSIO 40010 reveals the presence of a putative PTM BGC (*ptm'*, GenBank accession number MN234160) that displays high similarity to the *ptm* BGC in *S. pactum* SCSIO 02999 (Figure 2a) [20]. This *ptm'* BGC encodes six conserved enzymes, including the hybrid PKS/NRPS PtmA', the FAD-dependent oxidoreductase PtmB1', PtmB2', the alcohol dehydrogenase PtmC', the hydroxylase PtmD', and the P450 enzyme PtmE'. In addition to the scaffold constructing enzymes PtmA', PtmB1', PtmB2', and PtmC', two modifying enzymes PtmD' (resembling the C-25 hydroxylase FtdA [15], 63% identity) and PtmE' (resembling the P450 enzyme FtdF [15], 59% identity) were also found in the *ptm'* BGC in *Streptomyces* sp. SCSIO 40010.

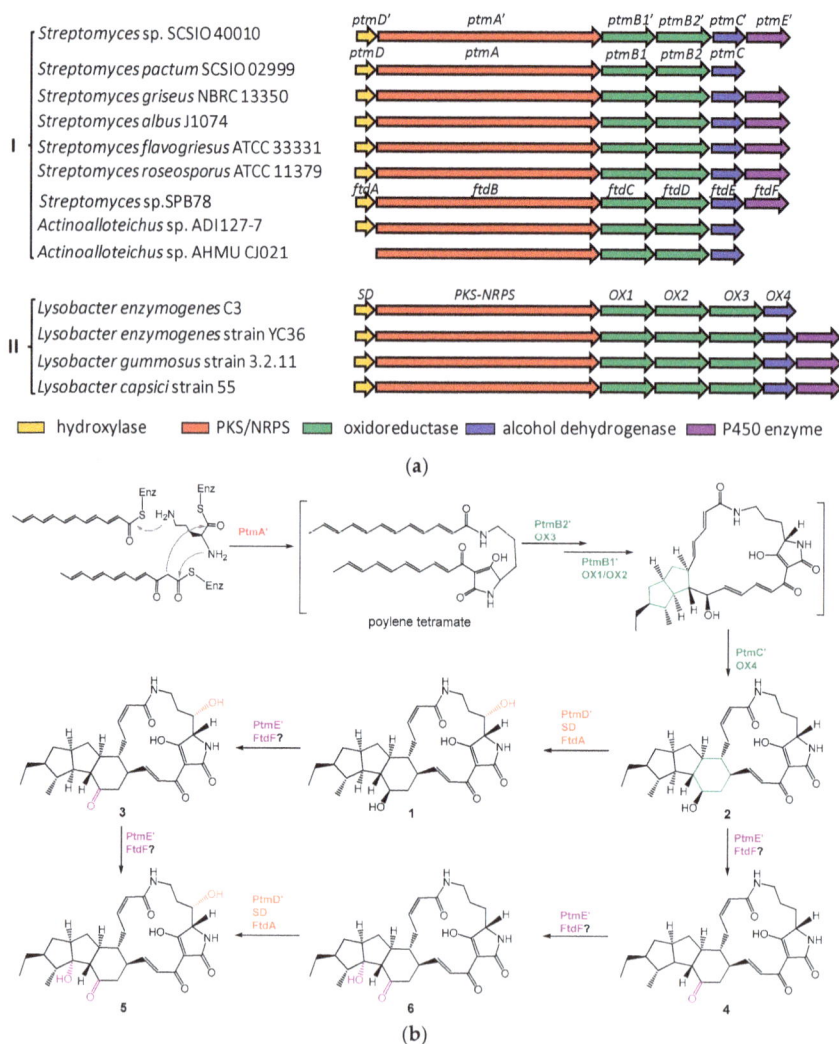

Figure 2. (a) Bioinformatics analysis of 5/5/6 type of PTM biosynthetic gene clusters (BGCs). (b) The proposed biosynthetic pathway for six new 5/5/6 type of PTMs.

Our preliminary genome mining of *Streptomyces* sp. SCSIO 40010 indicates that it should be a potential producer of PTMs with a 5/5/6 carbocyclic ring system [1,15]. Thus, we mined the available genome sequences for PTM BGCs and made a bioinformatics analysis of the PTM BGCs, typically for 5/5/6 type of PTMs [1,15]. Our analysis shows that the BGCs for 5/5/6 type of PTMs should fall into two categories (Figure 2a), depending on the number of oxidoreductases that are involved in the construction of the 5/5 ring system (two for Group I and three for Group II).

The PTM BGCs of Group I are mainly distributed in *Streptomyces* species (Figure 2a). Some of these *Streptomyces* strains have been demonstrated to produce 5/5/6 and/or 5/5 type of PTMs, such as pactamides in *S. pactum* SCSIO 02999 [20], compounds a–d in *S. griseus* NBRC 13350 [17], alteramides in *S. albus* J1074 [25], and frontalamides in *Streptomyces* sp. SPB78 [15]. In contrast, no PTMs have been reported from *S. flavogriseus* ATCC 33331 and *S. roseosporus* ATCC 11379 that contain Group I of

PTM BGCs (Figure 2a) [15]. In addition to *Streptomyces* species, *Actinoalloteichus* sp. ADI127-7 and AHMU CJ201 also contain Group I of PTM BGCs (Figure 2a), while no PTMs have been reported from them. *Actinoalloteichus cyanogriseus* WH1-2216-6 was reported to produce HSAF and its analogues [27]; however, its genome sequence is not yet publicly available.

Biosynthetically, it has been experimentally demonstrated in *S. pactum* SCSIO 02999 that two oxidoreductases PtmB1 and PtmB2 are responsible for the sequential formation of the 5/5 ring system in pactamide A, with the formation of the first 5-membered ring by PtmB2 and the second 5-membered ring by PtmB1 [20]. Subsequently, formation of the inner 6-membered is catalyzed by the alcohol dehydrogenase PtmC [20].

The PTM BGCs of Group II are mainly found in bacterial strains of *Lysobacter* (Figure 2a). The strains *Lysobacter enzymogenes* C3 and strain YC36 are validated to produce HSAF and related PTMs [7,16,19,23]. The strains *Lysobacter gummosus* strain 3.2.11 and *Lysobacter capsici* strain 55 are indicated as HSAF producers [28]. Unlike Group I PTM BGCs, three oxidoreductases (such as OX1, OX2, and OX3 in *Lysobacter enzymogenes*) from the PTM BGCs of Group II are involved in the formation of the 5/5 ring system [23]. It has been shown that OX3 is responsible for the first 5-membered ring formation in lysobacterene A, while OX1 and OX2 catalyze the formation of the second 5-membered ring, but with different stereo selectivity [23]. In a similar fashion to PtmC, OX4 catalyzes the formation of the inner 6-membered ring in HSAF (**7**, Figure 1) [23,24].

The metabolite profiles of *Streptomyces* sp. SCSIO 40010 were investigated by cultivation in four different media, including modifed-A1BFe+C [29], AM6, AM6-4, and modifed-ISP3 [26]. HPLC analyses showed that compounds exhibiting UV-visible absorption spectra that were similar to PTMs were better produced in the modifed-A1BFe+C medium (Supplementary Figure S1). Subsequently, a 20-L fermentation of *Streptomyces* sp. SCSIO 40010 was performed in the modifed-A1BFe+C. Butanone extracts of the 20-L fermentation cultures were subjected to multiple chromatographic methods to provide six new PTMs **1–6**.

Compound **1** was isolated as a white powder. The molecular formula of **1** was determined as $C_{29}H_{40}N_2O_6$ by HRESIMS ([M + H]$^+$, *m/z* 513.2960, calcd for 513.2965, Supplementary Figure S2). The planar structure of **1** was determined to be the same as that of HSAF (**7**, Figure 1) [27,30], by comparing NMR spectroscopic data of **1** (Tables 1 and 2; Supplementary Figures S3–S9) and HSAF (**7**) [27,30]. The geometries of double bonds in **1** were determined to be *trans* (*E*) or *cis* (*Z*) on the basis of their coupling constants (Z$\Delta^{2,3}$ $J_{2,3}$ 11.5 Hz; E$\Delta^{17,18}$ $J_{17,18}$ 15.5 Hz; Table 1). The relative configurations of **1** were assigned by NOESY correlations and then compared with pactamide A [20] and HSAF (**7**) [30] (Figure 3a). It should be noted that a *trans*-orientation of H-10 and H-11 was assigned for **1**, because of the obvious NOESY correlations of H-8/H-10, H10/H-12, and H-11/H-29b (Figure 3). Previously, a *trans*-orientation of H-10 and H-11 was reported for pactamide A [20], aburatubolactam A (X-ray crystallography structure available [31]), combamide D [32], deOH alteramides, and lysobacterene B [33]. In the recently reported 5/5 type of PTMs umezawamides, the relative orientation of H-10 and H-11 was not determined [34]. However, a *cis*-orientation of H-10 and H-11 was determined in **7** [30]. As described by Cao et al. [9] and Hoshino et al. [34], the small vicinal coupling constant between H-23 and H-25 strongly indicated the relative configuration between H-23 and H-25 to be (23*S**, 25*S**) in **1** ($^3J_{H-23/H-25}$ 1.2 Hz, Table 1). The configuration of C-23 was deduced to be 23*S* upon the proposed biogenesis from an ʟ-orinithine [7,15–21]. Recently, a crystallographic study unequivocally determined the absolute configuration as 23*S*, 25*S* in hydroxylikarugamycin A [26]. Thus, when considering the biosynthetic similarity between **1** and hydroxyikarugamycin A, **1** was deduced to also have the stereochemistry of 23*S*, 25*S*. Given that **1** and **7** (5*R*, 6*S*, 8*S*, 10*R*, 11*R*, 12*R*, 13*S*, 14*R*, 16*R*, 23*S*, 25*S* [30]) displayed an almost identical electronic circular dichroism (ECD) spectra (Supplementary Figure S10), the absolute configuration of **1** was deduced as 5*R*, 6*S*, 8*S*, 10*S*, 11*R*, 12*R*, 13*S*, 14*R*, 16*R*, 23*S*, 25*S*, which was only different from **7** by adopting an opposite configuration at C-10 (Figure 1). Thus, compound **1** was designated 10-*epi*-HSAF.

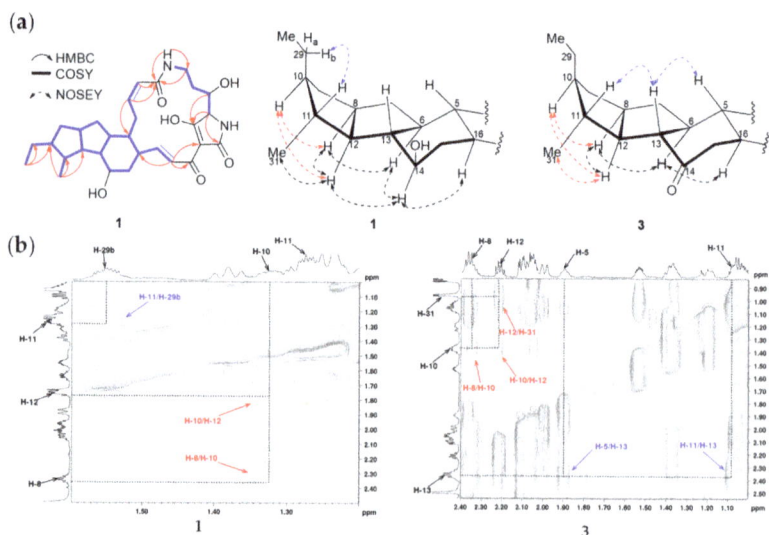

Figure 3. (a) Key COSY, HMBC correlations for **1**, and selected NOE correlations for **1** and **3**. (b) Key NOE correlations to support a *trans*-orientation of H-10/H-11 in **1** and **3**.

Table 1. ^{1}H NMR (700 MHz) Data for PTMs **1–6** in dimethyl sulfoxide (DMSO)-d_6 (δ_H, mult, *J* in Hz).

No.	1	2	3	4	5	6
2	5.70, dd, (2.3, 11.5)	5.73, dd, (1.9, 11.4)	5.72, dd, (2.2, 11.3)	5.75, dd, (2.0, 11.5)	5.75, dd, (2.1, 11.3)	5.77, dd, (2.0, 11.5)
3	5.90, td, (1.8, 11.2)	5.9, td, (2.2, 11.1)	5.92, td, (2.1, 11.0)	5.92, td, (2.1, 11.1)	5.96, td, (2.1, 11.1)	5.96, td, (2.2, 11.1)
4a	1.89, m	1.92, m	2.00, m	2.04, m	2.03, m	2.05, m
4b	3.52, m	3.49, m	3.62, m	3.62, m	3.65, m	3.64, m
5	1.28, m	1.27, m	1.90, m	1.90, m	1.87, m	1.88, m
6	1.64, m	1.27, m	2.09, m	2.10, m	2.56, m	2.57, m
7a	0.86, m	0.87, m	1.05, m	0.85, m	0.98, m	1.01, m
7b	1.96, m	1.97, m	2.05, m	2.06, m	2.11, m	2.14, m
8	2.35, m	2.35, m	2.35, m	2.35, m	2.03, m	2.03, m
9a	0.80, m	0.79, m	0.84, m	1.06, m	0.68, m	0.69, m
9b	2.01, m	2.01, m	2.05, m	2.06, m	2.02, m	2.03, m
10	1.33, m	1.32, m	1.37, m	1.37, m	1.53, m	1.53, m
11	1.27, m	1.64, m	1.09, m	1.09, m	1.21, m	1.22, m
12	1.75, m	1.77, m	2.21, m	2.21, m		
13	1.10, m	1.09, m	2.37, m	2.37, m	2.32, m	2.32, m
14	3.25, m	3.25, m				
15a	1.24, m	1.24, m	2.11, m	2.11, m	2.12, m	2.13, m
15b	1.74, m	1.75, m	2.59, m	2.59, m	2.59, m	2.60, m
16	2.06, m	2.07, m	2.38, m	2.40, m	2.44, m	2.47, m
17	6.57, dd, (10.5, 15.5)	6.55, dd, (10.5, 15.5)	6.63, dd, (10.3, 15.6)	6.61, t, (15.5)	6.60, dd, (10.3, 15.5)	6.59, dd, (11.5, 15.5)
18	6.86, d, (15.5)	6.95, d, (15.5)	6.87, d, (15.6)	6.96, d, (15.5)	6.88, d, (15.5)	6.96, d, (15.5)
22	NH, 8.68, s	NH, 8.68, s	NH, 8.95, s	NH, 8.73, brs	NH, 8.96, s	NH, 8.75, s
23	3.86, d, (1.2)	3.81, d, (5.7)	3.87, d, (1.4)	3.83, s	3.87, d, (1.1)	3.84, d, (6.1)
25a	3.81, dt, (1.5, 6.1)	1.74, m	3.81, dt, (1.4,6.2)	1.74, m	3.82, dt, (2.0, 6.4)	1.74, m
25b		1.84, m		1.86, m		1.86, m
26a	1.18, m	1.15, m	1.18, m	1.17, m	1.20, m	1.15, m
26b	1.38, m	1.32, m	1.39, m	1.35, m	1.39, m	1.34, m

Table 1. *Cont.*

No.	1	2	3	4	5	6
27a	2.57, m	2.39, m	2.59, m	3.23, m	2.58 m	2.40, m
27b	3.25, m	3.22, m	3.26, m	2.39, m	3.24, m	3.23, m
28	NH, 7.96, t, (5.7)	NH, 7.82, t, (5.3)	NH, 7.98, t, (5.6)	NH, 7.86, s	NH, 8.00, t, (5.6)	NH, 7.88, t, (5.6)
29a	1.04, m	1.04, m	1.03, m	1.03, m	0.99, m	1.02, m
29b	1.55, m	1.55, m	1.53, m	1.54, m	1.51, m	1.52, m
30	0.85, t (7.4)	0.85, t (7.4)	0.84, t (7.4)	0.84, t (7.4)	0.84, t (7.4)	0.84, t (7.4)
31	1.06, d (6.4)	1.06, d (6.4)	0.96, d (6.5)	0.96, d (6.5)	0.94, d (6.7)	0.94, d (6.8)

Table 2. ^{13}C NMR (176 MHz) Data for PTMs **1–6** in DMSO-d_6, (δ_C, type).

No.	1	2	3	4	5	6
1	165.5, C	165.5, C	165.5, C	165.6, C	165.5, C	165.6, C
2	124.1, CH	124.2, CH	124.4, CH	124.5, CH	124.5, CH	124.7, CH
3	139.1, CH	138.9, CH	138.4, CH	138.3, CH	138.2, CH	138.1, CH
4	28.0, CH$_2$	28.1, CH$_2$	27.7, CH$_2$	27.7, CH$_2$	27.4, CH$_2$	27.4, CH$_2$
5	43.5, CH	43.5, CH	43.2, CH	43.2, CH	43.0, CH	43.0, CH
6	47.4, CH	46.4, CH	51.1, CH	51.2, CH	47.9, CH	48.0, CH
7	37.2, CH$_2$	37.3, CH$_2$	38.4, CH$_2$	39.7, CH$_2$	36.5, CH$_2$	36.5, CH$_2$
8	41.5, CH	41.4, CH	40.4, CH	40.4, CH	51.3, CH	51.3, CH
9	40.3, CH$_2$	40.3, CH$_2$	39.6, CH$_2$	38.5, CH$_2$	38.1, CH$_2$	38.1, CH$_2$
10	53.5, CH	53.5, CH	53,2, CH	53,2, CH	50.4, CH	50.4, CH
11	46.5, CH	47.6, CH	46.7, CH	46.7, CH	49.3, CH	49.3, CH
12	58.1, CH	58.1, CH	50.4, CH	50.4, CH	89.7, CH	89.7, CH
13	59.1, CH	59.1, CH	63.0, CH	63.0, CH	64.2, CH	64.2, CH
14	72.7, CH	72.7, CH	207.4, C	207.4, C	210.1 C	210.2, C
15	41.9, CH$_2$	41.9, CH$_2$	45.6, CH$_2$	45.6, CH	46.0, CH$_2$	46.0, CH$_2$
16	45.7, CH	45.6, CH	47.7, CH	47.8, CH	47.1, CH	47.2, CH
17	150.1, CH	149.6, CH	147.8, CH	147.4, CH	147.5, CH	147.2, CH
18	121.3, CH	121.5, CH	122.0, CH	122.1, CH	122.1, CH	122.2, CH
19	172.2, C	171.8, C	171.9, C	175.1, C	171.8, C	171.2, C
20	100.4, C	100.7, C	100.7, C	101.1, C	100.7, C	101.0, C
21	175.7, C	175.3, C	175.6, C	171.4, C	175.6, C	175.1, C
23	68.6, CH	61.0, CH	68.5, CH	61.1, CH	68.6, CH	61.1, CH
24	193.0, C	195.8, C	193.0, C	195.8, C	193.0, C	195.9, C
25	70.1, CH	26.2, CH$_2$	70.1, CH	26.2, CH$_2$	70.1, CH	26.1, CH$_2$
26	31.1, CH$_2$	20.4, CH$_2$	31.0, CH$_2$	20.4, CH$_2$	31.0, CH$_2$	31.1, CH$_2$
27	36.4, CH$_2$	38.0, CH$_2$	36.4, CH$_2$	38.0, CH$_2$	36.4, CH$_2$	36.4, CH$_2$
29	25.8, CH$_2$	25.8, CH$_2$	25.5, CH$_2$	25.5, CH$_2$	25.3, CH$_2$	25.3, CH$_2$
30	12.6, CH$_3$	12.6, CH$_3$	12.4, CH$_3$	12.4, CH$_3$	12.0, CH$_3$	12.0, CH$_3$
31	18.4, CH$_3$	18.4, CH$_3$	17.6, CH$_3$	17.6, CH$_3$	11.5, CH$_3$	11.5, CH$_3$

Compound **2** was obtained as a white powder and it was assigned the molecular formula of $C_{29}H_{40}N_2O_5$ on the basis of HRESIMS ([M + H]$^+$, *m/z* 497.3010, calcd for 497.3015, Supplementary Figure S11). A detailed comparison of one-dimensional (1D) and two-dimensional (2D) NMR spectroscopic data of **2** (Tables 1 and 2, Supplementary Figures S12–S18) and deOH-HSAF (**8**, Figure 1) revealed the same planar structure for **2** and **8** [27,30]. The relative configuration of **2** was deduced by proton coupling constants ($Z\Delta^{2,3}$ $J_{2,3}$ 11.5 Hz; $E\Delta^{17,18}$ $J_{17,18}$ 15.5 Hz; Table 1) and careful analysis of NOESY correlations (Figure 4, Supplementary Figures S17 and S18). Similar to **1**, a *trans*-orientation of H-10/H-11 was determined in **2** (Figure 1), by deducing from NOESY correlations of H-8/H-10, H10/H-12, H12/Me-31, and H-11/H-29b. This was different from the *cis*-orientation of H-10/H-11 in **8** (Figure 1) [30]. **2** was deduced to have the absolute configuration of 5*R*, 6*S*, 8*S*, 10*S*, 11*R*, 12*R*, 14*R*, 13*S*, 16*R*, and 23*S* because of the almost identical ECD spectra of **1** and **8** (Supplementary Figure S10), and thus compound **2** was designated 10-*epi*-deOH-HSAF.

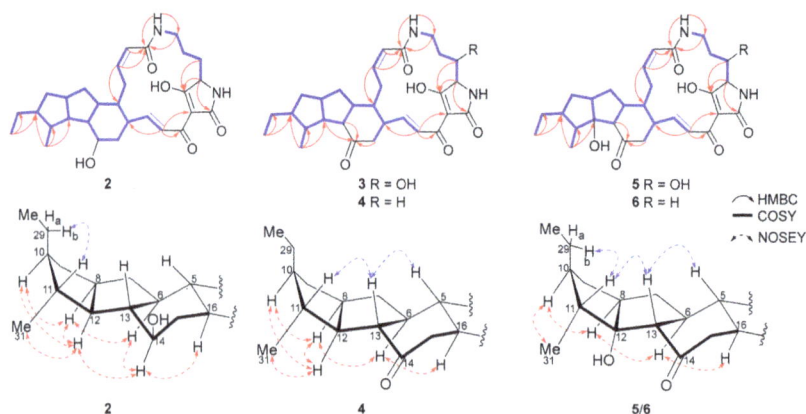

Figure 4. Key COSY, HMBC, and selected NOE correlations for **2–6**.

Compound **3** was isolated as a reddish and amorphous powder. The molecular formula of **3** was determined as $C_{29}H_{38}N_2O_6$ by HRESIMS ([M−H]⁻, *m/z* 509.2642, calcd for 509.2657, Supplementary Figure S19). Careful analysis of the 1D and 2D NMR data of **3** (Tables 1 and 2, Supplementary Figures S20–S26) revealed that **3** was an isomer of maltophilin (**9**, Figure 1) [27]. The *trans*-orientation of H-10/H-11 in **3**, which differed from the *cis*-orientation of H-10/H-11 in **9**, was supported by NOESY correlations of H-8/H-10, H-10/H-12, and H-12/Me-31 (Figure 3, Supplementary Figures S25 and S26). When considering the similar ECD spectra of **3** and **1** (Supplementary Figure S10), the absolute configuration of **3** was deduced as 5*R*, 6*S*, 8*S*, 10*S*, 11*R*, 12*R*, 13*S*, 16*R*, 23*S*, and 25*S*, and thus **3** was designated 10-*epi*-maltophilin.

Compound **4** was obtained as a white powder and it was assigned the molecular formula as $C_{29}H_{38}N_2O_5$ by HRESIMS ([M + H]⁺, *m/z* 495.2846, calcd for 495.2859, Supplementary Figure S27). Detailed comparison of NMR spectroscopic data of **4** (Tables 1 and 2, Supplementary Figures S28–S34) and xanthobaccin C (**10**, Figure 1) uncovered that **4** was an isomer of **10** [27]. The key NOESY correlations of H-8/H-10, H-10/H-12, and H-12/Me-31 in **4** (Figure 4, Supplementary Figure S34) supported a *trans*-orientation of H-10/H-11 in **4**. The absolute configuration of **4** was deduced as 5*R*, 6*S*, 8*S*, 10*S*, 11*R*, 12*R*, 13*S*, 16*R*, and 23*S* by comparing the ECD spectra of **4** and **2** (Supplementary Figure S10). Therefore, **4** was designated 10-*epi*-xanthobaccin C.

Compound **5** was obtained as a reddish powder. The molecular formula of **5** was assigned as $C_{29}H_{38}N_2O_7$ by HRESIMS ([M + H]⁺, *m/z* 527.2757, calcd for 527.2757, Supplementary Figure S35). A detailed comparison of NMR spectroscopic data of **5** and hydroxymaltophilin (**11**, Figure 1) suggested that both compounds should have the same planar structure (Tables 1 and 2, Supplementary Figures S36–S42) [27]. However, distinct from the *cis*-orientation of H-10/H-11 in **11** [27], a *trans*-orientation of H-10/H-11 was indicated in **5** by key NOE correlations of H-8/H-10, H10-/H-12, and H-11/H-29b (Figure 4, Supplementary Figures S41 and S42). Based on the similar ECD spectra of **5** and **11** (Supplementary Figure S10), **5** was suggested to have the configuration of 5*R*, 6*S*, 8*S*, 10*S*, 11*R*, 12*R*, 13*S*, 16*R*, 23*S*, and 25*S*, and it was thus designated 10-*epi*-hydroxymaltophilin.

Compound **6** was isolated as a yellowish solid. The molecular formula of **6** was determined to be $C_{29}H_{38}N_2O_6$ by HRESIMS ([M + H]⁺, *m/z* 511.2800, calcd for 511.2808, Supplementary Figure S43). An analysis of 1D, COSY, and HMBC correlations (Supplementary Figures S44–S48) showed that the planar structure of **6** was the same as that of FI-2 (**12**, Figure 1), an intermediate in frontalamide biosynthesis [15,27]. A *trans*-orientation of H-10/H-11 was indicated in **6** by key NOE correlations of H-8/H-10, H-10/H-12, and H-11/H-29b (Figure 4, Supplementary Figures S49 and S50), different from the *cis*-orientation of H-10/H-11 in **12** [27]. The high similarity in the ECD spectra of **6** and **12** indicated that **6** should be a 10-epimer of **12**, designated 10-*epi*-FI-2.

2.2. Biological Activities

The in vitro cytotoxicities of compounds **1–5** (compound **6** was not tested due to limited amount) were evaluated against four human cancer cell lines, including SF-268, MCF-7, A549, and HepG2, by the SRB method since most reported PTMs exhibits cytotoxic activities [1] (Table 3). Compounds **1–5** showed moderate activities against these four cancer cell lines with half inhibitory concentration (IC$_{50}$) values of 2.47–17.68 µM, which were comparable to those of the positive control cisplatin (Table 3). It should be noted that pactamide A, differing from **2** only by lacking C-14 OH, displayed much better cytotoxicities (IC$_{50}$ values ranging from 0.2–0.5 µM against these four cancer cell lines) than **2** [20].

Table 3. Cytotoxicities of PTMs **1–5**.

	IC$_{50}$ (µM)			
	SF-268	**MCF-7**	**A549**	**HepG2**
1	3.83 ± 0.13	2.47 ± 0.05	5.99 ± 0.15	3.48 ± 0.17
2	10.62 ± 0.45	3.84 ± 0.07	11.01 ± 1.09	10.34 ± 0.88
3	4.57 ± 0.18	3.18 ± 0.13	3.75 ± 0.62	6.30 ± 0.34
4	7.53 ± 0.27	3.54 ± 0.24	10.45 ± 0.46	17.86 ± 0.62
5	3.21 ± 0.18	6.83 ± 0.36	3.28 ± 0.04	3.12 ± 0.11
[a] CP	3.26 ± 0.29	3.19 ± 0.12	1.56 ± 0.08	2.42 ± 0.14

[a] Cisplatin, positive control.

2.3. Biosynthetic Implications

Based on bioinformatics analysis, the *ptm′* BGC in *Streptomyces* sp. SCSIO 40010 was highly similar to that of frontalamides (*ftd*) in *Streptomyces* sp. SPB78 and it should be classified into the Group I of 5/5/6 type of PTM BGCs (Figure 2a). Subsequently, six new PTM analogues with moderate antitumor activities were isolated from *Streptomyces* sp. SCSIO 40010 and the absolute configuration at C-10 in these PTMs was identified as being 10*S*, opposite to their known PTM congeners. These observations further highlight the importance of *Streptomyces* species as prolific sources for bioactive compounds and they indicate the worth of genome mining in marine-derived Streptomycetes [35]. Similar to the well-established biosynthetic pathway for 5/5/6 type of PTMs [20,23,33], PtmA′ catalyzes the formation of a common polyene tetramate precursor, which is sequentially cyclized by PtmB2′/PtmB1′ into an intermediate with the 5/5 carbocyclic ring system (Figure 2b). It has been hypothesized that OX3, which is a PtmB2′ homologous enzyme, is involved in controlling the formation of products with both *cis*- and *trans*-orientated H-10/H-11 in HSAF (**7**) biosynthesis [23]. However, it appears that PtmB2′ only generates products with *trans*-orientated H-10/H-11. Additionally, it has been proposed that C-14 oxidation occurs during the OX2 (PtmB1′ analogue)-catalyzed formation of the second five-membered ring [23], and a recent in vivo combinatorial study has confirmed that the second ring formation is coupled with the C-14 hydroxylation in the biosynthesis of HSAF and analogues [33]. However, the detailed biochemistry and enzymology responsible for such transformations have not been elucidated. Next, PtmC′ generates the inner six-membered ring in **2** (Figure 2b). Finally, different oxidations of **2** by PtmD′ (a C-25 hydroxlase, analogous to FtdA for frontalamides [15], SD for HSAF [36]) and PtmE′ (a putative C-12 hydroxylase and C-14 dehydrogenase, analogous to FtdF for frontalamides [15]) lead to the formation of products **1** and **3–6** due to the substrate promiscuity of PtmD′ and PtmE′ (Figure 2b).

2.4. Conclusion

Conclusively, on the basis of a genome mining approach, we isolated six new PTMs **1–6** from the marine-derived *Streptomyces* sp. SCSIO 40010. The 10*S* absolute configuration is the unique feature of these new PTM analogues, which is distinct from the 10*R* configuration in their known congeners. PTMs **1–5** display moderate cytotoxic activities toward four human cancer cell lines. Although a biosynthetic pathway for PTMs **1–6** is proposed, the precise biochemistry and enzymology involved in the polycyclic ring formation and the stereochemistry selectivity remains elusive and awaits further investigations.

3. Materials and Methods

3.1. General Experimental Procedures

Optical rotations were measured using a 341 Polarimeter (Perkin-kinelmer, Inc., Norwalk, CT, USA). The CD spectra were measured on a Chirascan circular dichroism spectrometer (Applied Photophysics, Ltd., Surrey, UK). UV spectra were measured on a U-2900 spectrophotometer (Hitachi, Tokyo, Japan). IR spectra were recorded on an Affinity-1 FT-IR spectrometer (Shimadzu, Tokyo, Japan). The 1D and 2D NMR spectra were recorded on a Bruker AV-700 MHz NMR spectrometer (Bruker Biospin GmbH, Rheinstetten, Germany) with tetramethylsilane (TMS) as the internal standard. Mass spectrometric data were obtained on a quadrupole-time-of-flight mass spectrometry (Bruker Maxis 4G) for HRESIMS. Column chromatography was performed while using silica gel (100–200 mesh, 300–400 mesh; Jiangyou Silica gel development, Inc., Yantai, China), Sephadex LH-20 (GE Healthcare Bio-Sciences AB, Uppsala, Sweden). HPLC was carried out while using a reversed-phase column (Phenomenex Gemini C18, 250 mm × 4.6 mm, 5 µm; Phenomenex, Torrance, CA, USA) with UV detection at 270 nm and 320 nm. Semi-preparative HPLC was performed on a Hitachi HPLC station (Hitachi-L2130) with a Diode Array Detector (Hitachi L-2455) using a Phenomenex ODS column (250 mm × 10.0 mm, 5 mm; Phenomenex, Torrance, CA, USA) with UV detection at 320 nm.

3.2. Strain, Screening and Culture Methods

Streptomyces sp. SCSIO 40010 was isolated from the Mangrove sediment obtained from Penang, Malaysia, and it was identified by 16S rDNA sequence analysis. The strain SCSIO 40010 was maintained in 40% glycerol aqueous solution at −80 °C in Research Center for Marine Microbiology Culture Collection Center of South China Sea Institute of Oceanology, Chinese Academy of Sciences. It was found that the strain SCSIO 40010 was best maintained on 38#-Agar medium containing 3% sea salt for optimal growth and sporulation. A single colony was inoculated into 50 mL of four different media, including modifed-A1BFe+C (soluble starch 1.0%, yeast extract 0.4%, tryptone 0.2%, $CaCO_3$ 0.2%, sea salts 3%, pH 7.2–7.4) [29], AM6 (soluble starch 2.0%, glucose 1.0%, tryptone 0.5%, yeast extract 0.5%, $CaCO_3$ 0.2%, sea salts 3%, pH 7.2–7.4) [37], AM6-4 (glycerol 0.1%, bacterial peptone 0.5%, glycine 0.01%, alanine 0.01%, $CaCO_3$ 0.5%, sea salts 3%, pH 7.2–7.4) [37], and modifed-ISP3 (oat meal 1.5%, $FeSO_4$ 0.0001%, $MnCl_2$ 0.0001%, $ZnSO_4$ 0.0001%, sea salts 3%, pH 7.2–7.4) [37], in 250 mL Erlenmeyer flasks, and then incubated on a rotary shaker (200 rpm) at 28 °C for seven days. The culture broths were extracted with an equal volume of n-butanol and the extracts were then monitored by HPLC-DAD. HPLC analyses were carried out under the following program: solvent system (solvent A, 10% acetonitrile in water supplemented with 0.08% formic acid; solvent B, 90% acetonitrile in water); 5% B to 100% B (linear gradient, 0–18 min.), 100% B (18–23 min.), 100% B to 5% B (23–27 min.), 5% B (27–32 min.); flow rate at 1 mL/min. A single colony was inoculated into 30 mL of modifed-A1BFe+C medium and incubated at 28 °C for 2–3 days. Then, a total of 20 L fermentation cultures were performed by inoculating 30 mL of the seed culture into a 1000 mL Erlenmeyer flask containing 200 mL of the modifed-A1BFe+C medium to cultivate on a rotary shaker (200 rpm) at 28 °C for 7 days.

3.3. Genome Mining and Bioinformatics Analysis

The strain SCSIO 40010 was inoculated into modifed-A1BFe+C medium and incubated at 28 °C for 48 h. Then the mycelia were collected by centrifugation. Genomic DNA was released from the mycelia by lysozyme and proteinase K digestion, which was extracted with Phenol-chloroform, followed by anhydrous ethanol precipitation. The draft genome of *Streptomyces* sp. 40010 was sequenced by using Illumina HiSeq 2500. The reads were de novo assembled by using SOAPdenovo ver 2.04 (http://soap.genomics.org.cn/soapdenovo.html). Gene sequences were predicted and annotated by the

Rapid Annotations using Subsystems Technology (RAST) server [38]. The putative PTM biosynthetic gene clusters in the genome were predicted with antiSMASH 4.0 [39]. The DNA sequences of the *ptm'* gene cluster were deposited under GenBank accession number MN234160. The function of gene products was predicted with protein blast and/or blastx program (https://blast.ncbi.nlm.nih.gov/Blast.cgi). The PTM BGCs were obtained from GenBank database for bioinformatics analysis: *Streptomyces* sp. SCSIO 40010 (MN234160); *Streptomyces pactum* SCSIO 02999 (KU569222); *Streptomyces griseus* NBRC 13350 (AP009493); *Streptomyces albus* J1074 (ABYC01000481); *Streptomyces flavogriseus* ATCC 33331 (NZ_ACZH01000010); *Streptomyces roseosporus* ATCC 11379 (ABYX01000252); *Streptomyces* sp. SPB78 (NZ_ACEU01000453 and NZ_ACEU01000454); *Actinoalloteichus* sp. ADI127-7 (CP016076); *Actinoalloteichus* sp. AHMU CJ021 (CP025990.1); *Lysobacter enzymogenes* C3 (EF028635.2); *Lysobacter enzymogenes* strain YC36 (CP040656.1); *Lysobacter gummosus* strain 3.2.11 (CP011131.1); *Lysobacter capsici* strain 55 (CP011130.1).

3.4. Extraction, Isolation and Purification

The 20 L of culture broth of *Streptomyces* sp. SCSIO 40010 were pooled and centrifuged at 3900 rpm for 15 min. at 25 °C. The mycelia were extracted three times, each with 2 L acetone. The acetone extracts were concentrated under reduced pressure to afford an aqueous residue, which was extracted four times with equal volume of *n*-butanone. The supernatants were extracted four times with equal volume of *n*-butanone. The butanone extracts were combined and concentrated under reduced pressure to afford the crude extracts (11.5 g). The crude extracts were subjected to the column chromatography over silica gel eluting with a gradient of CHCl$_3$/MeOH mixtures ranging from 100/0, 95/5, 90/10, 80/20,50/50 and 0/100 (*v*/*v*) yielded six fractions (Fr.1–Fr.6). Then Fr.2 (0.72 g) was further purified via MPLC (Medium Pressure Preparative Liquid Chromatography) with reverse phased C-18 column (14.5 × 2.5 cm i.d., 5 mm Agela Technologies) by eluting with a linear gradient of H$_2$O/MeOH (0–100%, 15 mL/min, 300 min) give fractions Fr.2.1–Fr.2.18. Fractions Fr.2.14–15 (170 mg) were further purified by semi-preparative HPLC while using a mobile phase of MeCN-H$_2$O (65:35, *v*/*v*) to give compounds **2** (3.4 mg), **3** (10.8 mg), and **4** (3.6 mg). The fraction Fr.3 (0.83 g) was purified by Sephadex LH-20 (120 × 3.5 cm i.d.), eluting with CHCl$_3$/MeOH (5:5, *v*/*v*) to give fractions Fr.3.1–Fr.3.25. Fractions Fr.3.5–9 (300 mg) were further purified by semi-preparative HPLC while using a mobile phase of MeCN-H$_2$O (45:55, *v*/*v*) to provide compounds **1** (4.1 mg), **5** (5.6 mg), and **6** (2.8 mg).

3.5. Physical and Chemical Properties of New Compounds 1–6

10-*epi*-HSAF (**1**): White powder; $[\alpha]_D^{25}$ + 50.7 (*c* 0.2, MeOH); UV (MeOH) λ_{max} (log ε) 322 (3.92) nm, 219 (4.18) nm; ECD (*c* 4.3 × 10^{-4} M, MeOH) λ_{max} ($\Delta\varepsilon$) 215 (+15.5), 241 (−18.1), 326 (+6.2) nm; IR ν_{max} 3356, 2951, 2918, 2369, 2341, 1653, 1541, 1471, 1020, 679 cm^{-1}; ^{1}H NMR (700 MHz, DMSO-d_6) and ^{13}C NMR (176 MHz, DMSO-d_6) data, see Tables 1 and 2; (+)-HRESIMS *m/z* [M + H]$^+$ 513.2960 (calcd for C$_{29}$H$_{41}$N$_2$O$_6$, 513.2965).

10-*epi*-deOH-HSAF (**2**): White powder; $[\alpha]_D^{25}$ + 53.7 (*c* 0.2, MeOH); UV (MeOH) λ_{max} (log ε) 322 (4.03) nm, 212 (4.37) nm; ECD (*c* 2.2 × 10^{-4} M, MeOH) λ_{max} ($\Delta\varepsilon$) 214 (+7.8), 244 (−9.9), 326 (+4.4) nm; IR ν_{max} 3356, 3334, 2953, 2868, 2358, 2341, 1647, 1541, 1506, 1203, 1024, 669 cm^{-1}; ^{1}H NMR (700 MHz, DMSO-d_6) and ^{13}C NMR (176 MHz, DMSO-d_6) data, see Tables 1 and 2; (+)-HRESIMS *m/z* [M + H]$^+$ 497.3010 (calcd for C$_{29}$H$_{41}$N$_2$O$_5$, 497.3015).

10-*epi*-maltophilin (**3**): Reddish solid; $[\alpha]_D^{25}$ + 42.4 (*c* 0.06, MeOH); UV (MeOH) λ_{max} (log ε) 322 (4.06) nm, 218 (4.31) nm; ECD (*c* 4.9 × 10^{-4} M, MeOH) λ_{max} ($\Delta\varepsilon$) 214 (+26.3), 238 (−23.1), 332 (+6.0) nm; IR ν_{max} 3336, 2953, 2920, 2358, 2341, 1647, 1456, 1022, 679 cm^{-1}; ^{1}H NMR (700 MHz, DMSO-d_6) and ^{13}C NMR (176 MHz, DMSO-d_6) data, see Tables 1 and 2; (-)-HRESIMS *m/z* [M − H]$^-$ 509.2642 (calcd for C$_{29}$H$_{37}$N$_2$O$_6$, 509.2952).

10-*epi*-xanthobaccin C (**4**): White powder; $[\alpha]_D^{25}$ + 8.31 (*c* 0.08, MeOH); UV (MeOH) λ_{max} (log ε) 322 (3.97) nm, 219 (4.27) nm; ECD (*c* 2.6 × 10^{-4} M, MeOH) λ_{max} ($\Delta\varepsilon$) 210 (+15.3), 247 (−16.9), 327 (+5.0) nm; IR v_{max} 3335, 2951, 2920, 2837, 2358, 2341, 1653, 1456, 1018, 758, 669 cm^{-1}; ^{1}H NMR (700 MHz, DMSO-d_6) and ^{13}C NMR (176 MHz, DMSO-d_6) data, see Tables 1 and 2; (+)-HRESIMS *m/z* [M + H]$^+$ 495.2846 (calcd for $C_{29}H_{39}N_2O_5$, 495.2859).

10-*epi*-hydroxymaltophilin (**5**): Reddish powder; $[\alpha]_D^{25}$ + 30.8 (*c* 0.06, MeOH); UV (MeOH) λ_{max} (log ε) 321 (4.02) nm, 216 (4.32) nm; ECD (*c* 4.0×10^{-4} M, MeOH) λ_{max} ($\Delta\varepsilon$) 214 (+24.1), 238 (−18.4), 326 (+4.4) nm; IR v_{max} 3334, 3327, 2955, 2927, 2359, 2342, 1697, 1653, 1541, 1471, 1217, 1024, 754, 678 cm^{-1}; ^{1}H NMR (700 MHz, DMSO-d_6) and ^{13}C NMR (176 MHz, DMSO-d_6) data, see Tables 1 and 2; (+)-HRESIMS *m/z* [M + H]$^+$ 527.2757 (calcd for $C_{29}H_{39}N_2O_7$, 527.2757).

10-*epi*-FI-2 (**6**): Yellowish solid; $[\alpha]_D^{25}$ + 41.4 (*c* 0.06, MeOH); UV (MeOH) λ_{max} (log ε) 322 (3.96) nm, 226 (4.39) nm; ECD (*c* 2.9×10^{-4} M, MeOH) λ_{max} ($\Delta\varepsilon$) 209 (+12.0), 239 (−7.4), 332 (+2.7) nm; IR v_{max} 3321, 2957, 2926, 2359, 2342, 1684, 1647, 1541, 1456, 1238, 669 cm^{-1}; ^{1}H NMR (700 MHz, DMSO-d_6) and ^{13}C NMR (176 MHz, DMSO-d_6) data, see Tables 1 and 2; (+)-HRESIMS *m/z* [M + H]$^+$ 511.2800 (calcd for $C_{29}H_{39}N_2O_6$, 511.2808).

3.6. Bioactivity Assays

The in vitro cytotoxic activities of PTMs **1–5** were evaluated against four tumor cell lines, SF-268 (human glioma cell line), HepG2 (human liver carcinoma cell line), and MCF-7 (human breast adenocarcinoma cell line), A549 (human lung adenocarcinoma cell) by the SRB method, according to a previously described protocol [40]. All of the cells were cultivated in RPMI 1640 medium [41]. Cells (180 μL) with a density of 3×10^4 cells/mL were seeded onto 96-well plates and incubated for 24 h at 37 °C, 5% CO$_2$. Subsequently, 20 μL of different concentrations of PTM compounds, ranging from 0 to 100 μM in dimethyl sulfoxide (DMSO), were added to each plate well. Equal volume of DMSO was used as a negative control. After a further incubation for 72 h, the cell monolayers were fixed with 50% (*wt/v*) trichloroacetic acid (50 μL) and then stained for 30 min. with 0.4% (*wt/v*) SRB dissolved in 1% acetic acid. Unbound dye was removed by repeatedly washing with 1% acetic acid. The protein-bound dye was dissolved in 10 mM Tris-base solution (200 μL) for the determination of optical density (OD) at 570 nm while using a microplate reader. The cytotoxic compound cisplatin was used as a positive control. All of the data were obtained in triplicate and presented as means ± S.D. IC$_{50}$ values were calculated with the SigmaPlot 14.0 software using the non-linear curve-fitting method.

Supplementary Materials: The following are available online at http://www.mdpi.com/1660-3397/17/12/663/s1, Figure S1: HPLC analysis of metabolite profiles of *Streptomyces* sp: SCSIO 40010 cultured in different media; Figure S2: Comparison of ECD spectra of compound **1–6** and the known compounds; Figure S3: HRESIMS (a) and IR (b) of compound **1**; Figure S4: ^{1}H NMR spectrum of compound **1** in DMSO-d_6; Figure S5: The ^{13}C NMR and DEPT 135 spectra of compound **1** in DMSO-d_6; Figure S6: The HSQC spectrum of compound **1** in DMSO-d_6; Figure S7: The HMBC spectrum of compound **1** in DMSO-d_6; Figure S8: The ^{1}H-^{1}H COSY spectrum of compound **1** in DMSO-d_6; Figure S9: The NOESY spectrum of compound **1** in DMSO-d_6; Figure S10: The key NOESY spectrum of compound **1** in DMSO-d_6; Figure S11: HRESIMS (a) and IR (b) of compound **2**; Figure S12: ^{1}H NMR spectrum of compound **2** in DMSO-d_6; Figure S13: The ^{13}C NMR and DEPT 135 spectra of compound **2** in DMSO-d_6; Figure S14: The HSQC spectrum of compound **2** in DMSO-d_6; Figure S15: The HMBC spectrum of compound **2** in DMSO-d_6; Figure S16: The ^{1}H-^{1}H COSY spectrum of compound **2** in DMSO-d_6; Figure S17: The NOESY spectrum of compound **2** in DMSO-d_6; Figure S18: The key NOESY spectrum of compound **2** in DMSO-d_6, Figure S19: HRESIMS (a) and IR (b) of compound **3**; Figure S20: ^{1}H NMR spectrum of compound **3** in DMSO-d_6; Figure S21: The ^{13}C NMR and DEPT 135 spectra of compound **3** in DMSO-d_6; Figure S22: The HSQC spectrum of compound **3** in DMSO-d_6; Figure S23: The HMBC spectrum of compound **3** in DMSO-d_6; Figure S24: The ^{1}H-^{1}HCOSY spectrum of compound **3** in DMSO-d_6; Figure S25: The NOESY spectrum of compound **3** in DMSO-d_6; Figure S26: The key NOESY spectrum of compound **3** in DMSO-d_6; Figure S27: HRESIMS (a) and IR (b) of compound **4**; Figure S28: ^{1}H NMR spectrum of compound **4** in DMSO-d_6; Figure S29: The ^{13}C NMR and DEPT 135 spectra of compound **4** in DMSO-d_6; Figure S30: The HSQC spectrum of compound **4** in DMSO-d_6; Figure S31: The HMBC spectrum of compound **4** in DMSO-d_6; Figure S32: The ^{1}H-^{1}HCOSY spectrum of compound **4** in DMSO-d_6; Figure S33: The NOESY spectrum of compound **4** in DMSO-d_6; Figure S34: The key NOESY spectrum of compound **4** in DMSO-d_6; Figure S35: HRESIMS (a) and IR (b) of compound **5**; Figure S36: ^{1}H NMR spectrum of compound **5** in DMSO-d_6; Figure S37: The ^{13}C NMR and DEPT 135 spectra of compound **5** in DMSO-d_6; Figure S38: The HSQC spectrum of compound **5** in DMSO-d_6; Figure S39: The HMBC spectrum of compound **5** in

Mar. Drugs **2019**, *17*, 663

DMSO-d_6; Figure S40: The ^{1}H-^{1}HCOSY spectrum of compound **5** in DMSO-d_6; Figure S41: The NOESY spectrum of compound **5** in DMSO-d_6; Figure S42: The key NOESY spectrum of compound **5** in DMSO-d_6; Figure S43: HRESIMS (**a**) and IR (**b**) of compound **6**; Figure S44: ^{1}H NMR spectrum of compound **6** in DMSO-d_6; Figure S45: The ^{13}C NMR and DEPT 135 spectra of compound **6** in DMSO-d_6; Figure S46: The HSQC spectrum of compound **6** in DMSO-d_6; Figure S47: The HMBC spectrum of compound **6** in DMSO-d_6; Figure S48: The ^{1}H-^{1}H COSY spectrum of compound **6** in DMSO-d_6; Figure S49: The NOESY spectrum of compound **6** in DMSO-d_6; Figure S50: The key NOESY spectrum of compound **6** in DMSO-d_6.

Author Contributions: W.L. contributed to compounds isolation and structure elucidation. Q.Z. and X.J. isolated and preserved the strain. H.J., L.Z. and G.Z. performed bioinformatics analysis. Y.C. and W.Z. (Weimin Zhang) performed cytotoxicity assays. W.L., W.Z. (Wenjun Zhang), and C.Z. analysed the data and wrote the manuscript. C.Z. and Z.S. supervised the study.

Funding: This work is supported in part by NSFC (31630004, 41606193), the Administration of Ocean and Fisheries of Guangdong Province (A201601C03), the Science and Technology Program of Guangzhou (201707010181), the Chinese Academy of Sciences (QYZDJ-SSW-DQC004), the Qingdao National Laboratory for Marine Science and Technology (QNLM2016ORP0304).

Acknowledgments: We thank Zhihui Xiao, Xiaohong Zheng, Chuanrong Li and Aijun Sun, Yun Zhang, Xuan Ma in the analytical facility center of the SCSIO for acquiring NMR data and MS data. We are grateful to Yongli Gao in the Equipment Public Service Center of the SCSIO for culturing strain.

Conflicts of Interest: The authors declare no conflict of interest.

References

1. Zhang, G.; Zhang, W.; Saha, S.; Zhang, C. Recent advances in discovery, biosynthesis and genome mining of medicinally relevant polycyclic tetramate macrolactams. *Curr. Top. Med. Chem.* **2016**, *16*, 1727–1739. [CrossRef] [PubMed]
2. Jomon, K.; Ajisaka, M.; Sakai, H.; Kuroda, Y. New antibiotic, ikarugamycin. *J. Antibiot.* **1972**, *25*, 271–280. [CrossRef] [PubMed]
3. Li, S.; Calvo, A.M.; Yuen, G.Y.; Du, L.; Harris, S.D. Induction of cell wall thickening by the antifungal compound dihydromaltophilin disrupts fungal growth and is mediated by sphingolipid biosynthesis. *J. Eukaryot. Microbiol.* **2009**, *56*, 182–187. [CrossRef] [PubMed]
4. Popescu, R.; Heiss, E.H.; Ferk, F.; Peschel, A.; Knasmueller, S.; Dirsch, V.M.; Krupitza, G.; Kopp, B. Ikarugamycin induces DNA damage, intracellular calcium increase, p38 MAP kinase activation and apoptosis in HL-60 human promyelocytic leukemia cells. *Mutat. Res.* **2011**, *709–710*, 60–66. [CrossRef] [PubMed]
5. Lacret, R.; Oves-Costales, D.; Gomez, C.; Diaz, C.; de la Cruz, M.; Perez-Victoria, I.; Vicente, F.; Genilloud, O.; Reyes, F. New ikarugamycin derivatives with antifungal and antibacterial properties from *Streptomyces zhaozhouensis*. *Mar. Drugs* **2015**, *13*, 128–140. [CrossRef]
6. Xie, Y.X.; Wright, S.; Shen, Y.M.; Du, L.C. Bioactive natural products from *Lysobacter*. *Nat. Prod. Rep.* **2012**, *29*, 1277–1287. [CrossRef]
7. Yu, F.; Zaleta-Rivera, K.; Zhu, X.; Huffman, J.; Millet, J.C.; Harris, S.D.; Yuen, G.; Li, X.-C.; Du, L. Structure and biosynthesis of deat-stable antifungal factor (HSAF), a broad-spectrum antimycotic with a novel mode of action. *Antimicrob. Agents Chemother.* **2007**, *51*, 64–72. [CrossRef]
8. Li, S.; Jochum, C.C.; Yu, F.; Zaleta-Rivera, K.; Du, L.; Harris, S.D.; Yuen, G.Y. An antibiotic complex from *Lysobacter enzymogenes* strain C3: Antimicrobial activity and role in plant disease control. *Phytopathology* **2008**, *98*, 695–701. [CrossRef]
9. Cao, S.G.; Blodgett, J.A.V.; Clardy, J. Targeted discovery of polycyclic tetramate macrolactams from an environmental *Streptomyces* strain. *Org. Lett.* **2010**, *12*, 4652–4654. [CrossRef]
10. Elkin, S.R.; Oswald, N.W.; Reed, D.K.; Mettlen, M.; MacMillan, J.B.; Schmid, S.L. Ikarugamycin: A natural product inhibitor of clathrin-mediated endocytosis. *Traffic* **2016**, *17*, 1139–1149. [CrossRef]
11. Paquette, L.A.; Macdonald, D.; Anderson, L.G.; Wright, J. A triply convergent enantioselective total synthesis of (+)-ikarugamycin. *J. Am. Chem. Soc.* **1989**, *111*, 8037–8039. [CrossRef]
12. Boeckman, R.K.; Weidner, C.H.; Perni, R.B.; Napier, J.J. An enantioselective and highly convergent synthesis of (+)-ikarugamycin. *J. Am. Chem. Soc.* **1989**, *111*, 8036–8037. [CrossRef]
13. Cramer, N.; Laschat, S.; Baro, A.; Schwalbe, H.; Richter, C. Enantioselective total synthesis of cylindramide. *Angew. Chem. Int. Ed.* **2005**, *44*, 820–822. [CrossRef] [PubMed]

14. Henderson, J.A.; Phillips, A.J. Total synthesis of aburatubolactam A. *Angew. Chem. Int. Ed.* **2008**, *47*, 8499–8501. [CrossRef] [PubMed]

15. Blodgett, J.A.V.; Oh, D.C.; Cao, S.G.; Currie, C.R.; Kolter, R.; Clardy, J. Common biosynthetic origins for polycyclic tetramate macrolactams from phylogenetically diverse bacteria. *Proc. Natl. Acad. Sci. USA* **2010**, *107*, 11692–11697. [CrossRef] [PubMed]

16. Lou, L.L.; Qian, G.L.; Xie, Y.X.; Hang, J.L.; Chen, H.T.; Zaleta-Riyera, K.; Li, Y.Y.; Shen, Y.M.; Dussault, P.H.; Liu, F.Q.; et al. Biosynthesis of HSAF, a tetramic acid-containing macrolactam from *Lysobacter enzymogenes*. *J. Am. Chem. Soc.* **2011**, *133*, 643–645. [CrossRef]

17. Luo, Y.; Huang, H.; Liang, J.; Wang, M.; Lu, L.; Shao, Z.; Cobb, R.E.; Zhao, H. Activation and characterization of a cryptic polycyclic tetramate macrolactam biosynthetic gene cluster. *Nat. Commun.* **2013**, *4*. [CrossRef]

18. Zhang, G.T.; Zhang, W.J.; Zhang, Q.B.; Shi, T.; Ma, L.; Zhu, Y.G.; Li, S.M.; Zhang, H.B.; Zhao, Y.L.; Shi, R.; et al. Mechanistic insights into polycycle formation by reductive cyclization in ikarugamycin biosynthesis. *Angew. Chem. Int. Ed.* **2014**, *53*, 4840–4844. [CrossRef]

19. Li, Y.Y.; Chen, H.T.; Ding, Y.J.; Xie, Y.X.; Wang, H.X.; Cerny, R.L.; Shen, Y.M.; Du, L.C. Iterative assembly of two separate polyketide chains by the same single-module bacterial polyketide synthase in the biosynthesis of HSAF. *Angew. Chem. Int. Ed.* **2014**, *53*, 7524–7530. [CrossRef]

20. Saha, S.; Zhang, W.J.; Zhang, G.T.; Zhu, Y.G.; Chen, Y.C.; Liu, W.; Yuan, C.S.; Zhang, Q.B.; Zhang, H.B.; Zhang, L.P.; et al. Activation and characterization of a cryptic gene cluster reveals a cyclization cascade for polycyclic tetramate macrolactams. *Chem. Sci.* **2017**, *8*, 1607–1612. [CrossRef]

21. Antosch, J.; Schaefers, F.; Gulder, T.A.M. Heterologous reconstitution of ikarugamycin biosynthesis in *E. coli*. *Angew. Chem. Int. Ed.* **2014**, *53*, 3011–3014. [CrossRef] [PubMed]

22. Greunke, C.; Glockle, A.; Antosch, J.; Gulder, T.A. Biocatalytic total synthesis of ikarugamycin. *Angew. Chem. Int. Ed.* **2017**, *56*, 4351–4355. [CrossRef] [PubMed]

23. Li, Y.; Wang, H.; Liu, Y.; Jiao, Y.; Li, S.; Shen, Y.; Du, L. Biosynthesis of the polycyclic system in the antifungal HSAF and analogues from *Lysobacter enzymogenes*. *Angew. Chem. Int. Ed.* **2018**, *57*, 6221–6225. [CrossRef] [PubMed]

24. Li, X.; Wang, H.; Shen, Y.; Li, Y.; Du, L. OX4 is an NADPH-dependent dehydrogenase catalyzing an extended Michael addition reaction to form the six-membered ring in the antifungal HSAF. *Biochemistry* **2019**. [CrossRef] [PubMed]

25. Olano, C.; Garcia, I.; Gonzalez, A.; Rodriguez, M.; Rozas, D.; Rubio, J.; Sanchez-Hidalgo, M.; Brana, A.F.; Mendez, C.; Salas, J.A. Activation and identification of five clusters for secondary metabolites in *Streptomyces albus* J1074. *Microb. Biotechnol.* **2014**, *7*, 242–256. [CrossRef] [PubMed]

26. Zhang, W.J.; Zhang, G.T.; Zhang, L.P.; Liu, W.; Jiang, X.D.; Jin, H.B.; Liu, Z.W.; Zhang, H.B.; Zhou, A.H.; Zhang, C.S. New polycyclic tetramate macrolactams from marine-derived *Streptomyces* sp. SCSIO 40060. *Tetrahedron* **2018**, *74*, 6839–6845. [CrossRef]

27. Mei, X.; Wang, L.; Wang, D.; Fan, J.; Zhu, W. Polycyclic tetramate macrolactams from the marine-derived *Actinoalloteichus cyanogriseus* WH1-2216-6. *Chin. J. Org. Chem.* **2017**, *37*, 2352–2360. [CrossRef]

28. De Bruijn, I.; Cheng, X.; de Jager, V.; Exposito, R.G.; Watrous, J.; Patel, N.; Postma, J.; Dorrestein, P.C.; Kobayashi, D.; Raaijmakers, J.M. Comparative genomics and metabolic profiling of the genus Lysobacter. *BMC Genom.* **2015**, *16*, 991. [CrossRef]

29. Zhang, W.J.; Ma, L.; Li, S.M.; Liu, Z.; Chen, Y.C.; Zhang, H.B.; Zhang, G.T.; Zhang, Q.B.; Tian, X.P.; Yuan, C.S.; et al. Indimicins A-E, Bisindole Alkaloids from the Deep-Sea-Derived Streptomyces sp SCSIO 03032. *J. Nat. Prod.* **2014**, *77*, 1887–1892. [CrossRef]

30. Xu, L.; Wu, P.; Wright, S.J.; Du, L.; Wei, X. Bioactive polycyclic tetramate macrolactams from *Lysobacter enzymogenes* and their absolute configurations by theoretical ECD calculations. *J. Nat. Prod.* **2015**, *78*, 1841–1847. [CrossRef]

31. Bae, M.A.; Yamada, K.; Ijuin, Y.; Tsuji, T.; Yazawa, K.; Tomono, Y.; Uemura, D. Aburatubolactam A, a novel inhibitor of superoxide anion generation from a marine microorganism. *Heterocycl. Commun.* **1996**, *2*, 315–318. [CrossRef]

32. Liu, Y.; Wang, H.; Song, R.; Chen, J.; Li, T.; Li, Y.; Du, L.; Shen, Y. Targeted Discovery and Combinatorial Biosynthesis of Polycyclic Tetramate Macrolactam Combamides A-E. *Org. Lett.* **2018**, *20*, 3504–3508. [CrossRef] [PubMed]

33. Li, X.; Wang, H.; Li, Y.; Du, L. Construction of a hybrid gene cluster to reveal coupled ring formation-hydroxylation in the biosynthesis of HSAF and analogues from *Lysobacter enzymogenes*. *MedChemComm* **2019**, *10*, 907–912. [CrossRef] [PubMed]

34. Hoshino, S.; Wong, C.P.; Ozeki, M.; Zhang, H.; Hayashi, F.; Awakawa, T.; Asamizu, S.; Onaka, H.; Abe, I. Umezawamides, new bioactive polycyclic tetramate macrolactams isolated from a combined-culture of Umezawaea sp. and mycolic acid-containing bacterium. *J. Antibiot. (Tokyo)* **2018**, *71*, 653–657. [CrossRef] [PubMed]

35. Ward, A.C.; Allenby, N.E. Genome mining for the search and discovery of bioactive compounds: The Streptomyces paradigm. *FEMS Microbiol. Lett.* **2018**, *365*. [CrossRef] [PubMed]

36. Li, Y.Y.; Huffman, J.; Li, Y.; Du, L.C.; Shen, Y.M. 3-Hydroxylation of the polycyclic tetramate macrolactam in the biosynthesis of antifungal HSAF from *Lysobacter enzymogenes* C3. *MedChemComm* **2012**, *3*, 982–986. [CrossRef]

37. Zhang, Q.; Li, H.; Li, S.; Zhu, Y.; Zhang, G.; Zhang, H.; Zhang, W.; Shi, R.; Zhang, C. Carboxyl formation from methyl via triple hydroxylations by XiaM in xiamycin A biosynthesis. *Org. Lett.* **2012**, *14*, 6142–6145. [CrossRef]

38. Aziz, R.K.; Bartels, D.; Best, A.A.; DeJongh, M.; Disz, T.; Edwards, R.A.; Formsma, K.; Gerdes, S.; Glass, E.M.; Kubal, M.; et al. The RAST Server: Rapid annotations using subsystems technology. *BMC Genom.* **2008**, *9*, 75. [CrossRef]

39. Blin, K.; Wolf, T.; Chevrette, M.G.; Lu, X.; Schwalen, C.J.; Kautsar, S.A.; Suarez Duran, H.G.; de los Santos, E.L.; Kim, H.U.; Nave, M.; et al. antiSMASH 4.0—Improvements in chemistry prediction and gene cluster boundary identification. *Nucleic Acids Res.* **2017**, *45*, W36–W41. [CrossRef]

40. Skehan, P.; Storeng, R.; Scudiero, D.; Monks, A.; McMahon, J.; Vistica, D.; Warren, J.T.; Bokesch, H.; Kenney, S.; Boyd, M.R. New colorimetric cytotoxicity assay for anticancer-drug screening. *J. Natl. Cancer Inst.* **1990**, *82*, 1107–1112. [CrossRef]

41. Nakabayashi, H.; Taketa, K.; Miyano, K.; Yamane, T.; Sato, J. Growth of Human Hepatoma-Cell Lines with Differentiated Functions in Chemically Defined Medium. *Cancer Res.* **1982**, *42*, 3858–3863. [PubMed]

marine drugs

MDPI

Article

Comparative Genomic Insights into Secondary Metabolism Biosynthetic Gene Cluster Distributions of Marine *Streptomyces*

Lin Xu [1,2,3,*], Kai-Xiong Ye [1,2], Wen-Hua Dai [1,2], Cong Sun [1,2,3], Lian-Hua Xu [1,2] and Bing-Nan Han [1,2,*]

[1]   Lab of Marine Functional Molecules, Zhejiang Sci-Tech University, Hangzhou 310018, China
[2]   College of Life Sciences and Medicine, Zhejiang Sci-Tech Univeristy, Hangzhou 310018, China
[3]   Key Laboratory of Marine Ecosystem and Biogeochemistry, State Oceanic Administration & Second Institute of Oceanography, Ministry of Natural Resources, Hangzhou 310012, China
*   Correspondence: linxu@zstu.edu.cn (L.X.); hanbingnan@zstu.edu.cn (B.-N.H.); Tel.: +86-13758243072 (L.X.); +86-15257165115 (B.-N.H.)

Received: 31 July 2019; Accepted: 21 August 2019; Published: 26 August 2019

Abstract: Bacterial secondary metabolites have huge application potential in multiple industries. Biosynthesis of bacterial secondary metabolites are commonly encoded in a set of genes that are organized in the secondary metabolism biosynthetic gene clusters (SMBGCs). The development of genome sequencing technology facilitates mining bacterial SMBGCs. Marine *Streptomyces* is a valuable resource of bacterial secondary metabolites. In this study, 87 marine *Streptomyces* genomes were obtained and carried out into comparative genomic analysis, which revealed their high genetic diversity due to pan-genomes owning 123,302 orthologous clusters. Phylogenomic analysis indicated that the majority of Marine *Streptomyces* were classified into three clades named Clade I, II, and III, containing 23, 38, and 22 strains, respectively. Genomic annotations revealed that SMBGCs in the genomes of marine *Streptomyces* ranged from 16 to 84. Statistical analysis pointed out that phylotypes and ecotypes were both associated with SMBGCs distribution patterns. The Clade I and marine sediment-derived *Streptomyces* harbored more specific SMBGCs, which consisted of several common ones; whereas the Clade II and marine invertebrate-derived *Streptomyces* have more SMBGCs, acting as more plentiful resources for mining secondary metabolites. This study is beneficial for broadening our knowledge about SMBGC distribution patterns in marine *Streptomyces* and developing their secondary metabolites in the future.

Keywords: *Streptomyces*; comparative genomics; secondary metabolites; biosynthetic gene clusters; phylotype; ecotype

---

1. Introduction

Bacterial secondary metabolites are defined as organic compounds that are not directly involved in the normal growth and proliferation of bacteria [1], and can be classified into several categories, such as alkaloids, antibiotics, carotenoids, pigments, and toxins [2]. Bacterial secondary metabolites play an important role in defending against adversities and increasing the survival of themselves, even their hosts, due to their antibacterial, antifungal, antitumor, and antiviral activities [3,4], meaning those organic compounds have considerable application potential in human and veterinary medicine as well as agriculture [5]. Since the initial discovery of bacterial secondary metabolites in the 1920s, they have shown a profound impact on human society [6]. Currently, marine-derived bacterial secondary metabolites with a broad range of complex structures are increasingly becoming sources of novel natural products for discovering and developing new drugs [7–10].

Genes involved in the biosynthesis of bacterial secondary metabolites are commonly organized in the secondary metabolism biosynthetic gene clusters (SMBGCs) [1,9]. The development of genomic sequencing technology facilitates the mining of marine bacterial SMBGCs [11–13]. Apart from core biosynthetic enzyme-encoding genes, SMBGCs generally also harbor genes encoding enzymes to synthesize specialized monomers, transporters, and regulatory elements as well as mediating host resistance [14]. Non-ribosomal peptide synthase (NRPS) and polyketide synthase (PKs) gene clusters are two main pathways for biosynthesizing bacterial secondary metabolites [15]. Those two core enzymes independently fold protein domains, operate in constructing polymeric chains, and tailor their functionalities [15]. In addition, another well-known class of SMBGCs is terpenes, which are derived biosynthetically from units of isopentenyl pyrophosphate through mevalonic acid pathway or 2-C-methyl-D-erythritol 4-phosphate pathway [16,17]. Because bacterial secondary metabolites improves fitness advantages of bacteria as well as their hosts and the frequency of horizontal gene transfer is high, some studies indicated that SMBGC distributions are related to the environment where bacteria live, called ecotype [18,19]. Meanwhile, recent studies demonstrated that bacterial secondary metabolite production is species-specific, which concerns phylogeny, called phylotype [20,21]. Therefore, what distribution patterns of SMBGCs are is still an open scientific question that is associated with phylotypes and ecotypes. The exploration of this question is beneficial for developing bacterial secondary metabolites.

The genus *Streptomyces* belongs to the family *Streptomycetaceae*, the order *Actinomycetales*, the class *Actinobacteria*, and the phylum *Actinobacteria* [22], and it is one of the largest group in this phylum with more than 600 species at the time of writing (http://www.bacterio.net/ streptomyces.html, [23]). The genus *Streptomyces* is well known for an important source of secondary metabolites, and the portion of recently novel antibiotics discovered from this genus can reach at about 20–30% [24,25]. Further, the genus *Streptomyces* inhabits a wide range of marine habitats, including seawater [26,27], marine sediments [28,29], alga [30,31], mangroves [32,33], sponges [34,35], corals [36,37], tunicates [38,39], mollusks [40,41], etc., resulting in the fact that this genus attracts continuous attentions of researchers to find valuable secondary metabolites. Furthermore, the genus *Streptomyces* is one of the earliest genome-sequenced prokaryotes, with the genome of *S. coelicolor* A3(2) sequenced and reported by Bentley et al. in 2002 [42]. Hundreds of *Streptomyces* genomes have been sequenced and deposited into public databases in the recent years [43,44], leading to the increases of comparative genomic studies about this genus. While comparative genomic studies of marine *Streptomyces* are mostly related to exploring their SMBGC resources as well as diversities [45–49] and investigating their marine adaptation mechanisms [50,51], there is still a lack of comprehensive study concerning SMBGC distribution patterns in marine *Streptomyces*. In this study, we proposed the hypothesis that both of phylotype- and ecotype-associated SMBGCs were in the genomes of marine *Streptomyces* and performed comparative genomic methods to test this hypothesis and analyze their distribution patterns. This study is beneficial for broadening our knowledge about SMBGC distribution patterns in marine *Streptomyces* and developing their secondary metabolites in the future.

2. Results and Discussion

2.1. Genomic Characteristics and Annotation Results of Marine Streptomyces

Eighty-seven marine *Streptomyces* genomes were screened into genomic analysis by confirming their high genomic qualities with the completeness >95% and the contamination <5% (Table S1). Those strains were isolated from various sources, including seawater ($n = 7$), marine sediments ($n = 38$), cyanobacteria ($n = 1$), algae ($n = 1$), mangroves ($n = 8$), sponges ($n = 22$), corals ($n = 3$), tunicates ($n = 2$), and mollusks ($n = 5$).

The G+C contents of marine *Streptomyces* were 69.9–73.8 mol% (Table S1), which was in accordance with high G+C content as a typical characteristic of the phylum *Actinobacteria* [52]. Genomic sizes and gene counts of those marine *Streptomyces* genomes varied remarkably, ranging from 5.77 to 11.50

Mbp and from 5363 to 10,776 (Figure 1 and Table S2), respectively. Furthermore, the number of genes was positively correlated with the genomic size of the marine *Streptomyces* ($y = 966.8x - 121.4$, $r^2 = 0.89$). Furthermore, it was found that 3978–8065 (71.15–78.9%) and 2005–3192 (27.6–38.0%) genes were assigned to Clusters of Orthologous Groups (COG) and Kyoto Encyclopedia of Genes and Genomes (KEGG) databases, respectively.

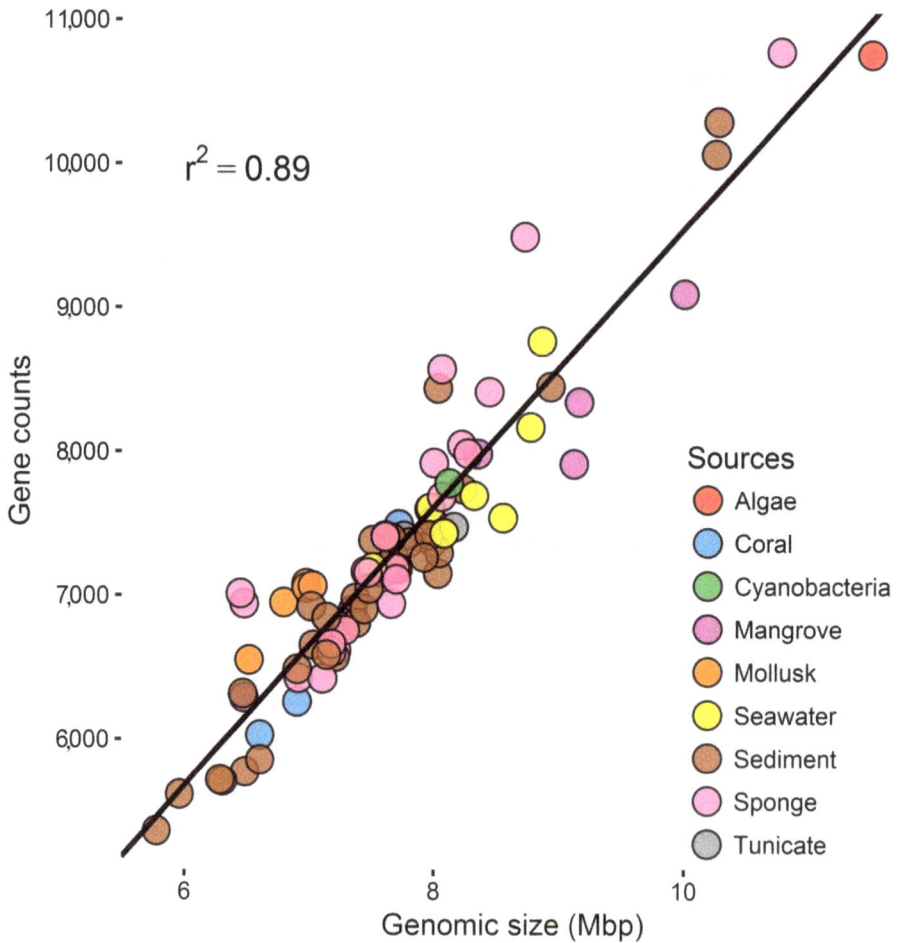

Figure 1. Genomic size and gene counts of *Streptomyces* derived from various marine environments.

It was detected that 16 to 84 SMBGCs (2 to 38 PKs, 1 to 15 NRPS, 0 to 8PKs/NRPS hybrid, 2 to 6 terpene, 2 to 17 other, and 2 to 25 hypothetical) were in the genomes of marine *Streptomyces* (Figure 2, Table S3) and the portion of SMBGCs in the genomes ranged from 1.94 to 9.21 Mbp^{-1}, revealing that SMBGC counts were not positively correlated with genomic sizes (Figure 2), which is different from the correlation between gene counts and genomic size. Hence, SMBGCs distributions in the genomes of marine might be associated with their phylotypes and ecotypes, which could be intrinsic factors.

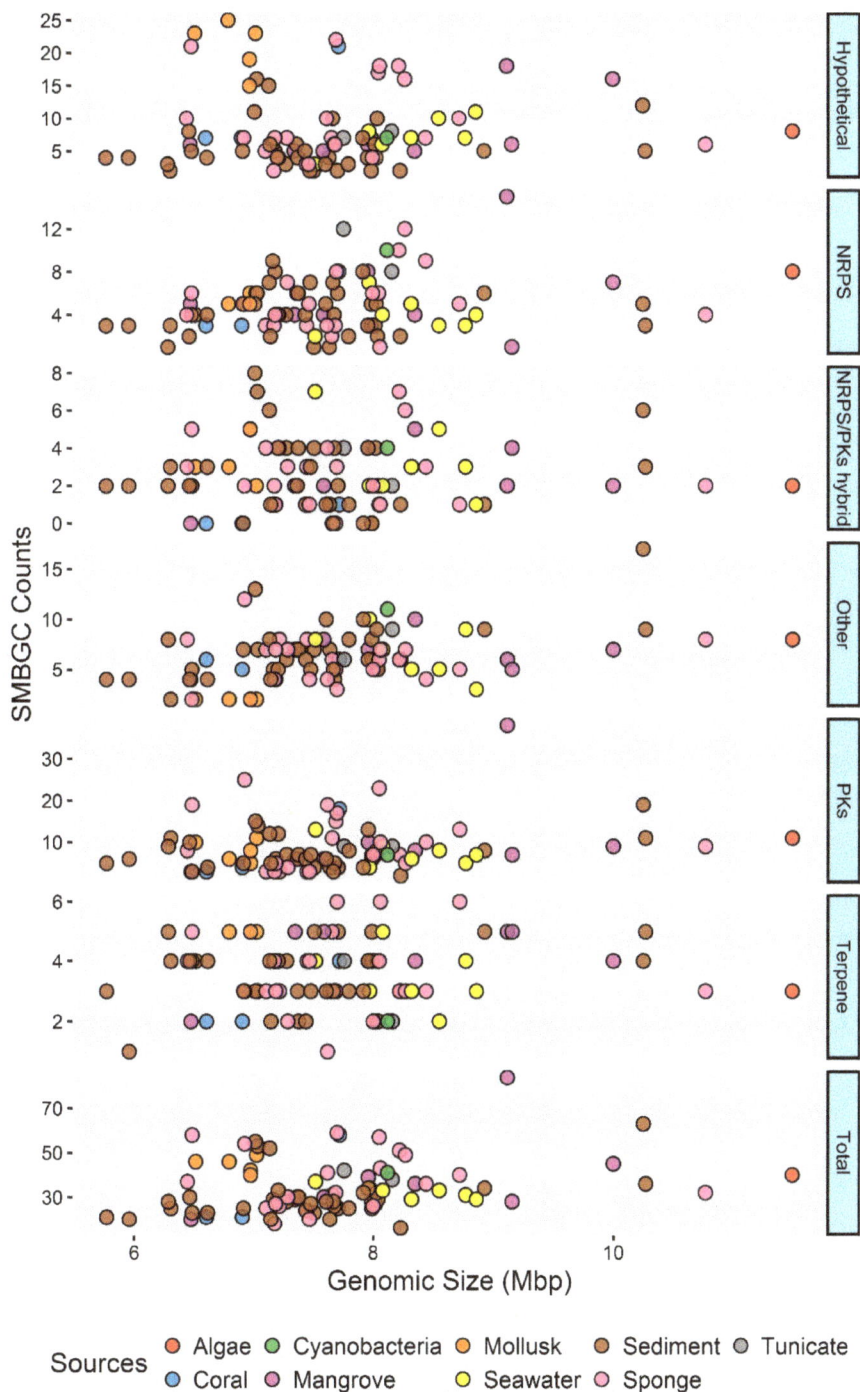

Figure 2. Secondary metabolism biosynthetic gene clusters (SMBGC) category counts identified in marine *Streptomyces* genomes.

2.2. Comparative Genomics and Phylogenomic Relationship of Marine Streptomyces

Comparative genomic analysis demonstrated that all of marine *Streptomyces* harbored 123,302 orthologous clusters (OCs) in their pan-genomes (Table S4), demonstrating their rich genetic diversities. Those strains contained 5258–10,376 OCs (average: 7116 ± 972, median: 6978) in their genomes, while they had 31–2793 (average: 861 ± 598, median: 714) exclusive OCs (Figure 3), also showing remarkable genetic diversities. It was detected that 996 OCs, of which 888 single-copy OCs were commonly in them and *Kitasatospora setae* KM-6054, were in their core-genomes (Table S4).

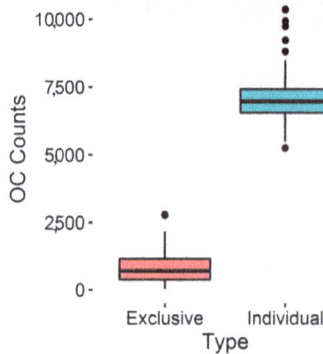

Figure 3. Individual and exclusive orthologous clusters (OCs) of marine *Streptomyces* genomes.

Based on the comparative genomic analysis, 888 single-copy OCs shared by all of marine *Streptomyces* and *Kitasatospora setae* KM-6054 (Table S4) were used to reconstruct a maximum-likelihood phylogenomic tree, revealing that the majority of marine *Streptomyces* were grouped into three clades (Clade I, II, and III) except for *S. antioxidans* MUSC 164, *S. xinghaiensis* S187, *Streptomyces* sp. NBRC 110027, and "*Streptomyces* sp. NRRL B-24484" (Figure 4). Further, "*Streptomyces* sp. NRRL B-24484" could not belong to the clade of the genus *Streptomyces*, which indicated that "*Streptomyces* sp. NRRL B-24484" was not a member of this genus, meaning it is excluded from further analysis.

Three major clades contain 23 (Clade I), 38 (Clade II), and 22 (Clade III) strains, respectively (Figure 4). Each clade consisted of strains derived from different sources, among which two majorities are marine sediment and sponge (Table 1). In addition, each clade had its own characteristic, which could be reflected by some ecotypes represented by ≥3 strains, such as coral- and mollusk-derived strains and only found in Clade I or II, and mangrove-derived strain only absent in the Clade III (Table 1).

Table 1. The percentage of sources in Clade I, II, and II.

Clade	1	2	3	4	5	6	7	8	9
I	0%	13%	0%	4%	0%	9%	56%	18%	0%
II	3%	0%	0%	16%	13%	5%	37%	23%	3%
III	0%	0%	4%	0%	0%	4%	46%	42%	4%

1, 2, 3, 4, 5, 6, 7, 8, and 9 represents algae, coral, cyanobacteria, mangrove, mollusk, seawater, marine sediment, sponge, and tunicate, respectively.

Phylogenomic analysis also indicated that numerous novel *Streptomyces* species are waiting for identifications. Moreover, average nucleotide identity (ANI) calculations pointed out that Clade I, II, and III contained at least 9, 13 ,and 15 novel species (Table S5), which had low ANI values (<95%, [53]) compared with validly published *Streptomyces* species in the phylogenomic tree.

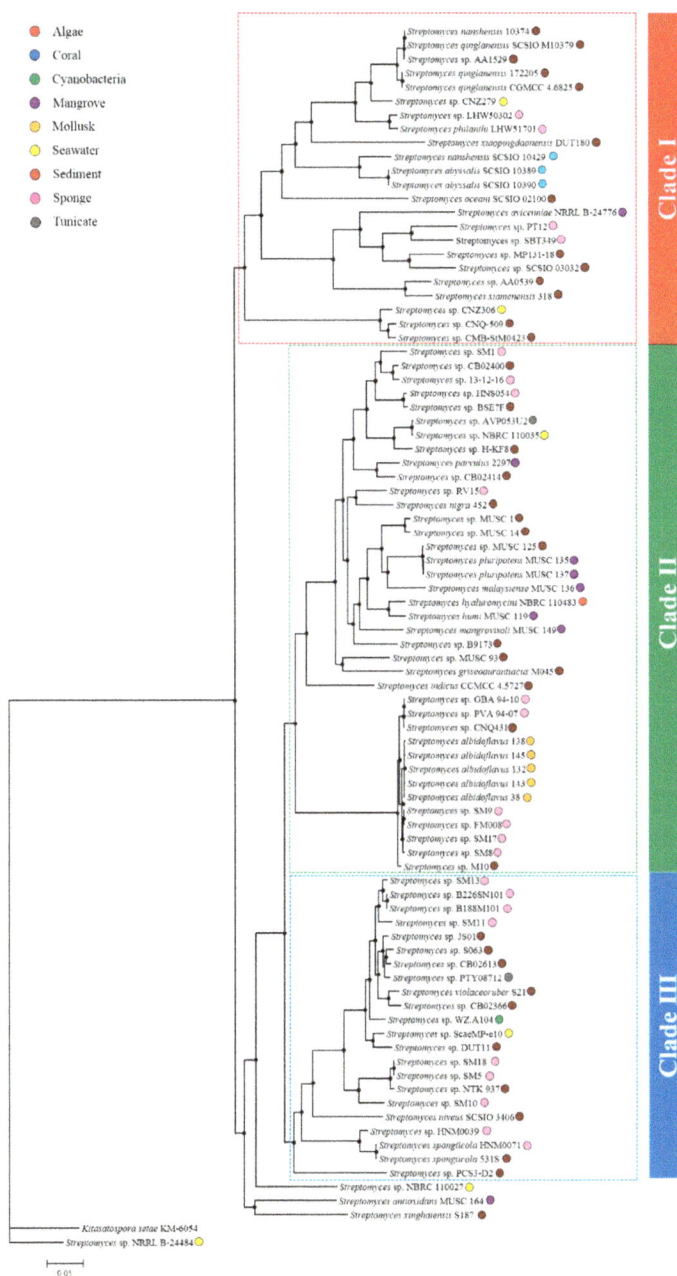

Figure 4. Maximum-likelihood phylogenomic tree based on the concatenation of 888 single-copy OC proteins shared by all of marine *Streptomyces*. Filled circle indicated nodes showing >85 of bootstrap values. *Kitasatospora setae* KM-6054 was used as an outgroup.

2.3. Phylotype-Associated SMBGCs

Except for hypothetical SMBGCs detected in the genomes of marine *Streptomyces*, significance tests among multi-clades revealed that 10 (6.0%) PKs, 7 (7.9%) NRPS, 3 (5.4%) PKs/NRPS hybrid, 2 (10.5%) terpene, and 7 (7.4%) other SMBGCs exhibited significant differences (Table 2).

Table 2. SMBGCs showing significant differences among three major clades of marine *Streptomyces*.

SMBGC	p Value	Activity
PKs		
Alkylresorcinol	0.003	Prevention of tumor [54] Antiprozoan [55]
Bafilomycin	0.003	Antitumor [56] Immunosuppressant [57]
Candicidin	0.005	Antifungus [58]
Coelimycin	1.60×10^{-6}	Yellow pigment [59]
Grincamycin	0.005	Antitumor [60]
Lactonamycin	0.005	Antibacteria [61]
Marineosin	0.004	Cytotoxicity [62]
Steffimycin	2.43×10^{-7}	Antitumor [63]
Spore pigment	0.0002	Regulate sporulation [64]
Xantholipin	0.0003	Antibacteria [65] Cytotoxicity [65]
FR-008	0.005	Antifungus [66]
NRPS		
Coelichelin	3.22×10^{-8}	Siderophore [67]
Daptomycin	2.84×10^{-5}	Antibacteria [68]
Echosides	0.006	Antivirus [69]
Ficellomycin	0.003	Antibacteria [70]
Griseobactin	4.72×10^{-5}	Siderophore [71]
Skyllamycin	0.003	Antitumor [72] Antibacteria [73]
Surugamide	0.0001	Antibacteria [74]
PKs/NRPS hybrid		
Antimycin	4.24×10^{-6}	Piscicide [75]
Oxazolomycin	0.005	Antibacteria [76]
SGR PTMs	0.003	Antifungal [77] Antioxidant [77]
Terpene		
Albaflavenone	2.20×10^{-16}	Antibacteria [78]
Pentalenolactone	0.002	Immunosuppressant [79]
Other		
2'-chloropentostatin	2.2×10^{-16}	Antivirus [80]
Informatipeptin	5.50×10^{-7}	Unknown
Keywimysin	1.72×10^{-7}	Unknown
Melanin	4.58×10^{-6}	Antioxidant [81]
Showdomycin	0.0002	Antitumor [82]
Spiroindimicin	0.004	Cytotoxicity [83]
AmfS	4.52×10^{-9}	Morphogen [84]
SCO-2138	0.006	Unknown

Among those SMBGCs, there were 23 SMBGCs showing clade-specificity (Figure 5), which is similar with previous studies regarding other genera [20,21]. (1) Marineosin, pentalenolactone, and spiroindimicin SMBGCs only appeared in the genomes of Clade I. (2) Albaflavenone, antimycin, candicidin, FR-008, grincamycin, informatipeptin, oxazolomycin, SCO-2138, as well as surugamide SMBGCs were exclusively present in the genomes of Clade II. (3) Amfs, daptomycin, and keywimysin SMBGCs was only found in the genomes of Clade III. (4) Bafilomycin, coelichelin, coelimycin, echosides, lactonamycin, SGR PTMs, skyllamycin, and xantholipin SMBGCs were present in the genomes of

Clade II and III, while absent in the genomes of Clade I. It was observed that Clade I encoded more specific SMBGCs, whereas Clade II had more various SMBGCs than other two clades, which indicated that Clade II should be more valued. SMBGCs exclusively present in the genomes of Clade II could be classified into PKs, NRPS, PKs/NRPS hybrid, terpene, and others, showing the category diversities. Furthermore, products of those SMBGCs could exhibit antibacterial, antifungal, antitumor, and piscicide activities [58,60,66,75,76,78], highlighting their functional diversities.

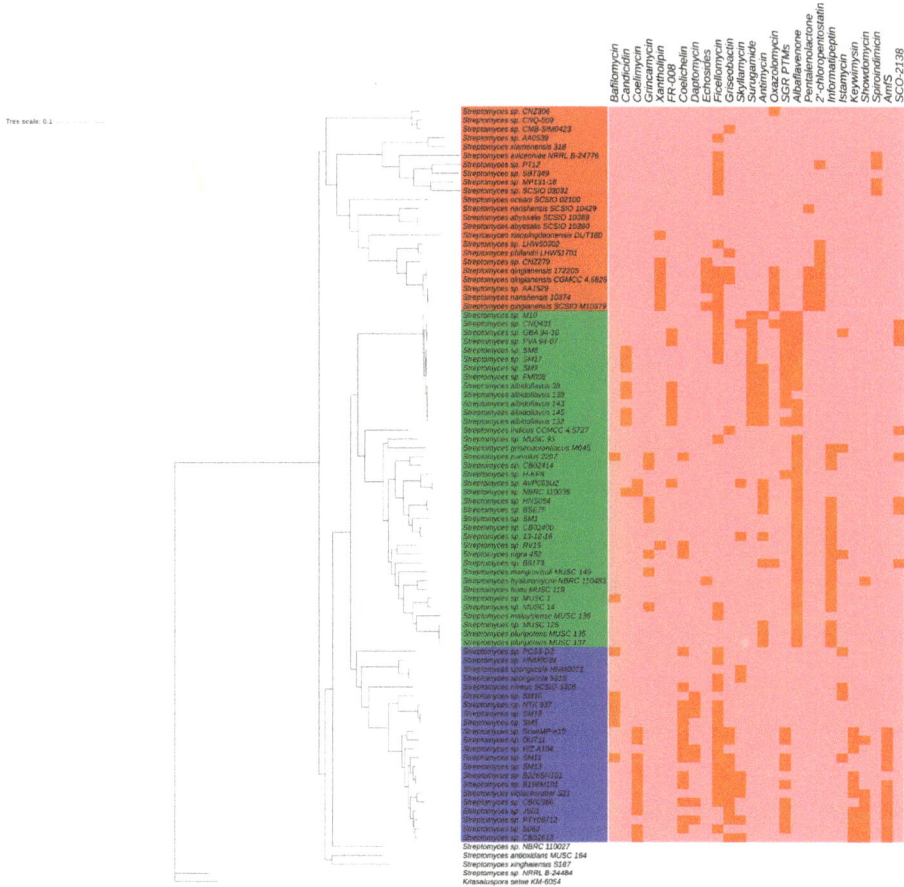

Figure 5. Heat maps of clade-specific SMBGCs. Dark brown and light brown indicate presence and absence of SMBGCs.

2.4. Ecotype-Associated SMBGCs

Because strains that were derived from algae, corals, cyanobacteria, and tunicates are few, those strains were excluded from ecotype-associated analysis. Significance tests among multi-clades revealed that 8 (4.8%) PKs, 11 (12.4%) NRPS, 5 (9.0%) PKs/NRPS hybrid, 2 (10.5%) terpene, and 8 (8.4%) other SMBGCs exhibited significant differences (Table 3). Among those SMBGCs, albaflavenone, 2'-chloropentostatin, daptomycin, echosides, FR-008, oxazolomycin, and pentalenolactone SMBGCs showed significant differences among both of phylotypes and ecotypes.

Table 3. SMBGCs showing significant differences among ecotypes of marine *Streptomyces*.

SMBGC		p Value	Activity
PKs			
	Brasilinolide	0.007	Antifungus [85]
	Butyrolactol A	0.0008	Antifungus [86]
	Incednine	0.007	Antiapotosis [87]
	Micromonolactam	0.0009	Unknown
	Napyradiomycin	0.007	Antibacteria [88] Cytotoxicity [88]
	Rifamycin	0.007	Antibacteria [89]
	Vicenistatin	0.007	Antitumor [90]
	FR-008	5.14×10^{-11}	Antifungus [66]
NRPS			
	Actinomycin	0.009	Antibacteria [91] Antitumor [91]
	Daptomycin	0.007	Antibacteria [68]
	Echosides	4.07×10^{-8}	Antivirus [69]
	Porothramycin	6.81×10^{-5}	Antibacteria [92] Antitumor [92]
	Rhizomide	0.008	Antibacteria [93]
	Scabichelin	0.009	Siderophore [94]
	Stenothricin	0.003	Antibacteria [95]
	Syringomycin	0.0005	Antifungus [96]
	Telomycin	0.0005	Antibacteria [97]
	Vazabitide A	0.002	Immunosuppressant [98]
	Yatakemycin	1.21×10^{-6}	Cytotoxicity [99]
PKs/NRPS hybrid			
	Ansatrienin	7.44×10^{-11}	Antibacteria [100]
	Kistamicin A	0.003	Antibacteira [101]
	Oxazolomycin	6.60×10^{-5}	Antibacteria [76]
	Rakicidin	1.07×10^{-5}	Antitumor [102]
	Zorbamycin	1.21×10^{-6}	Antitumor [103]
Terpene			
	Albaflavenone	0.003	Antibacteria [78]
	Pentalenolactone	8.94×10^{-7}	Immunosuppressant [79]
Other			
	2'-chloropentostatin	1.50×10^{-9}	Antivirus [80]
	Clavulanic acid	0.003	Antibacteria [104]
	Desferrioxamine	0.009	Antitumor [105]
	Lagmysin	6.81×10^{-5}	Unknown
	Legonaridin	1.21×10^{-6}	Cytotoxicity [106]
	Marinophenazines	0.005	Unknown
	Roseoflavin	0.0002	Antibacteria [107]

Among those SMBGCs, there were 11 SMBGCs showing clade-specificity (Figure 6). (1) Butyrolactol and FR-008 SMBGCs commonly appeared in the genomes of marine sediment-derived strains; (2) Albaflavenone SMBGC were mostly found in the strains isolated from seawater and marine sediments. (3) Daptomycin SMBGC were associated with strains living in sponges. (4) 2'-chloropentostatin, echosides, lagmysin, oxazolomycin pentalenolactone, porothramycin, and vazabitide A SMBGCs were usually detected in the mollusk-derived strains. Marine sediment-derived strains were mostly related to several SMBGCs, making those strains appear to be specific in SMBGCs distribution patterns. Compared with natural environments, strains isolated from marine invertebrates, particularly for mollusks, had more SMBGCs, showing symbiotic *Streptomyces* in marine invertebrates could be profitable resources of secondary metabolites.

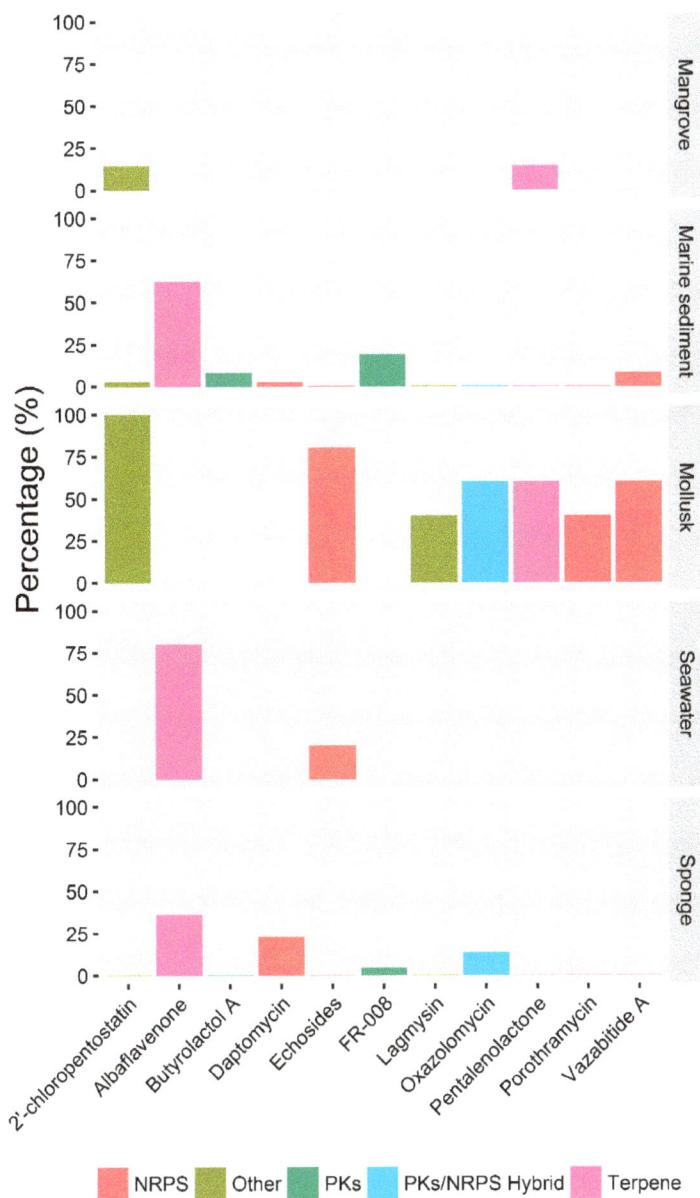

Figure 6. Distribution percentages of ecotype-specific SMBGCs.

3. Materials and Methods

3.1. Obtain, Assess, and Annotate Marine Streptomyces Genomes

Ninety-seven available marine *Streptomyces* genomes were obtained from NCBI GenBank database in January, 2019 (Table S1). Genomic qualities of those genomes were assessed by using CheckM software v1.0.7 (Australian Centre for Ecogenomics, The University of Queensland, Queensland,

Australia) [108] with the command "checkm lineage_wf -x fa bins/checkm/", and those genomes exhibiting the completeness >95% as well as the contamination <5% were screened to the further study.

rRNA genes were predicted by using with the command RNAmmer 1.2 package (Center for Biological Sequence Analysis, Technical University of Denmark, Lyngby, Denmark) "-S bac -m tsu,ssu,lsu" [109], while tRNA genes were annotated on the tRNAscan-SE On-line web server (http://trna.ucsc.edu/tRNAscan-SE/, [110]) with default mode except for the sequence source option set to "Bacterial". Open reading frames (ORFs) were predicted and annotated on the RAST webserver (http://rast.nmpdr.org/rast.cgi, [111]). SMBGCs were annotated using antiSMASH bacterial version with detection strictness set to "relaxed" and extra features sletected to "ActiveSiteFinder, KnownClusterBlast, SubClusterBlast" (https://antismash.secondarymetabolites.org/#!/start, [112]). Moreover, functional annotations based on COG and KEGG databases were carried out on WebMGA (http://weizhong-lab. ucsd.edu/webMGA/server/cog/) and KASS (https://www.genome.jp/tools/kaas/) webservers [113,114]. The GC content of those marine *Streptomyces* genomes were calculated by using OrthoANI [115].

3.2. Comparative Genomic Analysis of Marine Streptomyces Genomes

Kitasatospora setae KM-6054[T] was used as an outgroup in the further phylogenomic analysis based on recent polyphasic taxonomic studies [116–118], so its genome, which is under the NCBI GenBank assembly accession number of GCA_000269985.1, was also included in comparative genomic analysis. Comparative genomic analysis was modified based on the method described by Xu et al. [119]. Protein sequences translated from ORFs were compared pairwise using Proteinortho V5.16b (Interdisciplinary Center for Bioinformatics, University of Leipzig, Leipzig, Germany) with the command "-cov = 50 -identity = 50" [120] to identify OCs among genomes of marine *Streptomyces* and their outgroup. A set of OCs are defined as a class of genes transferred vertically from a common descent [121].

3.3. Phylogenomic Analysis and Genomic Similarity Calculation of Marine Streptomyces

Single-copy OCs shared by all of marine *Streptomyces* as well as *Kitasatospora setae* KM-6054[T] were screened by in-house perl script. Each single-copy OCs was aligned by using MAFFT version 7 (Computational Biology Research Center, The National Institute of Advanced Industrial Science and Technology, Tokyo, Japan) with the command "–auto" [122]. Then, aligned sequences were refined to remove poorly aligned regions by trimAL version 1.4.1 with the command "-automated1" [123], and concatenated manually. Subsequently, a maximum-likelihood phylogenomic tree based on concatenated protein sequences was reconstructed by using IQ-Tree 1.6.1 software (Center for Integrative Bioinformatics Vienna, Max F. Perutz Laboratories, University of Vienna, Medical University of Vienna, Vienna, Austria) with ultrafast bootstraps analysis set to 1000 replicates [124–126], following the best amino acid substitution model set as LG+F+R8 proposed by IQ-Tree 1.6.1 software with the command "-st AA -m MFP" [125].

Genome similarities of pairwise marine *Streptomyces* genomes were determined by calculating ANI values, which were obtained by using orthologous average nucleotide identity tool (OAT) 0.93.1 (Chunlab Inc., Seoul, Korea) [115] supplemented with basic local alignment search tool (BLAST) algorithm [127].

3.4. Statistical Analysis and Visualization

Unless stated, statistical analyses were performed by using R version 3.4.2 (R Foundation for Statistical Computing) [128]. Correlation of genomic size and gene counts were analyzed by using the function of *lm*. Significance test analyses of SMBGCs among phylotypes and ecotypes were performed by using the function of *kruskal.test*, with *p* values <0.01 showing the significant difference. Pan- and core-genomic analysis were carried out by summarizing OCs counts by using *"grep"* command in the CentOS 6 system (Red Hat, Inc., Raleigh, NC, USA).

The phylogenomic tree was visualized by using MEGA 7 software [129] and PowerPoint 2016 software (Microsoft Cooperation, Redmond, WA, USA). Unless heat maps were drawn by using Interactive Tree Of Life webserver (https://itol.embl.de/), other figures were constructed by using *ggplot2* and *Cairo* packages in R version 3.4.2 [128].

4. Conclusions

Marine *Streptomyces* is characterized by its rich species, genetic, and secondary metabolism diversities. Comparative genomics of Marine *Streptomyces* revealed that those group have a wide range of OCs showing high genetic diversity. Phylogenomic analysis in this study shows that enormous novel marine *Streptomyces* species needs to be identified and the majority can be classified into three clades. Phylotype and ecotype are both associated with SMBGCs distribution patterns. The Clade I and marine sediment-derived *Streptomyces* harbored more specific SMBGCs, which consisted of several common ones, such as butyrolactol, FR-008, marineosin, pentalenolactone, and spiroindimicin, whereas the Clade II and marine invertebrate-derived *Streptomyces* have more SMBGCs, such as 2′-chloropentostatin, albaflavenone, antimycin, candicidin, echosides, FR-008, grincamycin, informatipeptin, lagmysin, oxazolomycin, pentalenolactone, porothramycin, SCO-2138, and vazabitide A, indicating that those *Streptomyces* could act as plentiful resources for mining secondary metabolites. As stated above, our study is beneficial for broadening our knowledge about SMBGC distribution patterns in marine *Streptomyces* and developing their secondary metabolites in the future.

Supplementary Materials: The following are available online at http://www.mdpi.com/1660-3397/17/9/498/s1, Table S1. Genomic information and quality estimations of marine Streptomyces obtained from NCBI GenBank database. Table S2. Genomic annotations of marine *Streptomyces* by using RAST webserver. Table S3. SMBGC annotations of marine *Streptomyces* by using antiSMASH webserver. Table S4. OCs distributions in marine *Streptomyces*. Table S5. ANI values among marine *Streptomyces*.

Author Contributions: L.X. and B.-N.H. conceived and designed the study. K.-X.Y. and W.-H.D. obtained genomic sequences and checked their genomic qualities. L.X. and C.S. annotated genomic sequences and summarized annotation results. L.X. and L.-H.X. performed statistical analysis. L.X. and B.-N.H. wrote the manuscript.

Funding: This research was funded by the Special Fund for Agro-scientific Research in the Public Interest of Zhejiang Province (LGN18C190011), the Natural Science Foundation of Zhejiang Province (LQ19C010006), the National Natural Science Foundation of China (81973233), the China Postdoctoral Science Foundation (2019M652042), the Science Foundation of Zhejiang Sci-Tech University (17042058-Y, 17042187-Y), the Project for Jiaozhou Excellent Innovation Team (18-CX-1).

Conflicts of Interest: The authors declare no conflict of competing interests.

References

1. Sekurova, O.N.; Schneider, O.; Zotchev, S.B. Novel bioactive natural products from bacteria via bioprospecting, genome mining and metabolic engineering. *Microb. Biotechnol.* **2019**. [CrossRef] [PubMed]
2. Malik, V. Microbial secondary metabolism. *Trends Biochem. Sci.* **1980**, *5*, 68–72. [CrossRef]
3. Van der Meij, A.; Worsley, S.F.; Hutchings, M.I.; van Wezel, G.P. Chemical ecology of antibiotic production by actinomycetes. *FEMS Microbiol. Rev.* **2017**, *41*, 392–416. [CrossRef] [PubMed]
4. Mascuch, S.; Kubanek, J. A marine chemical defense partnership. *Science* **2019**, *364*, 1034–1035. [CrossRef] [PubMed]
5. Demain, A.L.; Sanchez, S. Microbial drug discovery: 80 years of progress. *J. Antibiot.* **2009**, *62*, 5–16. [CrossRef] [PubMed]
6. Purves, K.; Macintyre, L.; Brennan, D.; Hreggviethsson, G.O.; Kuttner, E.; Asgeirsdottir, M.E.; Young, L.C.; Green, D.H.; Edrada-Ebel, R.; Duncan, K.R. Using Molecular Networking for Microbial Secondary Metabolite Bioprospecting. *Metabolites* **2016**, *6*, 2. [CrossRef]
7. Machado, H.; Sonnenschein, E.C.; Melchiorsen, J.; Gram, L. Genome mining reveals unlocked bioactive potential of marine Gram-negative bacteria. *BMC Genom.* **2015**, *16*, 158. [CrossRef]
8. Kamjam, M.; Sivalingam, P.; Deng, Z.; Hong, K. Deep Sea Actinomycetes and Their Secondary Metabolites. *Front. Microbiol.* **2017**, *8*, 760. [CrossRef]

9.   Jackson, S.A.; Crossman, L.; Almeida, E.L.; Margassery, L.M.; Kennedy, J.; Dobson, A. Diverse and Abundant Secondary Metabolism Biosynthetic Gene Clusters in the Genomes of Marine Sponge Derived *Streptomyces* spp. Isolates. *Mar. Drugs* **2018**, *16*, 67. [CrossRef]
10.   Fenical, W.; Jensen, P.R. Developing a new resource for drug discovery: Marine actinomycete bacteria. *Nat. Chem. Bio.* **2006**, *2*, 666–673. [CrossRef]
11.   Ziemert, N.; Alanjary, M.; Weber, T. The evolution of genome mining in microbes–a review. *Nat. Prod. Rep.* **2016**, *33*, 988–1005. [CrossRef]
12.   Palazzotto, E.; Weber, T. Omics and multi-omics approaches to study the biosynthesis of secondary metabolites in microorganisms. *Curr. Opin. Microbiol.* **2018**, *45*, 109–116. [CrossRef]
13.   Foulston, L. Genome mining and prospects for antibiotic discovery. *Curr. Opin. Microbiol.* **2019**, *51*, 1–8. [CrossRef]
14.   Cimermancic, P.; Medema, M.H.; Claesen, J.; Kurita, K.; Wieland Brown, L.C.; Mavrommatis, K.; Pati, A.; Godfrey, P.A.; Koehrsen, M.; Clardy, J.; et al. Insights into secondary metabolism from a global analysis of prokaryotic biosynthetic gene clusters. *Cell* **2014**, *158*, 412–421. [CrossRef]
15.   Weissman, K.J. The structural biology of biosynthetic megaenzymes. *Nat. Chem. Bio.* **2015**, *11*, 660–670. [CrossRef]
16.   Eisenreich, W.; Bacher, A.; Arigoni, D.; Rohdich, F. Biosynthesis of isoprenoids via the non-mevalonate pathway. *Cell. Mol. Life Sci.* **2004**, *61*, 1401–1426. [CrossRef]
17.   Zurbriggen, A.; Kirst, H.; Melis, A. Isoprene production via the mevalonic acid pathway in *Escherichia coli* (Bacteria). *BioEnergy Res.* **2012**, *5*, 814–828. [CrossRef]
18.   Jensen, P.R.; Fenical, W. Strategies for the discovery of secondary metabolites from marine bacteria: Ecological perspectives. *Annu. Rev. Microbiol.* **1994**, *48*, 559–584. [CrossRef]
19.   O'Brien, J.; Wright, G.D. An ecological perspective of microbial secondary metabolism. *Curr. Opin. Biotechnol.* **2011**, *22*, 552–558. [CrossRef]
20.   Jensen, P.R.; Williams, P.G.; Oh, D.C.; Zeigler, L.; Fenical, W. Species-specific secondary metabolite production in marine actinomycetes of the genus *Salinispora*. *Appl. Environ. Microb.* **2007**, *73*, 1146–1152. [CrossRef]
21.   Adamek, M.; Alanjary, M.; Sales-Ortells, H.; Goodfellow, M.; Bull, A.T.; Winkler, A.; Wibberg, D.; Kalinowski, J.; Ziemert, N. Comparative genomics reveals phylogenetic distribution patterns of secondary metabolites in *Amycolatopsis* species. *BMC Genom.* **2018**, *19*, 426. [CrossRef]
22.   Kämpfer, P. *The Prokaryotes: Archaea. Bacteria: Firmicutes, Actinomycetes*; The family *Streptomycetaceae*, part I: Taxonomy; Rosenberg, E., DeLong, E.F., Lory, S., Stackbrandt, E., Thompson, F., Eds.; Springer: Berlin, Germany, 2006; Volume 3, pp. 538–604.
23.   Parte, A.C. LPSN–List of Prokaryotic names with Standing in Nomenclature (bacterio.net), 20 years on. *Int. J. Syst. Evol. Microbiol.* **2018**, *68*, 1825–1829. [CrossRef]
24.   Hwang, K.S.; Kim, H.U.; Charusanti, P.; Palsson, B.O.; Lee, S.Y. Systems biology and biotechnology of *Streptomyces* species for the production of secondary metabolites. *Biotechnol. Adv.* **2014**, *32*, 255–268. [CrossRef]
25.   Liu, R.; Deng, Z.; Liu, T. *Streptomyces* species: Ideal chassis for natural product discovery and overproduction. *Metab. Eng.* **2018**, *50*, 74–84. [CrossRef]
26.   Hakvag, S.; Fjaervik, E.; Josefsen, K.D.; Ian, E.; Ellingsen, T.E.; Zotchev, S.B. Characterization of *Streptomyces* spp. isolated from the sea surface microlayer in the Trondheim Fjord, Norway. *Mar. Drugs* **2008**, *6*, 620–635. [CrossRef]
27.   Zhu, H.; Jiang, S.; Yao, Q.; Wang, Y.; Chen, M.; Chen, Y.; Guo, J. *Streptomyces fenghuangensis* sp. nov., isolated from seawater. *Int. J Syst. Evol. Microbiol.* **2011**, *61*, 2811–2815. [CrossRef]
28.   Macherla, V.R.; Liu, J.; Bellows, C.; Teisan, S.; Nicholson, B.; Lam, K.S.; Potts, B.C. Glaciapyrroles A, B, and C, pyrrolosesquiterpenes from a *Streptomyces* sp. isolated from an Alaskan marine sediment. *J. Nat. Prod.* **2005**, *68*, 780–783. [CrossRef]
29.   Zhao, X.Q.; Li, W.J.; Jiao, W.C.; Li, Y.; Yuan, W.J.; Zhang, Y.Q.; Klenk, H.P.; Suh, J.W.; Bai, F.W. *Streptomyces xinghaiensis* sp. nov., isolated from marine sediment. *Int. J. Syst. Evol. Microbiol.* **2009**, *59*, 2870–2874. [CrossRef]
30.   Braña, A.F.; Sarmiento-Vizcaíno, A.; Pérez-Victoria, I.; Martín, J.; Otero, L.; Palacios-Gutiérrez, J.J.; Fernández, J.; Mohamedi, Y.; Fontanil, T.; Salmón, M. Desertomycin G, a New Antibiotic with Activity

against *Mycobacterium tuberculosis* and Human Breast Tumor Cell Lines Produced by *Streptomyces althioticus* MSM3, Isolated from the Cantabrian Sea Intertidal Macroalgae *Ulva* sp. *Mar. Drugs* **2019**, *17*, 114. [CrossRef]

31. Girão, M.; Ribeiro, I.; Ribeiro, T.; Azevedo, I.C.; Pereira, F.; Urbatzka, R.; Leão, P.N.; Carvalho, M.F. Actinobacteria isolated from *Laminaria ochroleuca*: A source of new bioactive compounds. *Front. Microbiol.* **2019**, *10*, 683. [CrossRef]

32. Xiao, J.; Wang, Y.; Luo, Y.; Xie, S.J.; Ruan, J.S.; Xu, J. *Streptomyces avicenniae* sp. nov., a novel actinomycete isolated from the rhizosphere of the mangrove plant *Avicennia mariana*. *Int. J. Syst. Evol. Microbiol.* **2009**, *59*, 2624–2628. [CrossRef]

33. Yan, L.-L.; Han, N.-N.; Zhang, Y.-Q.; Yu, L.-Y.; Chen, J.; Wei, Y.-Z.; Li, Q.-P.; Tao, L.; Zheng, G.-H.; Yang, S.-E. Antimycin A 18 produced by an endophytic *Streptomyces albidoflavus* isolated from a mangrove plant. *J. Antibiot.* **2010**, *63*, 259. [CrossRef]

34. Khan, S.T.; Komaki, H.; Motohashi, K.; Kozone, I.; Mukai, A.; Takagi, M.; Shin-ya, K. *Streptomyces* associated with a marine sponge *Haliclona* sp.; biosynthetic genes for secondary metabolites and products. *Env. Microbiol.* **2011**, *13*, 391–403. [CrossRef]

35. Huang, X.; Zhou, S.; Huang, D.; Chen, J.; Zhu, W. *Streptomyces spongiicola* sp. nov., an actinomycete derived from marine sponge. *Int. J. Syst. Evol. Microbiol.* **2016**, *66*, 738–743. [CrossRef]

36. Hodges, T.W.; Slattery, M.; Olson, J.B. Unique actinomycetes from marine caves and coral reef sediments provide novel PKS and NRPS biosynthetic gene clusters. *Mar. Biotechnol.* **2012**, *14*, 270–280. [CrossRef]

37. Braña, A.; Sarmiento-Vizcaíno, A.; Osset, M.; Pérez-Victoria, I.; Martín, J.; de Pedro, N.; de la Cruz, M.; Díaz, C.; Vicente, F.; Reyes, F. Lobophorin K, a new natural product with cytotoxic activity produced by *Streptomyces* sp. M-207 associated with the deep-sea coral *Lophelia pertusa*. *Mar. Drugs* **2017**, *15*, 144. [CrossRef]

38. Shaala, L.A.; Youssef, D.T.; Badr, J.M.; Harakeh, S.M. Bioactive 2(1H)-Pyrazinones and Diketopiperazine Alkaloids from a Tunicate-Derived Actinomycete *Streptomyces* sp. *Molecules* **2016**, *21*, 1116. [CrossRef]

39. Sung, A.A.; Gromek, S.M.; Balunas, M.J. Upregulation and Identification of Antibiotic Activity of a Marine-Derived *Streptomyces* sp. via Co-Cultures with Human Pathogens. *Mar. Drugs* **2017**, *15*, 250. [CrossRef]

40. Lin, Z.; Flores, M.; Forteza, I.; Henriksen, N.M.; Concepcion, G.P.; Rosenberg, G.; Haygood, M.G.; Olivera, B.M.; Light, A.R.; Cheatham III, T.E.; et al. Totopotensamides, polyketide-cyclic peptide hybrids from a mollusk-associated bacterium *Streptomyces* sp. *J. Nat. Prod.* **2012**, *75*, 644–649. [CrossRef]

41. Lin, Z.; Koch, M.; Pond, C.D.; Mabeza, G.; Seronay, R.A.; Concepcion, G.P.; Barrows, L.R.; Olivera, B.M.; Schmidt, E.W. Structure and activity of lobophorins from a turrid mollusk-associated *Streptomyces* sp. *J. Antibiot.* **2014**, *67*, 121. [CrossRef]

42. Bentley, S.D.; Chater, K.F.; Cerdeno-Tarraga, A.M.; Challis, G.L.; Thomson, N.R.; James, K.D.; Harris, D.E.; Quail, M.A.; Kieser, H.; Harper, D.; et al. Complete genome sequence of the model actinomycete *Streptomyces coelicolor* A3(2). *Nature* **2002**, *417*, 141–147. [CrossRef]

43. Studholme, D.J. Genome Update. Let the consumer beware: *Streptomyces* genome sequence quality. *Microb. Biotechnol.* **2016**, *9*, 3–7. [CrossRef]

44. Bush, M.J.; Chandra, G.; Bibb, M.J.; Findlay, K.C.; Buttner, M.J. Genome-Wide Chromatin Immunoprecipitation Sequencing Analysis Shows that WhiB Is a Transcription Factor That Cocontrols Its Regulon with WhiA To Initiate Developmental Cell Division in *Streptomyces*. *MBio* **2016**, *7*, e00523-e16. [CrossRef]

45. Doroghazi, J.R.; Metcalf, W.W. Comparative genomics of actinomycetes with a focus on natural product biosynthetic genes. *BMC Genom.* **2013**, *14*, 611. [CrossRef]

46. Zhang, H.; Wang, H.; Wang, Y.; Cui, H.; Xie, Z.; Pu, Y.; Pei, S.; Li, F.; Qin, S. Genomic sequence-based discovery of novel angucyclinone antibiotics from marine *Streptomyces* sp. W007. *FEMS Microbiol. Lett.* **2012**, *332*, 105–112. [CrossRef]

47. Seipke, R.F. Strain-level diversity of secondary metabolism in *Streptomyces albus*. *PLoS ONE* **2015**, *10*, e0116457. [CrossRef]

48. Zotchev, S.B. Marine actinomycetes as an emerging resource for the drug development pipelines. *J. Biotechnol.* **2012**, *158*, 168–175. [CrossRef]

49. Xu, M.J.; Wang, J.H.; Bu, X.L.; Yu, H.L.; Li, P.; Ou, H.Y.; He, Y.; Xu, F.D.; Hu, X.Y.; Zhu, X.M.; et al. Deciphering the streamlined genome of *Streptomyces xiamenensis* 318 as the producer of the anti-fibrotic drug candidate xiamenmycin. *Sci. Rep.* **2016**, *6*, 18977. [CrossRef]

50. Ian, E.; Malko, D.B.; Sekurova, O.N.; Bredholt, H.; Ruckert, C.; Borisova, M.E.; Albersmeier, A.; Kalinowski, J.; Gelfand, M.S.; Zotchev, S.B. Genomics of sponge-associated *Streptomyces* spp. closely related to *Streptomyces albus* J1074: Insights into marine adaptation and secondary metabolite biosynthesis potential. *PLoS ONE* **2014**, *9*, e96719. [CrossRef]

51. Tian, X.; Zhang, Z.; Yang, T.; Chen, M.; Li, J.; Chen, F.; Yang, J.; Li, W.; Zhang, B.; Zhang, Z.; et al. Comparative Genomics Analysis of *Streptomyces* Species Reveals Their Adaptation to the Marine Environment and Their Diversity at the Genomic Level. *Front. Microbiol.* **2016**, *7*, 998. [CrossRef]

52. Ventura, M.; Canchaya, C.; Tauch, A.; Chandra, G.; Fitzgerald, G.F.; Chater, K.F.; van Sinderen, D. Genomics of *Actinobacteria*: Tracing the evolutionary history of an ancient phylum. *Microbiol. Mol. Biol. Rev.* **2007**, *71*, 495–548. [CrossRef]

53. Kim, M.; Oh, H.S.; Park, S.C.; Chun, J. Towards a taxonomic coherence between average nucleotide identity and 16S rRNA gene sequence similarity for species demarcation of prokaryotes. *Int. J. Syst. Evol. Microbiol.* **2014**, *64*, 346–351. [CrossRef]

54. Ross, A.B.; Kamal-Eldin, A.; Aman, P. Dietary alkylresorcinols: Absorption, bioactivities, and possible use as biomarkers of whole-grain wheat- and rye-rich foods. *Nutr. Rev.* **2004**, *62*, 81–95. [CrossRef]

55. Hayashi, M.; Yamada, H.; Mitamura, T.; Horii, T.; Yamamoto, A.; Moriyama, Y. Vacuolar H$^+$-ATPase localized in plasma membranes of malaria parasite cells, *Plasmodium falciparum*, is involved in regional acidification of parasitized erythrocytes. *J. Biol. Chem.* **2000**, *275*, 34353–34358. [CrossRef]

56. Whitton, B.; Okamoto, H.; Packham, G.; Crabb, S.J. Vacuolar ATPase as a potential therapeutic target and mediator of treatment resistance in cancer. *Cancer Med.* **2018**, *7*, 3800–3811. [CrossRef]

57. Keller, C.W.; Schmidt, J.; Lunemann, J.D. Immune and myodegenerative pathomechanisms in inclusion body myositis. *Ann. Clin. Transl. Neur.* **2017**, *4*, 422–445. [CrossRef]

58. Szczeblewski, P.; Laskowski, T.; Kubacki, B.; Dziergowska, M.; Liczmanska, M.; Grynda, J.; Kubica, P.; Kot-Wasik, A.; Borowski, E. Analytical studies on ascosin, candicidin and levorin multicomponent antifungal antibiotic complexes. The stereostructure of ascosin A2. *Sci. Rep.* **2017**, *7*, 40158. [CrossRef]

59. Gomez-Escribano, J.P.; Song, L.; Fox, D.J.; Yeo, V.; Bibb, M.J.; Challis, G.L. Structure and biosynthesis of the unusual polyketide alkaloid coelimycin P1, a metabolic product of the cpk gene cluster of *Streptomyces coelicolor* M145. *Chem. Sci.* **2012**, *3*, 2716–2720. [CrossRef]

60. Lai, Z.; Yu, J.; Ling, H.; Song, Y.; Yuan, J.; Ju, J.; Tao, Y.; Huang, H. Grincamycins I–K, Cytotoxic Angucycline Glycosides Derived from Marine-Derived Actinomycete *Streptomyces lusitanus* SCSIO LR32. *Planta Med.* **2018**, *84*, 201–207. [CrossRef]

61. Matsumoto, N.; Tsuchida, T.; Maruyama, M.; Kinoshita, N.; Homma, Y.; Iinuma, H.; Sawa, T.; Hamada, M.; Takeuchi, T.; Heida, N.; et al. Lactonamycin, a new antimicrobial antibiotic produced by *Streptomyces rishiriensis* MJ773-88K4. I. Taxonomy, fermentation, isolation, physico-chemical properties and biological activities. *J. Antibiot.* **1999**, *52*, 269–275. [CrossRef]

62. Boonlarppradab, C.; Kauffman, C.A.; Jensen, P.R.; Fenical, W. Marineosins A and B, cytotoxic spiroaminals from a marine-derived actinomycete. *Org. Lett.* **2008**, *10*, 5505–5508. [CrossRef]

63. Gullon, S.; Olano, C.; Abdelfattah, M.S.; Brana, A.F.; Rohr, J.; Mendez, C.; Salas, J.A. Isolation, characterization, and heterologous expression of the biosynthesis gene cluster for the antitumor anthracycline steffimycin. *Appl. Environ. Microb.* **2006**, *72*, 4172–4183. [CrossRef]

64. Kelemen, G.H.; Brian, P.; Flardh, K.; Chamberlin, L.; Chater, K.F.; Buttner, M.J. Developmental regulation of transcription of *whiE*, a locus specifying the polyketide spore pigment in *Streptomyces coelicolor* A3(2). *J. Bacteriol.* **1998**, *180*, 2515–2521.

65. Wu, S.; Huang, T.; Xie, D.; Wo, J.; Wang, X.; Deng, Z.; Lin, S. Xantholipin B produced by the *stnR* inactivation mutant *Streptomyces flocculus* CGMCC 4.1223 WJN-1. *J. Antibiot.* **2017**, *70*, 90–95. [CrossRef]

66. Zhou, Y.; Li, J.; Zhu, J.; Chen, S.; Bai, L.; Zhou, X.; Wu, H.; Deng, Z. Incomplete beta-ketone processing as a mechanism for polyene structural variation in the FR-008/candicidin complex. *Chem. Biol.* **2008**, *15*, 629–638. [CrossRef]

67. Williams, J.C.; Sheldon, J.R.; Imlay, H.D.; Dutter, B.F.; Draelos, M.M.; Skaar, E.P.; Sulikowski, G.A. Synthesis of the Siderophore Coelichelin and Its Utility as a Probe in the Study of Bacterial Metal Sensing and Response. *Org. Lett.* **2019**, *21*, 679–682. [CrossRef]

68. Sader, H.S.; Flamm, R.K.; Farrell, D.J.; Jones, R.N. Daptomycin activity against uncommonly isolated streptococcal and other gram-positive species groups. *Antimicrob. Agents Chemother.* **2013**, *57*, 6378–6380. [CrossRef]

69. Zhu, J.; Chen, W.; Li, Y.Y.; Deng, J.J.; Zhu, D.Y.; Duan, J.; Liu, Y.; Shi, G.Y.; Xie, C.; Wang, H.X.; et al. Identification and catalytic characterization of a nonribosomal peptide synthetase-like (NRPS-like) enzyme involved in the biosynthesis of echosides from *Streptomyces* sp. LZ35. *Gene* **2014**, *546*, 352–358. [CrossRef]

70. He, X.; Li, M.; Song, S.; Wu, X.; Zhang, J.; Wu, G.; Yue, R.; Cui, H.; Song, S.; Ma, C.; et al. Ficellomycin: An aziridine alkaloid antibiotic with potential therapeutic capacity. *Appl. Microbiol. Biot.* **2018**, *102*, 4345–4354. [CrossRef]

71. Patzer, S.I.; Braun, V. Gene cluster involved in the biosynthesis of griseobactin, a catechol-peptide siderophore of *Streptomyces* sp. ATCC 700974. *J. Bacteriol.* **2010**, *192*, 426–435. [CrossRef]

72. Pohle, S.; Appelt, C.; Roux, M.; Fiedler, H.P.; Sussmuth, R.D. Biosynthetic gene cluster of the non-ribosomally synthesized cyclodepsipeptide skyllamycin: Deciphering unprecedented ways of unusual hydroxylation reactions. *J. Am. Chem. Soc.* **2011**, *133*, 6194–6205. [CrossRef]

73. Giltrap, A.M.; Haeckl, F.P.J.; Kurita, K.L.; Linington, R.G.; Payne, R.J. Synthetic Studies Toward the Skyllamycins: Total Synthesis and Generation of Simplified Analogues. *J. Org. Chem.* **2018**, *83*, 7250–7270. [CrossRef]

74. Wang, X.; Shaaban, K.A.; Elshahawi, S.I.; Ponomareva, L.V.; Sunkara, M.; Copley, G.C.; Hower, J.C.; Morris, A.J.; Kharel, M.K.; Thorson, J.S. Mullinamides A and B, new cyclopeptides produced by the Ruth Mullins coal mine fire isolate *Streptomyces* sp. RM-27-46. *J. Antibiot.* **2014**, *67*, 571–575. [CrossRef]

75. Hamilton, B.T.; Moore, S.E.; Williams, T.B.; Darby, N.; Vinson, M.R. Comparative effects of rotenone and antimycin on macroinvertebrate diversity in two streams in Great Basin National Park, Nevada. *N. Am. J. Fish. Manag.* **2009**, *29*, 1620–1635. [CrossRef]

76. Angelov, P.; Chau, Y.K.; Fryer, P.J.; Moloney, M.G.; Thompson, A.L.; Trippier, P.C. Biomimetic synthesis, antibacterial activity and structure-activity properties of the pyroglutamate core of oxazolomycin. *Org. Biomol. Chem.* **2012**, *10*, 3472–3485. [CrossRef]

77. Luo, Y.; Huang, H.; Liang, J.; Wang, M.; Lu, L.; Shao, Z.; Cobb, R.E.; Zhao, H. Activation and characterization of a cryptic polycyclic tetramate macrolactam biosynthetic gene cluster. *Nat. Commun.* **2013**, *4*, 2894. [CrossRef]

78. Lin, X.; Cane, D.E. Biosynthesis of the sesquiterpene antibiotic albaflavenone in Streptomyces coelicolor. Mechanism and stereochemistry of the enzymatic formation of epi-isozizaene. *J. Am. Chem. Soc.* **2009**, *131*, 6332–6333. [CrossRef]

79. Uyeda, M.; Mizukami, M.; Yokomizo, K.; Suzuki, K. Pentalenolactone I and hygromycin A, immunosuppressants produced by Streptomyces filipinensis and *Streptomyces hygroscopicus*. *Biosci. Biotech. Bioch.* **2001**, *65*, 1252–1254. [CrossRef]

80. Gao, Y.; Xu, G.; Wu, P.; Liu, J.; Cai, Y.S.; Deng, Z.; Chen, W. Biosynthesis of 2′-Chloropentostatin and 2′-Amino-2′-Deoxyadenosine Highlights a Single Gene Cluster Responsible for Two Independent Pathways in *Actinomadura* sp. Strain ATCC 39365. *Appl. Environ. Microbiol.* **2017**, *83*, e00078-17. [CrossRef]

81. Riley, P.A. Melanin. *Int. J. Biochem. Cell Biol.* **1997**, *29*, 1235–1239. [CrossRef]

82. Numao, N.; Hemmi, H.; Naujokaitis, S.A.; Rabinovitz, M.; Beisler, J.A. Showdomycin analogues: Synthesis and antitumor evaluation. *J. Med. Chem.* **1981**, *24*, 515–520. [CrossRef]

83. Kasanah, N.; Triyanto, T. Bioactivities of Halometabolites from Marine Actinobacteria. *Biomolecules* **2019**, *9*, 225. [CrossRef]

84. Ueda, K.; Oinuma, K.; Ikeda, G.; Hosono, K.; Ohnishi, Y.; Horinouchi, S.; Beppu, T. AmfS, an extracellular peptidic morphogen in *Streptomyces* griseus. *J. Bacteriol.* **2002**, *184*, 1488–1492. [CrossRef]

85. Tanaka, Y.; Komaki, H.; Yazawa, K.; Mikami, Y.; Nemoto, A.; Tojyo, T.; Kadowaki, K.; Shigemori, H.; Kobayashi, J. Brasilinolide A, a new macrolide antibiotic produced by *Nocardia brasiliensis*: Producing strain, isolation and biological activity. *J. Antibiot.* **1997**, *50*, 1036–1041. [CrossRef]

86. Kotake, C.; Yamasaki, T.; Moriyama, T.; Shinoda, M.; Komiyama, N.; Furumai, T.; Konishi, M.; Oki, T. Butyrolactols A and B, new antifungal antibiotics. *J. Antibiot.* **1992**, *45*, 1442–1450. [CrossRef]

87. Futamura, Y.; Sawa, R.; Umezawa, Y.; Igarashi, M.; Nakamura, H.; Hasegawa, K.; Yamasaki, M.; Tashiro, E.; Takahashi, Y.; Akamatsu, Y. Discovery of incednine as a potent modulator of the anti-apoptotic function of Bcl-xL from microbial origin. *J. Am. Chem. Soc.* **2008**, *130*, 1822–1823. [CrossRef]

88. Cheng, Y.B.; Jensen, P.R.; Fenical, W. Cytotoxic and Antimicrobial Napyradiomycins from Two Marine-Derived, MAR 4 *Streptomyces* Strains. *Eur. J. Org. Chem.* **2013**, *2013*, 3751–3757. [CrossRef]

89. Xia, M.; Suchland, R.J.; Carswell, J.A.; Van Duzer, J.; Buxton, D.K.; Brown, K.; Rothstein, D.M.; Stamm, W.E. Activities of rifamycin derivatives against wild-type and *rpoB* mutants of *Chlamydia trachomatis*. *Antimicrob. Agents Chemother.* **2005**, *49*, 3974–3976. [CrossRef]

90. Shindo, K.; Kamishohara, M.; Odagawa, A.; Matsuoka, M.; Kawai, H. Vicenistatin, a novel 20-membered macrocyclic lactam antitumor antibiotic. *J. Antibiot.* **1993**, *46*, 1076–1081. [CrossRef]

91. Koba, M.; Konopa, J. Actinomycin D and its mechanisms of action. *Postepy Hig. Med. Dosw.* **2005**, *59*, 290–298.

92. Tsunakawa, M.; Kamei, H.; Konishi, M.; Miyaki, T.; Oki, T.; Kawaguchi, H. Porothramycin, a new antibiotic of the anthramycin group: Production, isolation, structure and biological activity. *J. Antibiot.* **1988**, *41*, 1366–1373. [CrossRef]

93. Revathi, S.; Malathy, N.S. Antibacterial Activity of Rhizome of *Curcuma aromatica* and Partial Purification of Active Compounds. *Indian J. Pharm. Sci.* **2013**, *75*, 732–735.

94. Kodani, S.; Bicz, J.; Song, L.; Deeth, R.J.; Ohnishi-Kameyama, M.; Yoshida, M.; Ochi, K.; Challis, G.L. Structure and biosynthesis of scabichelin, a novel tris-hydroxamate siderophore produced by the plant pathogen *Streptomyces scabies* 87.22. *Org. Biomol. Chem.* **2013**, *11*, 4686–4694. [CrossRef]

95. Hasenbohler, A.; Kneifel, H.; Konig, W.A.; Zahner, H.; Zeiler, H.J. Metabolic products of microorganisms. 134. Stenothricin, a new inhibitor of the bacterial cell wall synthesis (author's transl). *Arch. Microbiol.* **1974**, *99*, 307–321.

96. Hu, F.P.; Young, J.M.; Fletcher, M.J. Preliminary description of biocidal (syringomycin) activity in fluorescent plant pathogenic *Pseudomonas* species. *J. Appl. Microbiol.* **1998**, *85*, 365–371. [CrossRef]

97. Fu, C.; Keller, L.; Bauer, A.; Bronstrup, M.; Froidbise, A.; Hammann, P.; Herrmann, J.; Mondesert, G.; Kurz, M.; Schiell, M.; et al. Biosynthetic Studies of Telomycin Reveal New Lipopeptides with Enhanced Activity. *J. Am. Chem. Soc.* **2015**, *137*, 7692–7705. [CrossRef]

98. Yashiro, T.; Sakata, F.; Sekimoto, T.; Shirai, T.; Hasebe, F.; Matsuda, K.; Kurosawa, S.; Suzuki, S.; Nagata, K.; Kasakura, K.; et al. Immunosuppressive effect of a non-proteinogenic amino acid from *Streptomyces* through inhibiting allogeneic T cell proliferation. *Biosci. Biotechnol. Biochem.* **2019**, *83*, 1111–1116. [CrossRef]

99. Tichenor, M.S.; MacMillan, K.S.; Trzupek, J.D.; Rayl, T.J.; Hwang, I.; Boger, D.L. Systematic exploration of the structural features of yatakemycin impacting DNA alkylation and biological activity. *J. Am. Chem. Soc.* **2007**, *129*, 10858–10869. [CrossRef]

100. Moore, B.S.; Floss, H.G. Biosynthetic studies on the origin of the cyclohexanecarboxylic acid moiety of ansatrienin A and omega-cyclohexyl fatty acids. *J. Nat. Prod.* **1994**, *57*, 382–386. [CrossRef]

101. Greule, A.; Izore, T.; Iftime, D.; Tailhades, J.; Schoppet, M.; Zhao, Y.; Peschke, M.; Ahmed, I.; Kulik, A.; Adamek, M.; et al. Kistamicin biosynthesis reveals the biosynthetic requirements for production of highly crosslinked glycopeptide antibiotics. *Nat. Commun.* **2019**, *10*, 2613. [CrossRef]

102. Sang, F.; Li, D.; Sun, X.; Cao, X.; Wang, L.; Sun, J.; Sun, B.; Wu, L.; Yang, G.; Chu, X.; et al. Total synthesis and determination of the absolute configuration of rakicidin A. *J. Am. Chem. Soc.* **2014**, *136*, 15787–15791. [CrossRef]

103. Wang, L.; Yun, B.S.; George, N.P.; Wendt-Pienkowski, E.; Galm, U.; Oh, T.J.; Coughlin, J.M.; Zhang, G.; Tao, M.; Shen, B. Glycopeptide antitumor antibiotic zorbamycin from *Streptomyces flavoviridis* ATCC 21892: Strain improvement and structure elucidation. *J. Nat. Prod.* **2007**, *70*, 402–406. [CrossRef]

104. Finlay, J.; Miller, L.; Poupard, J.A. A review of the antimicrobial activity of clavulanate. *J. Antimicrob. Ch.* **2003**, *52*, 18–23. [CrossRef]

105. Umemura, M.; Kim, J.H.; Aoyama, H.; Hoshino, Y.; Fukumura, H.; Nakakaji, R.; Sato, I.; Ohtake, M.; Akimoto, T.; Narikawa, M.; et al. The iron chelating agent, deferoxamine detoxifies Fe(Salen)-induced cytotoxicity. *J. Pharm. Sci.* **2017**, *134*, 203–210. [CrossRef]

106. Rateb, M.E.; Zhai, Y.; Ehrner, E.; Rath, C.M.; Wang, X.; Tabudravu, J.; Ebel, R.; Bibb, M.; Kyeremeh, K.; Dorrestein, P.C.; et al. Legonaridin, a new member of linaridin RiPP from a Ghanaian *Streptomyces* isolate. *Org. Biomol. Chem.* **2015**, *13*, 9585–9592. [CrossRef]

107. Lee, E.R.; Blount, K.F.; Breaker, R.R. Roseoflavin is a natural antibacterial compound that binds to FMN riboswitches and regulates gene expression. *RNA Biol.* **2009**, *6*, 187–194. [CrossRef]

108. Parks, D.H.; Imelfort, M.; Skennerton, C.T.; Hugenholtz, P.; Tyson, G.W. CheckM: Assessing the quality of microbial genomes recovered from isolates, single cells, and metagenomes. *Genome Res.* **2015**, *25*, 1043–1055. [CrossRef]

109. Lagesen, K.; Hallin, P.; Rodland, E.A.; Staerfeldt, H.H.; Rognes, T.; Ussery, D.W. RNAmmer: Consistent and rapid annotation of ribosomal RNA genes. *Nucleic Acids Res.* **2007**, *35*, 3100–3108. [CrossRef]

110. Lowe, T.M.; Chan, P.P. tRNAscan-SE On-line: Integrating search and context for analysis of transfer RNA genes. *Nucleic Acids Res.* **2016**, *44*, W54–W57. [CrossRef]

111. Overbeek, R.; Olson, R.; Pusch, G.D.; Olsen, G.J.; Davis, J.J.; Disz, T.; Edwards, R.A.; Gerdes, S.; Parrello, B.; Shukla, M.; et al. The SEED and the Rapid Annotation of microbial genomes using Subsystems Technology (RAST). *Nucleic Acids Res.* **2014**, *42*, D206–D214. [CrossRef]

112. Blin, K.; Shaw, S.; Steinke, K.; Villebro, R.; Ziemert, N.; Lee, S.Y.; Medema, M.H.; Weber, T. antiSMASH 5.0: Updates to the secondary metabolite genome mining pipeline. *Nucleic Acids Res.* **2019**, *47*, W81–W87. [CrossRef]

113. Moriya, Y.; Itoh, M.; Okuda, S.; Yoshizawa, A.C.; Kanehisa, M. KAAS: An automatic genome annotation and pathway reconstruction server. *Nucleic Acids Res.* **2007**, *35*, W182–W185. [CrossRef]

114. Wu, S.; Zhu, Z.; Fu, L.; Niu, B.; Li, W. WebMGA: A customizable web server for fast metagenomic sequence analysis. *BMC Genom.* **2011**, *12*, 444. [CrossRef]

115. Lee, I.; Ouk Kim, Y.; Park, S.C.; Chun, J. OrthoANI: An improved algorithm and software for calculating average nucleotide identity. *Int. J. Syst. Evol. Microbiol.* **2016**, *66*, 1100–1103. [CrossRef]

116. Amin, A.; Ahmed, I.; Khalid, N.; Osman, G.; Khan, I.U.; Xiao, M.; Li, W.J. *Streptomyces caldifontis* sp. nov., isolated from a hot water spring of Tatta Pani, Kotli, Pakistan. *Antonie van Leeuwenhoek* **2017**, *110*, 77–86. [CrossRef]

117. Cao, T.; Mu, S.; Lu, C.; Zhao, S.; Li, D.; Yan, K.; Xiang, W.; Liu, C. *Streptomyces amphotericinicus* sp. nov., an amphotericin-producing actinomycete isolated from the head of an ant (*Camponotus japonicus* Mayr). *Int. J. Syst. Evol. Microbiol.* **2017**, *67*, 4967–4973. [CrossRef]

118. Take, A.; Inahashi, Y.; Omura, S.; Takahashi, Y.; Matsumoto, A. *Streptomyces boninensis* sp. nov., isolated from soil from a limestone cave in the Ogasawara Islands. *Int. J. Syst. Evol. Microbiol.* **2018**, *68*, 1795–1799. [CrossRef]

119. Xu, L.; Wu, Y.H.; Zhou, P.; Cheng, H.; Liu, Q.; Xu, X.W. Investigation of the thermophilic mechanism in the genus *Porphyrobacter* by comparative genomic analysis. *BMC Genom.* **2018**, *19*, 385. [CrossRef]

120. Lechner, M.; Findeiss, S.; Steiner, L.; Marz, M.; Stadler, P.F.; Prohaska, S.J. Proteinortho: Detection of (co-)orthologs in large-scale analysis. *BMC Bioinform.* **2011**, *12*, 124. [CrossRef]

121. Jordan, I.K.; Rogozin, I.B.; Wolf, Y.I.; Koonin, E.V. Essential genes are more evolutionarily conserved than are nonessential genes in bacteria. *Genome Res.* **2002**, *12*, 962–968. [CrossRef]

122. Katoh, K.; Standley, D.M. MAFFT multiple sequence alignment software version 7: Improvements in performance and usability. *Mol. Biol. Evol.* **2013**, *30*, 772–780. [CrossRef]

123. Capella-Gutiérrez, S.; Silla-Martínez, J.M.; Gabaldón, T. trimAl: A tool for automated alignment trimming in large-scale phylogenetic analyses. *Bioinformatics* **2009**, *25*, 1972–1973. [CrossRef]

124. Felsenstein, J. Evolutionary trees from DNA sequences: A maximum likelihood approach. *J. Mol. Evol.* **1981**, *17*, 368–376. [CrossRef]

125. Nguyen, L.-T.; Schmidt, H.A.; von Haeseler, A.; Minh, B.Q. IQ-TREE: A fast and effective stochastic algorithm for estimating maximum-likelihood phylogenies. *Mol. Biol. Evol.* **2014**, *32*, 268–274. [CrossRef]

126. Hoang, D.T.; Chernomor, O.; von Haeseler, A.; Minh, B.Q.; Vinh, L.S. UFBoot2: Improving the Ultrafast Bootstrap Approximation. *Mol. Biol. Evol.* **2018**, *35*, 518–522. [CrossRef]

127. Altschul, S.F.; Gish, W.; Miller, W.; Myers, E.W.; Lipman, D.J. Basic local alignment search tool. *J. Mol. Biol.* **1990**, *215*, 403–410. [CrossRef]

128. Ihaka, R.; Gentleman, R. R: A language for data analysis and graphics. *J. Comput. Graph. Stat.* **1996**, *5*, 299–314.

129. Kumar, S.; Stecher, G.; Tamura, K. MEGA7: Molecular Evolutionary Genetics Analysis Version 7.0 for Bigger Datasets. *Mol. Biol. Evol.* **2016**, *33*, 1870–1874. [CrossRef]

marine drugs

MDPI

Article

Producing Novel Fibrinolytic Isoindolinone Derivatives in Marine Fungus *Stachybotrys longispora* FG216 by the Rational Supply of Amino Compounds According to Its Biosynthesis Pathway

Ying Yin [1,4], Qiang Fu [2], Wenhui Wu [2], Menghao Cai [1], Xiangshan Zhou [1] and Yuanxing Zhang [1,3,*]

1   State Key Laboratory of Bioreactor Engineering, East China University of Science and Technology, 130 Meilong Road, Shanghai 200237, China; yingy@sibs.ac.cn (Y.Y.); cmh022199@ecust.edu.cn (M.C.); xszhou@ecust.edu.cn (X.Z.)
2   College of Food Biotechnology, Shanghai Ocean University, 999 Huchenghuan Road, Shanghai 201306, China; m140208432@st.shou.edu.cn (Q.F.); whwu@shou.edu.cn (W.W.)
3   Shanghai Collaborative Innovation Center for Biomanufacturing, 130 Meilong Road, Shanghai 200237, China
4   Institute of Plant Physiology and Ecology, Chinese Academy of Sciences, 300 Feng Lin Road, Shanghai 200032, China
*   Correspondence: yxzhang@ecust.edu.cn; Tel.: +86-21-6425-3025

Academic Editor: Orazio Taglialatela-Scafati
Received: 22 March 2017; Accepted: 3 July 2017; Published: 5 July 2017

Abstract: Many fungi in the *Stachybotrys* genus can produce various isoindolinone derivatives. These compounds are formed by a spontaneous reaction between a phthalic aldehyde precursor and an ammonium ion or amino compounds. In this study, we suggested the isoindolinone biosynthetic gene cluster in *Stachybotrys* by genome mining based on three reported core genes. Remarkably, there is an additional nitrate reductase (NR) gene copy in the proposed cluster. NR is the rate-limiting enzyme of nitrate reduction. Accordingly, this cluster was speculated to play a role in the balance of ammonium ion concentration in *Stachybotrys*. Ammonium ions can be replaced by different amino compounds to create structural diversity in the biosynthetic process of isoindolinone. We tested a rational supply of amino compounds ((±)-3-amino-2-piperidinone, glycine, and L-threonine) in the culture of an isoindolinone high-producing marine fungus, *Stachybotrys longispora* FG216. As a result, we obtained four new kinds of isoindolinone derivatives (FGFC4–GFC7) by this method. Furthermore, high yields of FGFC4–FGFC7 confirmed the outstanding production capacity of FG216. Among the four new isoindolinone derivatives, FGFC6 and FGFC7 showed promising fibrinolytic activities. The knowledge of biosynthesis pathways may be an important attribute for the discovery of novel bioactive marine natural products.

Keywords: *Stachybotrys*; isoindolinone biosynthesis; genome mining; amino compound; fibrinolytic activity

---

1. Introduction

"Supply problem" is a major bottleneck in the discovery of marine drugs [1], and many research efforts on marine natural products (MNPs) are currently directed towards microorganisms. Marine microorganisms became the main source of novel MNPs (677 out of 1378) in 2014. Among them, marine fungi are the most talented MNP producers (426 out of 677) [2]. Considering their great ability to produce new compounds, marine fungi should be investigated more carefully. For instance, an increasing amount of information on fungi genomes is now available, which facilitates the discovery of biosynthesis pathways by genome mining. The knowledge of biosynthesis pathways is beneficial for MNP production.

Since a high fat and high calorie diet has been a part of modern life, stroke becomes a serious health risk for humans, especially when they get old. The plasminogen/plasmin system plays a key role in thrombolysis [3]. Therefore, small-molecule modulators of plasminogen activation have attracted increasing attention in cardiovascular drug development [4]. Many isoindolinone derivatives isolated from *Stachybotrys microspora* IFO 30018 (known as *Stachybotrys microspora* triprenyl phenols, SMTPs) are plasminogen activators [5–8]. They share a common triprenyl isoindolinone core unit. Among them, SMTP-7 (see Figure 1) has shown the best fibrinolytic effect [7,9]. Most attractively, it was effective in treating thrombotic stroke in primates [10]. Marine fungus *S. longispora* FG216 can produce fungi fibrinolytic compound 1 (FGFC1), which has the same structure as SMTP-7 [11]. In addition, the production capacity of FG216 is outstanding. It has been shown to produce FGFC1 on a 10 g/L level under an optimized glucose/ornithine replenishment strategy [12]. FGFC1 has also shown positive effects on thrombolysis and hemorrhagic activities, both in vitro and in vivo [13]. Thus, FG216 has the potential for manufacturing cardiovascular drugs.

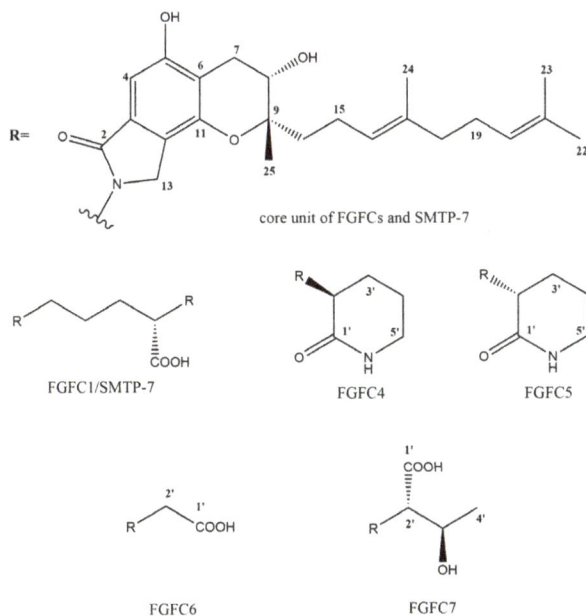

Figure 1. The structures of FGFCs and SMTP-7.

The precursor of all SMTPs has been reported to be pre-SMTP, which is a kind of prenylated phthalic aldehyde, and pre-SMTP is derived from a meroterpenoid ilicicolin B (also called LL-Z1272β) [14]. Recently, three genes responsible for the biosynthesis of ilicicolin B have been discovered in *S. bisbyi* [15]. In this study, we tried to ascertain the complete biosynthetic gene cluster for FGFCs or SMTPs by genome mining based on the three reported core genes. According to the biosynthesis pathway, a rational supply strategy of amino compounds was developed for the production of other novel isoindolinone derivatives. Since FGFC1 is a high yield compound, abundant new derivatives could also be easily obtained. Additionally, their fibrinolytic activities were also tested to evaluate the potential of FG216 for manufacturing cardiovascular drugs more comprehensively.

2. Results and Discussion

2.1. Isoindolinone Biosynthesis Pathway in Stachybotrys

As mentioned above, the biosynthesis pathway of ilicicolin B, the common precursor of all isoindolinone derivatives in *Stachybotrys*, has recently been discovered in *S. bisbyi* PYH05-7 [15]. A non-reducing polyketide synthase (NR-PKS) StbA, a putative UbiA-like prenyltransferase (PT) StbC, and a non-ribosomal peptide synthase (NRPS)-like enzyme StbB were reported to be successively involved. Although the genome information of *S. bisbyi* was not available, homologous genes of *stbA-C* could be found in the genomes of *S. chartarum* and *S. chlorohalonata* [15,16]. Therefore, we tried to identify the complete gene cluster by genome mining.

As a result, we found a cluster containing 10 genes in their genomes (see Table 1). The gene cluster in *S. chlorohalonata* IBT 40285 was named as *idlA-I, R*. Nine genes in this cluster showed a high protein identity (above 94%) to their homologues in *S. chartarum* IBT 7711. The identity between *idlA-C* and *stbA-C* was a bit lower, at about 75%. The *idl* gene cluster and the proposed biosynthesis processes of isoindolinone derivatives in *Stachybotrys* are shown in Figure 2. The three steps catalyzed by the three core genes are shown in an orange block. Orsellinic acid is synthesized by NR-PKS StbA/IdlA in the first step. Then, a molecule of farnesyl pyrophosphate (FPP) is transferred by PT StbC/IdlC to form grifolic acid. The carboxyl group in grifolic acid is reduced by the reducing (R) domain of the NRPS-like enzyme StbB/IdlB [15]. According to the structures of the reported isoindolinone derivatives in *Stachybotrys*, we predict that two modes of cyclization occur afterwards, which may result from epoxidation on different double bonds of the farnesyl group. For example, stachybotrins [17,18] and chartarutines [19] share the same cyclization mode with FGFC1 and SMTPs, while stachybotrylactams [20], spirodihydrobenzofuranlactams [21], phenylspirodrimanes [22], spirocyclic drimanes [23], and chartarlactams [24] share the other mode. No matter which mode of cyclization occurs, the methyl group of orsellinic acid will be oxidized to an aldehyde group. The phthalic aldehyde precursor then combines with an ammonium ion or amino compounds to form different isoindolinone derivatives.

Table 1. Predicted gene cluster responsible for isoindolinone biosynthesis in *Stachybotrys*. The gene cluster in *S. chlorohalonata* IBT 40285 was compared to its homologues in *S. chartarum* IBT 7711 and *S. bisbyi* PYH05-7.

Gene	Locus Tag	*S. chartarum* IBT 7711 Homologue	Protein Identity (%)	*S. bisbyi* PYH05-7 Homologue	Protein Identity (%)	Putative Function
idlR	S40285_07604	S7711_05923	98			Transcriptional regulator
idlI	S40285_07605	S7711_05924	78 (34% coverage)			Nitrate reductase (partial in *S. chartarum* IBT 7711)
		S7711_05925				Carboxylesterase
idlH	S40285_07606	S7711_05926	97			Esterase
idlG	S40285_07607	S7711_05927	100			Short-chain dehydrogenase
idlF	S40285_07608	S7711_05928	94			Isomerase/epimerase
idlE	S40285_07609	S7711_05929	97			Copper dependent oxidase
idlD	S40285_07610	S7711_05930	96			Short-chain dehydrogenase
idlC	S40285_10521	S7711_10996	98	*stbC*	76	PT
idlB	S40285_07611	S7711_05931	98	*stbB*	73	NRPS-like
idlA	S40285_07612	S7711_05932	98	*stbA*	75	NR-PKS

Homologous genes of NR-PKS, NRPS-like enzyme, and two putative short-chain dehydrogenases can be found in the gene cluster responsible for cichorine biosynthesis in the well-studied fungus *Aspergillus nidulans* [25,26], confirming their involvement in the biosynthesis of an isoindolinone skeleton. There is no prenylated structure in cichorine, so there are no homologues of the PT gene in this cluster. On the other hand, the homologous gene of PT can be found in another cluster responsible for the biosynthesis of meroterpenoids austinol and dehydroaustinol in *A. nidulans* [27]. A similar gene can also be found in the terretonin cluster in *A. terreus* [28]. These genes are related to the formation of the prenylated structure.

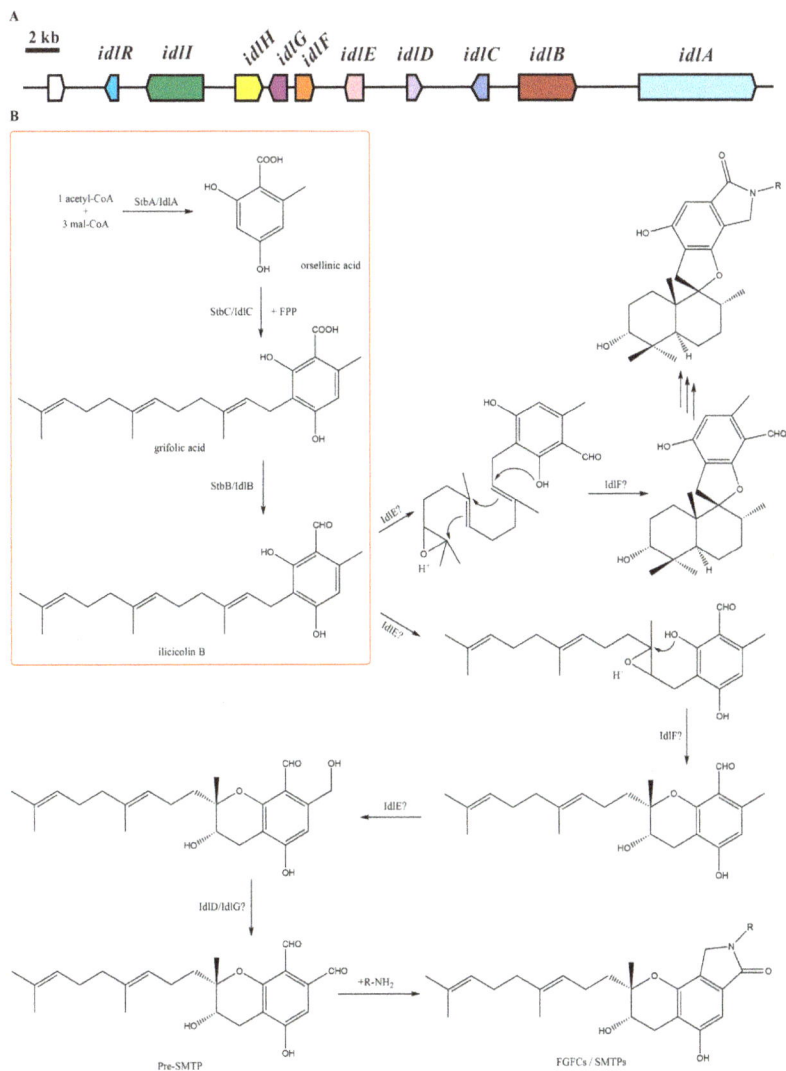

Figure 2. Proposed gene cluster and biosynthesis processes of isoindolinone derivatives in *Stachybotrys*. (**A**) Predicted gene cluster responsible for isoindolinone biosynthesis in *S. chlorohalonata* IBT 40285. (**B**) Proposed biosynthesis processes of isoindolinone derivatives in *Stachybotrys*. The three steps catalyzed by the three core genes are shown in an orange block.

There is a nitrate reductase (NR) gene in the cluster of *S. chlorohalonata* IBT 40285, while its homologue in the cluster of *S. chartarum* IBT 7711 lacks a molybdopterin binding domain. Futhermore, there is a carboxylesterase gene in the cluster of IBT 7711 that does not occur in IBT 40285. We speculated that mutations might occur in this locus in the evolution process of *S. chartarum*. On the other hand, there is another intact NR gene (not in this cluster) in both genomes of IBT 40285 and IBT 7711. This suggests that there is or used to be an additional copy of the NR gene in the isoindolinone biosynthesis cluster in *Stachybotrys*.

An additional gene is probably a resistance gene. In the case of *A. terreus*, there is an additional copy of the 3-hydroxy-3-methylglutaryl-coenzyme A (HMG-CoA) reductase gene in the lovastatin (HMG-CoA reductase inhibitor) cluster, and there is an additional copy of the ATP synthase β-chain gene in the citreoviridin (ATP synthase β-chain inhibitor) cluster [29]. As we mentioned, an isoindolinone skeleton is formed by the reaction between the phthalic aldehyde precursor produced by the cluster and ammonium ion or amino compounds. This reaction is reported to happen spontaneously [14], so ammonium ions will be wiped off by the product of this cluster. In contrast, NR is the rate-limiting enzyme of nitrate reduction, which is an important step in the formation of ammonium. Consequently, we speculated that this cluster might play a role in the balance of ammonium ion concentration in *Stachybotrys*.

2.2. Production of New FGFCs by Rational Amino Compounds Supply

To generate novel isoindolinone derivatives, the spontaneous reaction between the phthalic aldehyde and an ammonim ion can be manipulated by substituting various amino compounds. In other words, it is likely to control the structures of the isoindolinone derivatives by the supply of amino compounds. In the study of SMTPs, amino acids [6], glucosamine, sulfanilic acid [7], phenylamine [8], phenylglycine [30], and naphthenic amine [31] were used as amino precursors. Likewise, a high level production of FGFC1 was achieved in FG216 by a sufficient L-ornithine supply [12]. So novel isoindolinone derivatives were expected to be obtained through the rational supply of amino compounds. In this study, we tested four amino compound precursors (3-amino-2-piperidinone, (3S)-3-amino-2-piperidinone hydrochloride, glycine, and L-threonine). 3-Amino-2-piperidinone is the lactone of L-ornithine. Glycine and L-threonine are amino acids. They are all analogues of L-ornithine, so the isoindolinone derivatives derived from them are comparable with FGFC1 in fibrinolytic activities.

The fermentation medium used in this study was the optimized medium for FGFC1 production [12] without an L-ornithine hydrochloride supply. The meroterpenoid precursor for FGFCs could be produced on a high level in this medium, and new isoindolinone derivatives could be easily detected when other amino compounds were supplied. As shown in Figure 3, obvious new product peaks can be observed compared with the no supply control (Figure 3e).

3-Amino-2-piperidinone is a racemic mixture, so two adjacent peaks occur in Figure 3a. FGFC4 was recognized as the product of (3S)-3-amino-2-piperidinone, and FGFC5 was recognized as the product of (3R)-3-amino-2-piperidinone by comparing it with the optical product derived by (3S)-3-amino-2-piperidinone hydrochloride supply (Figure 3b). Considering the possible toxicity, 3-amino-2-piperidinone and its hydrochloride were only fed on 0.2% concentration. Consequently, we collected a 238 mg mixture of FGFC4 and FGFC5 (feeding 3-amino-2-piperidinone) and 163 mg FGFC4 (feeding (3S)-3-amino-2-piperidinone hydrochloride) from 100 mL broth, respectively. In our former study, 913 mg FGFC1 could be derived from 100 mL broth, when 1.3% L-ornithine hydrochloride was fed. Therefore, the feeding concentration of 3-amino-2-piperidinone could be further increased to achieve a higher production under the premise of not affecting the growth of FG216. Some mixture (FGFC4 and FGFC5) was further separated, and 40 mg FGFC5 was collected.

Glycine (Figure 3c) and L-threonine (Figure 3d) were fed on 0.5% concentration, but only 62 mg FGFC6 and 72 mg FGFC7 were derived from 100 mL broth, respectively. We predict that large amounts of these two amino acids are probably utilized for cell growth and energy metabolism, since they are essential amino acids. Therefore, the way to feed essential amino acids may need further modification, e.g., feeding in the beginning of the stable growth phase.

Overall, the gram level per liter production of FGFC4–FGFC7 derived by a non-optimized method was an inspiring start, which confirmed the outstanding production capacity of FG216.

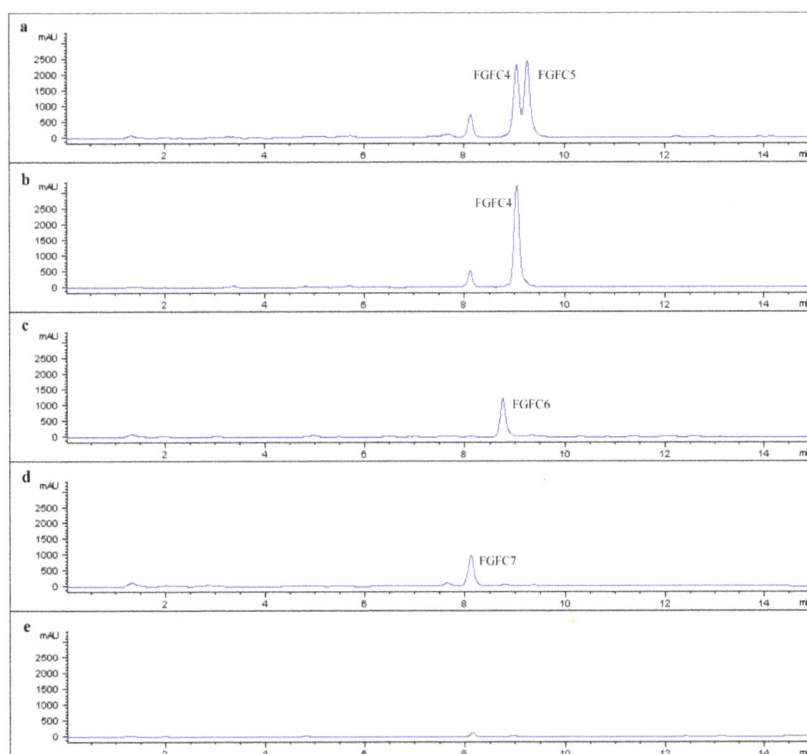

Figure 3. HPLC assay of *S. longispora* FG216 metabolites with different amino compounds supply. ((**a**) 3-amino-2-piperidinone, (**b**) (3*S*)-3-amino-2-piperidinone hydrochloride, (**c**) glycine, (**d**) L-threonine, and (**e**) no supply control.)

2.3. Structure Determination of New FGFCs

FGFC4–FGFC7 all had the same UV absorption feature (λ_{max} 214, 258, 302 nm) as FGFC1, so we speculated that they had the same core unit as FGFC1 (Figure 1). As mentioned above, the *N*-linked side chains of FGFC4–FGFC7 were derived from (3*S/R*)-3-amino-2-piperidinone, glycine and L-threonine, respectively. Thus, we suggested the structures of FGFC4–FGFC7 in Figure 1. Then, we verified this speculation by ESI-TOF-MS and NMR. The NMR data of FGFC4–FGFC7 is shown in Table 2 and Figures S1–S14, and the HMBC correlations are shown in Figure 4. The NMR data also inferred that the four compounds shared a common core unit, which was also consistent with that of FGFC1 [11] or SMTP-7 [32].

The molecular formula of FGFC4 was determined as $C_{28}H_{38}O_5N_2$ according to the ESI-TOF-MS peaks at *m/z* 483.2813 [M + H]$^+$ and the NMR data. Six aromatic singlets in the ^{13}C NMR spectrum at δ_C 132.4 (C-3), 100.9 (C-4), 158.0 (C-5), 113.5 (C-6), 150.0 (C-11), and 121.9 (C-12), along with an aromatic singlet in the ^1H NMR spectrum at δ_H 6.77 (1H, s, H-4), suggested the signal of a benzene ring, which was consistent with the UV absorption at 258 and 302 nm. HMBC interactions from H-4 to C-5, C-6, and C-11 also confirmed the existence of a benzene ring. Among these aromatic carbons, C-5 and C-11 were supposed to link with a hydroxy group or oxygen atom, according to their higher chemical shift. H-4 showed a correlation to a carbonyl carbon C-2 (δ_C 171.6). Two methylene protons, H-13-1 (δ_H 4.36, 1H, d) and H-13-2 (δ_H 4.24, 1H, d), showed correlations to C-2 and C-12. Furthermore, the methylene carbon C-13 (δ_C 46.8) was likely to link with the nitrogen atom because of its high chemical shift. Thus, an isoindolinone nucleus was recognized. Two methylene protons, H-7-1 (δ_H 3.00, 1H, dd) and H-7-2 (δ_H 2.67, 1H, dd), showed correlations to two aromatic carbons, C-6 and C-11, and a methylidyne carbon, C-8 (δ_C 68.3). The methylidyne proton H-8 (δ_H 3.90, 1H, dd) had a correlation to a quaternary carbon, C-9 (δ_C 80.2). Both C-8 and C-9 exhibited a high chemical shift, indicating that they linked with a hydroxy group or oxygen atom. According to the core unit structure of FGFC1, C-9 and C-11 were linked by an oxygen atom, constituting a pyran ring. Two methylene protons H$_2$-14 (δ_H 1.69, 2H, m) and three methyl protons H$_3$-25 (δ_H 1.29, 3H, s) showed correlations to the pyran ring at C-9. These two methylene protons H$_2$-14 also showed correlations to a geranyl, which had an ^{1}H NMR spectrum of δ_H 2.20 (2H, m, H$_2$-15), 5.16 (1H, t, H-16), 1.99 (2H, m, H$_2$-18), 2.07 (2H, m, H$_2$-19), 5.10 (1H, t, H-20), 1.67 (3H, s, H$_3$-22), 1.59 (3H, s, H$_3$-23), and 1.60 (3H, s, H$_3$-24), and an ^{13}C NMR spectrum of δ_C 22.6 (C-15), 125.5 (C-16), 136.3 (C-17), 40.8 (C-18), 27.7 (C-19), 125.4 (C-20), 132.2 (C-21), 25.9 (C-22), 17.8 (C-23), 16.0 (C-24). Hence, geranyl linked to the pyran ring through a methylene. As for the structure of the *N*-linked side chain, it showed a high consistency with 3-amino-2-piperidinone. The central methylidyne proton H-2′ (δ_H 4.80, 1H, dd) showed correlations to C-2 and C-13, confirming the linkage between the *N*-linked side chain and isoindolinone nucleus. It also showed a correlation to a carbonyl carbon, C-1′ (δ_C 171.7), which had a similar chemical shift to C-2. Among the three methylenes, C-5′ (δ_C 43.0) had the highest chemical shift, suggesting that it was the one linked with acylamino. FGFC5 showed the same molecular weight and NMR spectrum as FGFC4, which confirmed that they were chiral isomers.

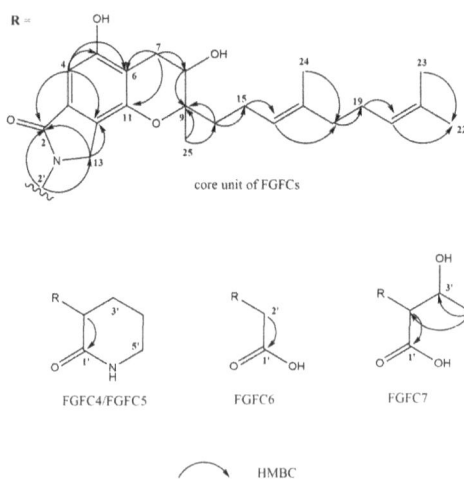

Figure 4. The HMBC correlation data of FGFC4–FGFC7.

The molecular formula of FGFC6 was determined as $C_{25}H_{33}O_6N$ according to the ESI-TOF-MS peak at *m/z* 442.2233 [M − H]$^-$ and the NMR data. FGFC6 had the same core unit as FGFC4 and FGFC5. The structure of its *N*-linked side chain was very simple, since the amino compound precursor was glycine. The methylene protons H$_2$-2′ (δ_H 4.34, 2H, s) showed correlations to C-2, C-13, and carboxyl carbon C-1′ (δ_C 171.7).

The molecular formula of FGFC7 was determined as $C_{27}H_{37}O_7N$ according to the ESI-TOF-MS peak at *m/z* 486.2544 [M − H]$^-$ and the NMR data. FGFC7 also had the same isoindolinone core unit. Its *N*-linked side chain was derived from L-threonine. The central methylidyne proton H-2′ (δ_H 4.89, 1H, d) also showed correlations to C-2, C-13, and C-1′ (δ_C 173.0). The methylidyne carbon C-3′ (δ_C 68.9) linked with a hydroxy group. The methyl protons H$_3$-4′ (δ_H 1.20, 3H, d) showed correlations to both C-2′ (δ_C 61.4) and C-3′.

Table 2. ^{1}H (400 MHz) and ^{13}C NMR (100 MHz) data for new FGFC congeners (in MeOH-d_4, J in Hz).

No.	FGFC4 δC	FGFC4 δH	FGFC5 δC	FGFC5 δH	FGFC6 δC	FGFC6 δH	FGFC7 δC	FGFC7 δH
2	171.6, C		171.5, C		171.7, C		172.6, C	
3	132.4, C		132.4, C		132.2, C		131.7, C	
4	100.9, CH	6.77 s	101.0, CH	6.77 s	101.1, CH	6.78 s	101.0, CH	6.80 s
5	158.0, C		158.2, C		158.0, C		157.8, C	
6	113.5, C		113.6, C		113.6, C		113.6, C	
7	27.7, CH$_2$	3.00 dd (17.6, 5.4), 2.67 dd (17.6, 7.1)	27.8, CH$_2$	3.01 dd (17.7, 5.5), 2.66 dd (17.7, 7.4)	27.8, CH$_2$	3.00 dd (17.7, 5.4), 2.68 dd (17.7, 7.0)	27.8, CH$_2$	3.01 dd (17.7, 5.3), 2.70 dd (17.7, 6.8)
8	68.3, CH	3.90 dd (7.0, 5.6)	68.5, CH	3.89 dd (7.2, 5.7)	68.4, CH	3.90 t (6.2)	68.4, CH	3.91 t (6.0)
9	80.2, C		80.2, C		80.3, C		80.2, C	
11	150.0, C		150.0, C		150.0, C		150.0, C	
12	121.9, C		121.9, C		122.2, C		123.1, C	
13	46.8, CH$_2$	4.36 d (16.7), 4.24 d (16.7)	46.9, CH$_2$	4.34 d (16.7), 4.25 d (16.7)	49.6, CH$_2$	4.39 s	48.6, CH$_2$	4.65 m
14	38.6, CH$_2$	1.69 m	38.6, CH$_2$	1.71 m	38.6, CH$_2$	1.69 m	38.5, CH$_2$	1.70 m
15	22.6, CH$_2$	2.20 m	22.6, CH$_2$	2.21 m	22.6, CH$_2$	2.20 m	22.6, CH$_2$	2.19 m
16	125.5, CH	5.16 t (6.8)	125.6, CH	5.17 t (6.7)	125.5, CH	5.16 t (6.9)	125.4, CH	5.16 t (6.9)
17	136.3, C		136.2, C		136.3, C		136.3, C	
18	40.8, CH$_2$	1.99 m	40.9, CH$_2$	1.99 m	40.8, CH$_2$	1.98 m	40.8, CH$_2$	1.97 m
19	27.7, CH$_2$	2.07 m	27.7, CH$_2$	2.07 m	27.7, CH$_2$	2.07 m	27.7, CH$_2$	2.07 m
20	125.4, CH	5.10 t (7.0)	125.4, CH	5.09 t (6.9)	125.4, CH	5.08 t (6.9)	125.4, CH	5.07 t (6.9)
21	132.2, C		132.2, C		132.2, C		132.2, C	
22	25.9, CH$_3$	1.67 s	25.9, CH$_3$	1.67 s	25.9, CH$_3$	1.66 s	25.9, CH$_3$	1.65 s
23	17.8, CH$_3$	1.59 s	17.8, CH$_3$	1.59 s	17.8, CH$_3$	1.58 s	17.8, CH$_3$	1.57 s
24	16.0, CH$_3$	1.60 s	16.0, CH$_3$	1.61 s	16.0, CH$_3$	1.60 s	16.0, CH$_3$	1.58 s
25	18.8, CH$_3$	1.29 m	18.6, CH$_3$	1.28 s	18.8, CH$_3$	1.30 s	19.0, CH$_3$	1.31 s
1'	171.7, C		171.8, C		171.7, C		173.0, C	
2'	53.6, CH	4.80 dd (10.6, 7.1)	53.6, CH	4.78 dd (10.6, 6.9)	45.0, CH$_2$	4.34 s	61.4, CH	4.89 d (3.8)
3'	27.3, CH$_2$	2.13 m	27.3, CH$_2$	2.14 m			68.2, CH	4.62 m
4'	22.9, CH$_2$	2.03 m	22.9, CH$_2$	2.03 m			20.7, CH$_3$	1.20 d (6.3)
5'	43.0, CH$_2$	3.37 brs	43.0, CH$_2$	3.37 brs				

2.4. Fibrinolytic Activities of New FGFCs

The reciprocal activation of prourokinase (pro-uPA) and plasminogen (plg) is believed to play a key role in tissue fibrinolysis. Compared with urokinase-type plasminogen activator (uPA), pro-uPA has a weak intrinsic activity, which can convert plg into plasmin (plm) [33,34]. Plasmin then converts pro-uPA into activeuPA. Accordingly, more plg can be converted into plm by uPA. A fibrinolytic reaction occurs.

This reciprocal activation process can be sped up when more pro-uPA is supplied. In addition, a small molecule fibrinolytic compound can further promote the reciprocal activation of pro-uPA and plg. In this study, we tested the fibrinolytic activities of FGFC4–FGFC7. The rate of urokinase catalyzing chromogenic substrate S-2444 to form nitroaniline was set as a standard to measure the activation degree of the enzymatic reaction system. The relative reaction rate of the negative control (enzymatic reaction system with no isoindolinone derivatives or extra pro-uPA) in the first 50 min was set as 1. The folds of the reaction rate enhanced by positive controls (extra 30 nmol/L pro-uPA and 0.1 g/L FGFC1, which were the most suitable working concentrations for them respectively) and test samples (different concentrations of FGFC4–FGFC7) were determined as relative activities.

It was a pity that the supplies of FGFC4 and FGFC5 did not promote the enzyme reaction rate. The relative activities of FGFC6 and FGFC7 under different concentrations are shown in Figure 5. The best working concentrations for FGFC6 and FGFC7 were both 0.025 g/L, which was much lower than that of FGFC1 (0.1 g/L), suggesting good application prospects for them. The relative activity of 0.025 g/L FGFC7 was 6.90, which was much higher than that of 0.1 g/L FGFC1 (3.50, $p < 0.05$), and roughly equivalent to that of extra 30 nmol/L pro-uPA (6.46). Additionally, the relative activity of 0.025 g/L FGFC6 was 3.86, which was roughly equivalent to that of 0.1 g/L FGFC1. Moreover, 0.1 g/L FGFC7 also showed a higher relative activity (5.41) than 0.1 g/L FGFC1 ($p < 0.05$). Some other concentrations (0.01 g/L FGFC6 and FGFC7, 0.1 g/L FGFC6 and 0.4 g/L FGFC7) showed similar relative activities to 0.1 g/L FGFC1. In contrast, high concentrations of FGFC6 and FGFC7 did not show good effects, especially 1 g/L FGFC6 and FGFC7, which did not exhibit a significant difference to the negative control.

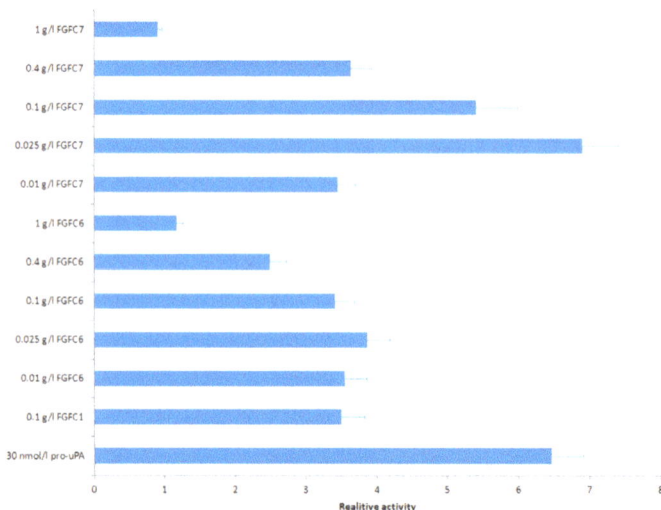

Figure 5. The relative activities of FGFC6 and FGFC7 in different concentrations. (Extra 30 nmol/L pro-uPA and 0.1 g/L FGFC1 were set as positive controls, respectively. The relative activity of no supply negative control was set as 1.)

Diversity of structure resulted in the diversity of bioactivity. Hasegawa et al. [7] summarized that negative ionization groups such as the carboxyl or sulfonic acid group in the *N*-linked side chain of SMTPs was a necessary basis for their plasminogen activation activity. Our results agreed with this conclusion. When ornithine was lactonized to 3-amino-2-piperidinone, the loss of the carboxyl acid group in the *N*-linked side chain resulted in inactive FGFC4 and FGFC5. However, FGFC6 and FGFC7, which were gained from amino acids, showed excellent activities that were even better than FGFC1, exhibiting potential application prospects.

Besides plasminogen activation activity, isoindolinone derivatives also showed other important bioactivities. For example, many SMTPs showed anti-inflammatory activities, which were attributable to their abilities to inhibit soluble epoxide hydrolase (sEH) [31]. Stachybotrin C was a neuritogenic compound [35]. Chartarutines B, G, and H exhibited significant inhibitory effects against HIV-1 virus [18]. Stachybotrylactams were immunosuppressant [19]. Spirodihydrobenzofuranlactams showed antagonistic effects in the endothelin receptor binding assay [36]. Spirocyclic drimanes exhibited antibacterial activity against the clinically relevant methicillin-resistant *Staphylococcus aureus* (MRSA) [22]. Chartarlactams showed antihyperlipidemic effects in HepG2 cells [23]. Therefore, these new FGFCs will be detected for other bioactivities in the future in order to understand their application prospects more adequately.

In summary, we explored the isoindolinone biosynthesis pathway in *Stachybotrys*, and obtained four new isoindolinone derivatives (FGFC4–FGFC7) from marine fungus *S. longispora* FG216 by the rational supply of amino compounds. Enantiomers FGFC4 and FGFC5 were achieved by feeding racemic mixture 3-amino-2-piperidinone. However, they did not show fibrinolytic activities for their lack of negative ionization groups in the *N*-linked side chains. FGFC6 and FGFC7 were achieved by feeding glycine and L-threonine, respectively, and they showed competitive fibrinolytic activities. Especially, FGFC7 had a comparable effect to pro-uPA, although fermentation optimization was needed to improve the yield.

3. Materials and Methods

3.1. General Experimental Procedures

Extraction was performed on an ultrasonicator (KQ-800, Kunshan Ultrasonic Instruments Co., Ltd., Kunshan, China). High performance liquid chromatography (HPLC) was performed on an Agilent 1200 HPLC system (Agilent Technologies, Santa Clara, CA, USA) with a C18 column (ZORBAX Eclipse XDB, 150 mm × 4.6 mm, 5 μm, 100 Å-spherical silica) and a semipreparation C18 column (ZORBAX Eclipse XDB, 250 mm × 9.4 mm, 5 μm, 100 Å-spherical silica). The molecular weight was determined by the Agilent electrospray ionization time of flight mass spectrometry (ESI-TOF-MS) 6230. A nuclear magnetic resonance (NMR) assay was performed on Bruker AM-400 spectrometers (Bruker Daltonics Inc., Billerica, MA, USA). Fibrinolytic activity was detected by a microplate reader (SH-1000, CORONA, Ibarakiken, Japan).

3.2. Strain, Medium, and Cultural Conditions

Stachybotrys longispora FG216 (CCTCCM 2012272) was kindly provided by Shanghai Ocean University.

The sporiparous medium was potato sucrose agar (PSA), which contained 200 g/L potato, 20 g/L sucrose, and 20 g/L agar. The seed medium contained 35 g/L glucose, 10 g/L soluble starch, 20 g/L defatted soybean flour (Sigma, St. Louis, MO, USA), 5 g/L bacteriological peptone (Oxoid, Basingstoke, UK), 5 g/L beef extract paste (Oxoid, Basingstoke, UK), 3 g/L yeast extract (Oxoid, Basingstoke, UK), 2 g/L NaCl, 0.5 g/L K_2HPO_4, and 0.05 g/L $MgSO_4$. The fermentation medium contained 125 g/L sucrose, 3.3 g/L $NaNO_3$, 0.7 g/L yeast extract (Oxoid, Basingstoke, UK), 0.625 g/L KCl, 0.4 g/L $MgSO_4 \cdot 7H_2O$, 0.07 g/L $K_2HPO_4 \cdot 3H_2O$, 18.75 mg/L $FeSO_4 \cdot 7H_2O$, 6.5 mg/L $CaCl_2$, and 3.125 mg/L

CoCl$_2$·6H$_2$O. The seed and fermentation media were adjusted to pH 5.8 before sterilization. All media were sterilized at 121 °C for 20 min.

FG216 was statically incubated on PSA at 25 °C for 7 d to induce spore formation. Then, the spores were washed with sterile water, and 1×10^6 spores were inoculated into 30 mL seed medium in a 250 mL Erlenmeyer flask and incubated at 25 °C, 180 rpm for 2.5 d to obtain the first-stage seed. Second-stage seed was gained by 1.5 mL of first-stage seed inoculated into 30 mL seed medium and incubated under the same conditions for 1 d. Finally, 1.5 mL second-stage seed was inoculated into 30 mL fermentation medium and incubated under the same conditions for 10 d.

3.3. Amino Compouds Supply

3-Amino-2-piperidinone was purchased from Ark Pharm, Inc. (Libertyville, IL, USA). (3S)-3-Amino-2-piperidinone hydrochloride was purchased from Shanghai Macklin Biochemical Co., Ltd. (Shanghai, China). Glycine and L-threonine were purchased from Shanghai Sangon Biotechnology Co., Ltd. (Shanghai, China).

Each of them was dissolved in distilled water (3-amino-2-piperidinone in ethanol), sterilized with a millipore filter, and fed into the fermentation medium at a concentration of 0.2% (3-amino-2- piperidinone and (3S)-3-amino-2-piperidinone hydrochloride) or 0.5% (glycine and L-threonine) on day 0.

3.4. Metabolites Detection and Isolation

One hundred milliliter of fermentation broth was centrifuged at 5000× g for 20 min, and the precipitate was extracted with 100 mL methanol. The mixture was placed in an ultrasonicator at a frequency of 40 kHz and a nominal power of 800 W for 30 min. The extraction was repeated three times. The crude extract was obtained by concentration in vacuo.

The crude extract was detected by an HPLC system and then separated with a semipreparation column. The operating temperature was 40 °C, the flow rate was 1 mL/min and 4 mL/min for detection and separation, respectively, and UV detection was set at 258 nm. The mobile phase was composed of acetonitrile and 0.1% formic acid solution. For detection, the ratio acetonitrile increased from 10 to 100% in 15 min. For separation, the ratio of acetonitrile was set as 50%. FGFC4, FGFC5, FGFC6, and FGFC7 were collected at 12.3 min, 13.3 min, 10.2 min, and 16.0 min, respectively.

3.5. Bioactivity Assays

Single chain urokinase-type plasminogen activator (pro-uPA), plasminogen (plg), and bovine serum albumin (BSA) were purchased from Sigma-Aldrich Co. (St. Louis, MO, USA). Chromogenic substrate S-2444 of urokinase was purchased from HYPHEN BioMed Co. (Neuville-sur-Oise, France).

These reagents were respectively dissolved in Tris-HCl buffer solution (50 mmol/L Tris-HCl, 100 mmol/L NaCl, pH 7.4). Their concentrations were as follows: 20 nmol/L pro-uPA, 10 nmol/L plg, 10 g/L BSA, and 2 mmol/L S-2444. FGFC4 and FGFC5 were dissolved in methanol. FGFC6 and FGFC7 were dissolved in sodium bicarbonate solution (pH 7.5). Ten microlitre of each solution was mixed in a 96 hole round bottom plate to form a 50 μL enzymatic reaction system. The nitroaniline was formed at 37 °C for 150 min, and detected in a microplate reader at 405 nm.

FGFC4–FGFC7 were tested on five concentrations (0.01, 0.025, 0.1, 0.4, 1 g/L). An enzymatic reaction system with no isoindolinone derivatives was set as the negative control. Enzymatic reaction systems with an extra 30 nmol/L pro-uPA (the best concentration for pro-uPA) or 0.1 g/L FGFC1 (the best concentration for FGFC1) were set as positive controls, respectively. The assay was repeated three times.

Supplementary Materials: The following are available online at www.mdpi.com/1660-3397/15/7/214/s1, Figure S1: ^{1}H NMR Spectrum of FGFC4 in MeOH-d_4, Figure S2: ^{13}C NMR Spectrum of FGFC4 in MeOH-d_4, Figure S3: HSQC Spectrum of FGFC4 in MeOH-d_4, Figure S4: HMBC Spectrum of FGFC4 in MeOH-d_4, Figure S5: ^{1}H NMR Spectrum of FGFC5 in MeOH-d_4, Figure S6: ^{13}C NMR Spectrum of FGFC5 in MeOH-d4, Figure S7: ^{1}H NMR Spectrum of FGFC6 in MeOH-d_4, Figure S8: ^{13}C NMR Spectrum of FGFC6 in MeOH-d_4, Figure S9: HSQC Spectrum of FGFC6 in MeOH-d_4, Figure S10: HMBC Spectrum of FGFC6 in MeOH-d_4, Figure S11: ^{1}H NMR

Spectrum of FGFC7 in MeOH-d_4, Figure S12: ^{13}C NMR Spectrum of FGFC7 in MeOH-d_4, Figure S13: HSQC Spectrum of FGFC7 in MeOH-d_4, Figure S14: HMBC Spectrum of FGFC7 in MeOH-d_4.

Acknowledgments: This work was financially supported by the Open Funding Project of the State Key Laboratory of Bioreactor Engineering, China.

Author Contributions: Ying Yin and Yuanxing Zhang conceived and designed the experiments; Ying Yin and Qiang Fu performed the experiments; Wenhui Wu and Menghao Cai analyzed the data; Xiangshan Zhou contributed reagents/materials/analysis tools; Ying Yin wrote the paper.

Conflicts of Interest: The authors declare no conflict of interest.

References

1. Leal, M.C.; Sheridan, C.; Osinga, R.; Dioníasio, G.; Rocha, R.J.M.; Silva, B.; Rosa, R.; Calado, R. Marine microorganism-invertebrate assemblages: Perspectives to solve the "supply problem" in the initial steps of drug discovery. *Mar. Drugs* **2014**, *12*, 3929–3952. [CrossRef] [PubMed]

2. Blunt, J.W.; Copp, B.R.; Keyzers, R.A.; Munro, M.H.G.; Prinsep, M.R. Marine natural products. *Nat. Prod. Rep.* **2016**, *33*, 382–431. [CrossRef] [PubMed]

3. Rijken, D.C.; Lijnen, H.R. New insights into the molecular mechanisms of the fibrinolytic system. *J. Thromb. Haemost.* **2009**, *7*, 4–13. [CrossRef] [PubMed]

4. Hasumi, K.; Yamamichi, S.; Harada, T. Small-molecule modulators of zymogen activation in the fibrinolytic and coagulation systems. *FEBS J.* **2010**, *277*, 3675–3687. [CrossRef] [PubMed]

5. Shinohara, C.; Hasumi, K.; Hatsumi, W.; Endo, A. Staplabin, a novel fungal triprenyl phenol which stimulates the binding of plasminogen to fibrin and U937 cells. *J. Antibiot.* **1996**, *49*, 961–966. [CrossRef] [PubMed]

6. Hu, W.-M.; Narasaki, R.; Ohyama, S.; Hasumi, K. Selective production of staplabin and SMTPs in cultures of *Stachybotrys microspora* fed with precursor amines. *J. Antibiot.* **2001**, *54*, 962–966. [CrossRef] [PubMed]

7. Hasegawa, K.; Koide, H.; Hu, W.-M.; Nishimura, N.; Narasaki, R.; Kitano, Y.; Hasumi, K. Structure–activity relationships of 11 new congeners of the SMTP plasminogen modulator. *J. Antibiot.* **2010**, *63*, 589–593. [CrossRef] [PubMed]

8. Koide, H.; Hasegawa, K.; Nishimura, N.; Narasaki, R.; Hasumi, K. A new series of the SMTP plasminogen modulators with a phenylamine-based side chain. *J. Antibiot.* **2012**, *65*, 361–367. [CrossRef] [PubMed]

9. Hu, W.-M.; Narasaki, R.; Nishimura, N.; Hasumi, K. SMTP (*Stachybotrys microspora* triprenyl phenol) enhances clot clearance in a pulmonary embolism model in rats. *Thromb. J.* **2012**, *10*, 1–9. [CrossRef] [PubMed]

10. Sawada, H.; Nishimura, N.; Suzuki, E.; Zhuang, J.; Hasegawa, K.; Takamatsu, H.; Honda, K.; Hasumi, K. SMTP-7, a novel small-molecule thrombolytic for ischemic stroke: A study in rodents and primates. *J. Cereb. Blood Flow Metab.* **2014**, *34*, 235–241. [CrossRef] [PubMed]

11. Wang, G.; Wu, W.-H.; Zhu, Q.-G.; Fu, S.-Q.; Wang, X.-Y.; Hong, S.-T.; Guo, R.-H.; Bao, B. Identification and fibrinolytic evaluation of an isoindolone derivative isolated from a rare marine fungus *Stachybotrys longispora* FG216. *Chin. J. Chem.* **2015**, *33*, 1089–1095. [CrossRef]

12. Wang, M.-X.; He, H.; Na, K.; Cai, M.-H.; Zhou, X.-S.; Zhao, W.-J.; Zhang, Y.-X. Designing novel glucose/ornithine replenishment strategies by biosynthetic and bioprocess analysis to improve fibrinolytic FGFC1 production by the marine fungus *Stachybotrys longispora*. *Process. Biochem.* **2015**, *50*, 2012–2018. [CrossRef]

13. Yan, T.; Wu, W.-H.; Su, T.-W.; Chen, J.-J.; Zhu, Q.-G.; Zhang, C.-Y.; Wang, X.-Y.; Bao, B. Effects of a novel marine natural product: Pyrano indolone alkaloid fibrinolytic compound on thrombolysis and hemorrhagic activities in vitro and in vivo. *Arch. Pharm. Res.* **2015**, *38*, 1530–1540. [CrossRef] [PubMed]

14. Nishimura, Y.; Suzuki, E.; Hasegawa, K.; Nishimura, N.; Kitano, Y.; Hasumi, K. Pre-SMTP, a key precursor for the biosynthesis of the SMTP plasminogen modulators. *J. Antibiot.* **2012**, *65*, 483–485. [CrossRef] [PubMed]

15. Li, C.; Matsuda, Y.; Gao, H.; Hu, D.; Yao, X.-S.; Abe, I. Biosynthesis of LL-Z1272β: Discovery of a new member of NRPS-like enzymes for aryl-aldehyde formation. *ChemBioChem* **2016**, *17*, 904–907. [CrossRef] [PubMed]

16. Semeiks, J.; Borek, D.; Otwinowski, Z.; Grishin, N.V. Comparative genome sequencing reveals chemotype-specific gene clusters in the toxigenic black mold *Stachybotrys*. *BMC Genom.* **2014**, *15*, 590. [CrossRef] [PubMed]

17. Xu, X.-M.; De Guzman, F.S.; Gloer, J.B.; Shearer, C.A. Stachybotrins A and B: Novel bioactive metabolites from a brackish water isolate of the fungus *Stachybotrys* sp. *J. Org. Chem.* **1992**, *57*, 6700–6703. [CrossRef]

18. Nozawa, Y.; Ito, M.; Sugawara, K.; Hanada, K.; Mizoue, K. Stachybotrin C and parvisporin, novel neuritogenic compounds. II. Structure determination. *J. Antibiot.* **1997**, *50*, 641–645. [CrossRef] [PubMed]
19. Li, Y.; Liu, D.; Cen, S.; Proksch, P.; Lin, W.-H. Isoindolinone-type alkaloids from the sponge-derived fungus *Stachybotrys chartarum*. *Tetrahedron* **2014**, *70*, 7010–7015. [CrossRef]
20. Jarvis, B.B.; Salemme, J.; Morals, A. *Stachybotrys* toxins. 1. *Nat. Toxins* **1995**, *3*, 10–16. [CrossRef] [PubMed]
21. Roggo, B.E.; Hug, P.; Moss, S.; Stämpfli, A.; Kriemler, H.P.; Peter, H.H. Novel spirodihydrobenzofuranlactams as antagonists of endothelin and as inhibitors of HIV-1 protease produced by *Stachybotrys* sp. II. Structure determination. *J. Antibiot.* **1996**, *49*, 374–379. [CrossRef] [PubMed]
22. Ma, X.-H.; Li, L.-T.; Zhu, T.-J.; Ba, M.-Y.; Li, G.-Q.; Gu, Q.-Q.; Guo, Y.; Li, D.-H. Phenylspirodrimanes with anti-HIV activity from the sponge-derived fungus *Stachybotrys chartarum* MXH-X73. *J. Nat. Prod.* **2013**, *76*, 2298–2306. [CrossRef] [PubMed]
23. Wu, B.; Oesker, V.; Wiese, J.; Malien, S.; Schmaljohann, R.; Imhoff, J.F. Spirocyclic drimanes from the marine fungus *Stachybotrys* sp. strain MF347. *Mar. Drugs* **2014**, *12*, 1924–1938. [CrossRef] [PubMed]
24. Li, Y.; Wu, C.-M.; Liu, D.; Proksch, P.; Guo, P.; Lin, W.-H. Chartarlactams A-P, phenylspirodrimanes from the sponge-associated fungus *Stachybotrys chartarum* with antihyperlipidemic activities. *J. Nat. Prod.* **2014**, *77*, 138–147. [CrossRef] [PubMed]
25. Sanchez, J.F.; Entwistle, R.; Corcoran, D.; Oakley, B.R.; Wang, C.C.C. Identification and molecular genetic analysis of the cichorine gene cluster in *Aspergillus nidulans*. *Med. Chem. Comm.* **2012**, *3*, 997–1002. [CrossRef] [PubMed]
26. Yin, Y.; Cai, M.-H.; Zhou, X.-S.; Li, Z.-Y.; Zhang, Y.-X. Polyketides in *Aspergillus terreus*: Biosynthesis pathway discovery and application. *Appl. Microbiol. Biotechnol.* **2016**, *100*, 7787–7798. [CrossRef] [PubMed]
27. Lo, H.-C.; Entwistle, R.; Guo, C.-J.; Ahuja, M.; Szewczyk, E.; Hung, J.-H.; Chiang, Y.-M.; Oakley, B.R.; Wang, C.C.C. Two separate gene clusters encode the biosynthetic pathway for the meroterpenoids austinol and dehydroaustinol in *Aspergillus nidulans*. *J. Am. Chem. Soc.* **2012**, *134*, 4709–4720. [CrossRef] [PubMed]
28. Guo, C.-J.; Knox, B.P.; Chiang, Y.-M.; Lo, H.-C.; Sanchez, J.F.; Lee, K.-H.; Oakley, B.R.; Bruno, K.S.; Wang, C.C.C. Molecular genetic characterization of a cluster in *A. terreus* for biosynthesis of the meroterpenoid terretonin. *Org. Lett.* **2012**, *14*, 5684–5687. [CrossRef] [PubMed]
29. Lin, T.-S.; Chiang, Y.-M.; Wang, C.C.C. Biosynthetic pathway of the reduced polyketide product citreoviridin in *Aspergillus terreus* var. *aureus* revealed by heterologous expression in Aspergillus nidulans. *Org. Lett.* **2016**, *18*, 1366–1369. [CrossRef] [PubMed]
30. Koide, H.; Narasaki, R.; Hasegawa, K.; Nishimura, N.; Hasumi, K. A new series of the SMTP plasminogen modulator with a phenylglycine-based side chain. *J. Antibiot.* **2012**, *65*, 91–93. [CrossRef] [PubMed]
31. Matsumoto, N.; Suzuki, E.; Tsujihara, K.; Nishimura, Y.; Hasumi, K. Structure–activity relationships of the plasminogen modulator SMTP with respect to the inhibition of soluble epoxide hydrolase. *J. Antibiot.* **2015**, *68*, 685–690. [CrossRef] [PubMed]
32. Hu, W.-M.; Ohyama, S.; Hasumi, K. Activation of fibrinolysis by SMTP-7 and -8, novel staplabin analogs with a pseudosymmetric structure. *J. Antibiot.* **2000**, *53*, 241–247. [CrossRef] [PubMed]
33. Pannel, R.; Gurewich, V. Activation of plasminogen by single-chain urokinase or by tow-chain urokinase—A demonstration that single-chain urokinase has a low catalytic activity. *Blood* **1987**, *69*, 22–26.
34. Petersen, L.C.; Lund, L.R.; Nielsen, L.S.; Dano, K.; Skriver, L. One-chain urokinase-type plasminogen activator from human sarcoma cell is a proenzyme with little or no intrinsic activity. *J. Biol. Chem.* **1988**, *263*, 11189–11195. [PubMed]
35. Nozawa, Y.; Yamamoto, K.; Ito, M.; Sakai, N.; Mizoue, K.; Mizobe, F.; Hanada, K. Stachybotrin C and parvisporin, novel neuritogenic compounds. I. Taxonomy, isolation, physico-chemical and biological properties. *J. Antibiot.* **1997**, *50*, 635–640. [CrossRef] [PubMed]
36. Roggo, B.E.; Petersen, F.; Sills, M.; Roesel, J.L.; Moerker, T.; Peter, H.H. Novel spirodihydrobenzofuranlactams as antagonists of endothelin and as inhibitors of HIV-1 protease produced by *Stachybotrys* sp. I. Fermentation, isolation and biological activity. *J. Antibiot.* **1996**, *49*, 13–19. [CrossRef] [PubMed]

marine drugs

MDPI

Article

Chrysoxanthones A–C, Three New Xanthone–Chromanone Heterdimers from Sponge-Associated *Penicillium chrysogenum* HLS111 Treated with Histone Deacetylase Inhibitor

Xin Zhen [†], Ting Gong [†], Yan-Hua Wen, Dao-Jiang Yan, Jing-Jing Chen and Ping Zhu *

State Key Laboratory of Bioactive Substance and Function of Natural Medicines, Key Laboratory of Biosynthesis of Natural Products of National Health Committee, Institute of Materia Medica, Chinese Academy of Medical Sciences and Peking Union Medical College, 1 Xian Nong Tan Street, Beijing 100050, China; zhenxin@imb.pumc.edu.cn (X.Z.); gongting@imm.ac.cn (T.G.); wenyanhua@imm.ac.cn (Y.-H.W.); yandaojiang@imm.ac.cn (D.-J.Y.); chenjingjing@imm.ac.cn (J.-J.C.)
* Correspondence: zhuping@imm.ac.cn; Tel.: +86-10-6316-5197
† These authors contributed equally to this work.

Received: 11 September 2018; Accepted: 28 September 2018; Published: 1 October 2018

Abstract: By treating with histone-deacetylase inhibitor valproate sodium, three new heterodimeric tetrahydroxanthone–chromanone lactones chrysoxanthones A–C (**1–3**), along with 17 known compounds were isolated from a sponge-associated *Penicillium chrysogenum* HLS111. The planar structures of chrysoxanthones A–C were elucidated by means of spectroscopic analyses, including MS, 1D, and 2D NMR. Their absolute configurations were established by electronic circular dichroism (ECD) calculations. Chrysoxanthones A–C exhibited moderate antibacterial activities against *Bacillus subtilis* with minimum inhibitory concentration (MIC) values of 5–10 µg/mL.

Keywords: *Penicillium chrysogenum*; secondary metabolites; histone-deacetylase inhibitor; antibacterial activity; polyketide synthase

---

1. Introduction

The *Penicillium chrysogenum* species is commonly used in the industrial production of β-lactam antibiotics. This species is present in various habitats including plants, gorgonians, marine sediments, algae, sponges, and mangroves. The whole genome of the strain *P. chrysogenum* Wisconsin 54-1255 was sequenced in 2008, and up to 33 secondary metabolic gene clusters were predicted, showing huge potential to produce various types of compounds in addition to β-lactam antibiotics [1]. However, limited products with six chemical types were obtained from *P. chrysogenum*, much fewer than the estimated number that the secondary metabolic gene clusters could generate. Therefore, it is believed that many of the secondary metabolic gene clusters may be silent in laboratory culture conditions. In recent years, strategies related to epigenetic modifications, such as the addition of a histone-deacetylase inhibitor and disruption of gene-encoding histone deacetylase (HDAC) have been proven to be able to effectively activate the silent biosynthetic gene clusters and lead to the formation of diverse secondary metabolites in fungi [2–5].

The effects of HDAC inhibitors (valproic acid (VPA), suberoylanilide hydroxamic acid (SAHA), sodium butyrate (NaBu)) on the production of secondary metabolites of a sponge-associated fungus, *P. chrysogenum* HLS111, have been examined. As a result, the strain cultured with VPA produced more compounds than those of the control, and the contents of some compounds were much higher than those of the control (Figure 1). In this study, the differential parts of a crude extract of the test group attracted our attention, and its subsequent separation and purification provided three new tetrahydroxanthone—chromanone lactone heterdimers, designated chrysoxanthones A–C (**1**–**3**, Figure 2), and 17 known compounds (**4**–**20**, Supplementary Materials Figure S1). Herein, we report the isolation, structure elucidation, and biological activities of these compounds. Finally, the biosynthetic pathway of chrysoxanthones A–C was proposed.

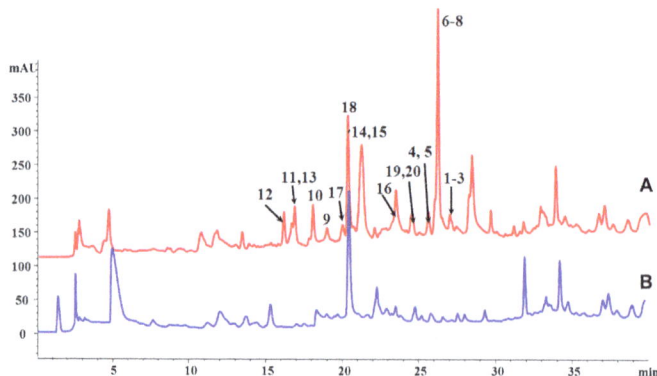

Figure 1. High Performance Liquid Chromatography (HPLC) analysis of the secondary metabolites of *P. chrysogenum* HLS111. (**A**) Test group (with valproic acid (VPA)). Peak numbers match the compound numbers in Figure 2 and Supplementary Materials Figure S1. (**B**) Control group.

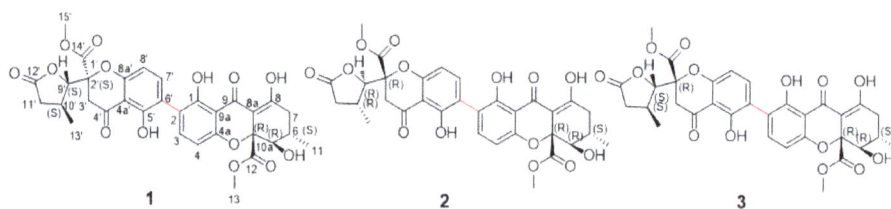

Figure 2. Chemical structures of compounds **1**–**3**.

2. Results and Discussion

2.1. Structure Elucidation of Compounds

The organic extract of *P. chrysogenum* HLS111 treated by VPA was purified by repeated silica-gel chromatography, reversed-phase chromatography, as well as semipreparative HPLC to yield compounds **1**–**20**. Chrysoxanthones A–C (**1**–**3**) were new isomerides that shared the common molecular formula of $C_{32}H_{30}O_{14}$, determined by HRESIMS {m/z 639.1716 [M + H]$^+$ (compound **1**), m/z 639.1712 [M + H]$^+$ (compound **2**), m/z 639.1713 [M + H]$^+$ (compound **3**), calcd for $C_{32}H_{31}O_{14}$ 639.1714}, requiring 18 degrees of unsaturation. Compounds **1**–**3** have similar ultraviolet spectral properties λ_{max} (log ε) 248.2 (3.6), 335.5 (3.6) nm, indicating that they share related skeletons.

Chrysoxanthones A (**1**) and B (**2**) were attained as a pale yellow powder. They showed identical ^{1}H and ^{13}C NMR spectra (Table 1, Supplementary Materials Figures S4, S5, S10, and S11), despite they were separated using a nonenantioselective method, presuming they were isomers but not enantiomers. By comparison of the NMR data (Table 1) with the structurally related compounds, they were identified as the tetrahydroxanthone–chromanone lactone heterdimers [6–8]. The HMBC correlations of H-3 to C-6′ and H-7 to C-2 indicated the monomers of compounds **1** and **2** were connected by a 2-6′ linkage (Figure 3, Supplementary Materials, Figures S6 and S12). The ^{1}H and ^{13}C NMR data (Table 1) of **1** and **2** were in agreement with those for blennolide J [6], which suggested they might have the same planar structures. The key coupling constants in ^{1}H NMR spectra (Table 1, Supplementary Materials, Figures S4 and S10), the key Rotating Frame Overhauser Effect Spectroscopy

(ROESY) correlations (Figure 4) and NMR data rules could prove the relative stereochemistry of **1** and **2**. The *trans*-configuration of H-5 and H-6 of the tetrahydroxanthone monomer was established by chemical shifts (compounds **1** and **2** $\delta_{5\text{-H/C}}$ 3.93/76.9, $\delta_{6\text{-H/C}}$ 2.42/29.2 vs. blennolide I $\delta_{5\text{-H/C}}$ 4.12/71.3, $\delta_{6\text{-H/C}}$ 2.11/28.5; *cis*-configuration) and the coupling constant ($^3J_{\text{H-5/H-6}}$ = 11.5 Hz) [6], as well as the ROESY correlation between H-5 (δ_H 3.93) and H-11 (δ_H 1.18). The ROESY correlation between H-13 (δ_H 3.68) and H-6 (δ_H 2.38) revealed the relative configurations of chiral carbons 5, 6, and 10a to be (5S, 6R, 10aS, or 5R, 6S, 10aR) (Figure 4C). Coupling constant $^3J_{\text{H-9'/H-10'}}$ = 4.0 Hz established the *trans*-configuration of H-9' and H-10' of the β-methyl-γ-lactone moiety. Since the biaryl bond could rotate, it was ambiguous to determine the relative stereochemistry of carbon 2' based on ROESY correlations. However, a summary of the NMR data [6–8] could provide some clues to identify the relative configurations of C-2' and C-10'. When the difference between the H-3a' and H-3b' values was greater than 0.41 ppm, the relative configurations were proposed to be 2'S, 10'R or 2'R, 10'S, while when it was less than 0.19 ppm, the relative configuration was 2'S, 10'S or 2'R, 10'R. Thus, the relative configurations of **1** and **2** were the same as those of blennolide J [6], but the optical rotation data and CD spectra of compounds **1** and **2** were opposite to those of blennolide J (compound **1** $[\alpha]_D^{20}$ +30.8 (*c* 0.05, MeOH), compound **2** $[\alpha]_D^{20}$ +58.4 (*c* 0.15, MeOH), blennolide J $[\alpha]_D^{20}$ −93.5 (*c* 0.08, CHCl$_3$)) [6], indicating the absolute configurations of two subunits of compounds **1** and **2** were the same or enantiotropic to those of blennolide J. Then, the absolute structures of compounds **1** and **2** were suggested to be (1a/2a (5R,6S,10aR,2'S,9'S,10'S); 1b/2b (5R,6S,10aR,2'R,9'R,10'R); 1d/2d (5S,6R,10aS,2'R,9'R,10'R)). The absolute configurations of compounds **1** and **2** were determined by comparing experimentally measured CD spectra with those calculated with the Time-Dependent Density Functional Theory (TD-DFT) method. The experimental electronic circular dichroism (ECD) spectrum of compounds **1** and **2** matched very well with the theoretical ECD spectrum for 1a(5R,6S,10aR,2'S,9'S,10'S) and 2b(5R,6S,10aR,2'R,9'R,10'R) (Figure 5A,B), respectively, confirming the absolute structures of **1** and **2**.

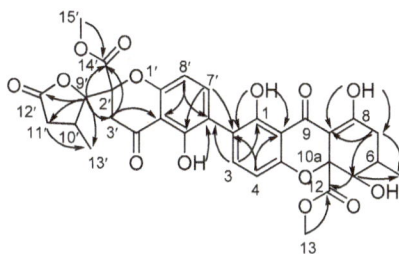

Figure 3. Key HMBC correlations of compounds **1**–**3**.

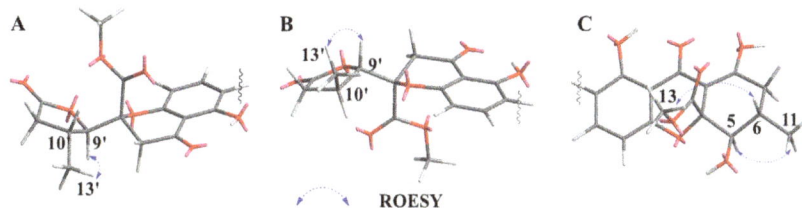

Figure 4. (A) Key Rotating Frame Overhauser Effect Spectroscopy (ROESY) correlations of β-methyl-γ-lactone moiety in compounds **1** and **2**; (B) key ROESY correlations of β-methyl-γ-lactone moiety in compound **3**; (C) key ROESY correlations of tetrahydroxanthone moiety in compounds **1**–**3**.

Table 1. ^{1}H (500 MHz) and ^{13}C NMR (125 MHz) data of compounds **1–3** in CDCl$_3$.

Position	1 and 2		3	
	δ_H, mult, J in Hz	δ_C	δ_H, mult, J in Hz	δ_C
1		159.3		159.3
2		117.7		117.8
3	7.46, d, 8.5	140.1	7.46, d, 8.5	140.2
4	6.63, d, 8.5	107.5	6.62, d, 8.5	107.6
4a		158.4		158.3
5	3.93, dd, 11.5, 2.5	76.9	3.93, dd, 11.0, 1.5	76.9
6	2.42, m	29.2	2.42, m	29.2
7a	2.32, dd, 19.0, 10.5	36.2	2.31, dd, 19.0, 10.5	36.3
7b	2.74, dd, 19.0, 6.5		2.74, dd, 19.0, 6.5	
8		177.6		177.6
8a		101.5		101.5
9		187.1		187.1
9a		106.8		106.8
10a		84.7		84.7
11	1.18, d, 6.5	18.0	1.17, d, 6.5	18.0
12		170.2		170.2
13	3.73, s	53.3	3.73, s	53.3
1-OH	11.72, s		11.72, s	
5-OH	2.78, d, 2.5		2.80, br s	
8-OH	13.76, s		13.78, s	
2'		84.2		84.3
3a'	3.05, d, 17.0	39.6	3.10, d, 17.0	40.4
3b'	3.21, d, 17.0		3.51, d, 17.0	
4'		194.0		195.0
4a'		107.6		107.3
5'		159.2		159.1
6'		118.1		118.0
7'	7.52, d, 8.5	141.3	7.51, d, 8.5	141.1
8'	6.62, d, 8.5	107.3	6.57, d, 8.5	107.0
8a'		158.6		158.6
9'	4.45, d, 4.0	87.5	4.38, d, 3.5	86.4
10'	2.84, m	30.0	2.88, m	29.6
11a'	2.92, dd, 17.5, 9.0	36.0	3.03, dd, 18.0,9.0	36.2
11b'	2.23, dd, 17.5, 4.0		2.23, dd, 18.0,4.0	
12'		175.2		175.5
13'	1.29, d, 7.0	20.9	1.19, d, 7.0	20.6
14'		168.8		169.0
15'	3.77, s	53.7	3.78, s	53.7
5'-OH	11.89, s		11.92, s	

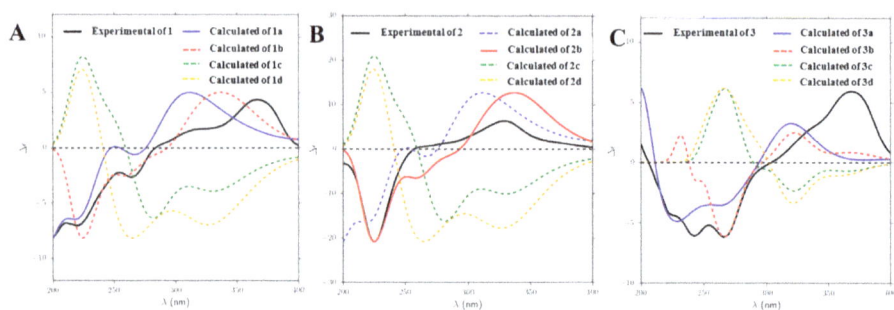

Figure 5. Experimentally measured and calculated electronic circular dichroism (ECD) spectra for compounds **1–3**.

Compound **3** was also isolated as a pale yellow powder. Its NMR spectroscopic data were similar with that of compounds **1** and **2**, except the chemical shifts [H-3a' (δ_H 3.05), H-3b' (δ_H 3.21), H-9' (δ_H 4.45); C-3'(δ_C 39.6), C-9'(δ_C 87.5)] moved to [H-3a' (δ_H 3.10), H-3b' (δ_H 3.51), H-9' (δ_H 4.38); C-3'(δ_C 40.4), C-9'(δ_C 86.4)] (Table 1). The key HMBC correlations of compound **3** suggested the planar structure of **3** was identical with structures **1** and **2** (Figure 3). The obvious NMR differences might be induced by the change of the stereogenic centers of 2', 9', and 10' in β-methyl-γ-lactone moiety. The *trans*-configuration of H-9' and H-10' (9'S,10'S or 9'R,10'R) of the β-methyl-γ-lactone moiety was uniform with compounds **1** and **2** by the chemical shifts (Table 1) and coupling constant ($^3J_{H-9'/H-10'}$ = 3.5 Hz), together with the NOE correlations between H-13' (δ_H 1.19) and H-9' (δ_H 4.38) (Figure 4B). Therefore, epimerization at C-2' led to the above NMR differences. The absolute configurations of chiral carbons 2', 9', and 10' were assigned as (2'R,9'S,10'S or 2'S,9'R,10'R), which were in agreement with paecilin B [6]. Above all, the accordance of the experimentally measured CD spectra with the calculated ECD of 3a confirmed the absolute configuration of compound **3** was 5R,6S,10aR,2'R,9'S,10'S (Figure 5C).

The known compounds were identified as andrastins A (keto form, **4** and enol form, **5**) [9,10], secalonic acid A (**6**) [11–14], secalonic acid D (**7**) [14], secalonic acid F (**8**) [14,15], griseoxanthone C (**9**) [16], norlichexanthone (**10**) [16], 7-hydroxy-3-(2-hydroxypropyl)-5-methyl-isochromen-1-one (**11**) [17], 7-hydroxy-3,5-dimethyl-isochromen-1-one (**12**) [17], penicisimpin B (**13**) [18], meleagrin (**14**) [19], oxaline (**15**) [20], rcqueforcine C (**16**) [21], citreoindole (**17**) [22], haenamindole (**18**) [22], conidiogenone C (**19**) [23], and conidiogenone D (**20**) [23] by comparing their spectroscopic data with those reported in the literature (chemical structures in Supplementary Materials Figure S1). Moreover, the NMR data of andrastins A in keto form were reported the first time (Supplementary Materials Table S3).

2.2. Biological Assays

Chrysoxanthones A–C were assayed for antibacterial activities against *Staphylococcus epidermidis* (ATCC 12228, MSSE), *Staphylocccus aureus* (ATCC 29213, MSSA), *Bacillus subtilis* (ATCC 63501), *Enterococcus faecalis* (ATCC 29212, VSE), and *Escherichia coli* (ATCC 25922). Chrysoxanthones A–C exhibited the highest antibacterial activities against *B. subtilis* with minimum inhibitory concentration (MIC) values of 5–10 μg/mL and moderate activities against *S. epidermidis* and *S. aureus* with MICs of 10–80 μg/mL (Table 2). Meanwhile, the antitumor activities in vitro of chrysoxanthones A–C were evaluated against human multiform glioblastoma (U87 MG), human nonsmall cell lung cancer (NCI-H1650), human colonic carcinoma (HT29), human renal carcinoma (A498), and human leukemia (HL60) cell lines using the MTT method. Compared with secalonic acid D, the antitumor activities of chrysoxanthones A–C decreased significantly (Supplementary Materials Table S5), indicating that the β-methyl-γ-lactone ring reduced cytotoxicity, supporting that the tricyclic core was essential for bioactivities [6].

Table 2. Minimum inhibitory concentrations (MIC, μg/mL) of compounds **1–3**.

Compounds	MICs (μg/mL)				
	Staphylococcus epidermidis ATCC 12228	*Staphylococcus aureus* ATCC 29213	*Bacillus subtilis* ATCC 63501	*Enterococcus faecalis* ATCC 29212	*Escherichia coli* ATCC25922
1	—	—	5	—	>100
2	10	20	5	>100	>100
3	20	80	10	>100	>100
Ampicillin	6.25	6.25	3.12	6.25	25

Note: "—" represents untested.

2.3. Plausible Biosynthetic Gene Cluster and Pathway of Chrysoxanthones A–C

Based on the common tetrahydroxanthones moiety, we proposed that chrysoxanthones A–C shared the same biosynthetic gene cluster and pathway with those of secalonic acids. The biosynthetic gene cluster of secalonic acids in *Claviceps purpurea* has been confirmed by deleting gene *CPUR_05437*, which led to the abolishment of secalonic acid production [24]. In our effort to find the biosynthetic gene cluster, the gene *pc16g10750* of *P. chrysogenum* wisconsin 54-1255 showed the highest identity with the gene *CPUR_05437* (71%) by the NCBI BLAST. *Pc16g10750* putatively encodes an iterative nonreducing PKS (NR-PKS) and has a domain architecture of KS-AT-PT-ACP, as ascertained by NCBI Conserved Domains. In these special types of NR-PKSs lacking the thioesterase (TE) domain, polyketide is released from the PKS by a metallo-β-lactamase-type thioesterase (MβL-TE) [24]. The functions of the upstream and downstream genes were also analyzed by BLAST analysis, indicating the integrated gene cluster composed of 23 Open Reading Frames (ORFs) (Supplementary Materials, Figure S24). The gene cluster included one NR-PKS (*pc16g10750*), one O-methyltransferase (*pc16g10740*), two decarboxylases (*pc16g10730* and *pc16g10800*), two major facilitator superfamily proteins (*pc16g10710* and *pc16g10850*), two fungal specific transcription factors (*pc16g10640* and *pc16g10860*), ten oxidases or reductases, and five function-unknown genes (Supplementary Materials, Table S4). Finally, the plausible biosynthetic pathway of chrysoxanthones was depicted in Scheme 1. The route followed the sequence octaketide–anthraquinone–benzophenone–tetrahydroxanthone–β-methyl-γ-lactone–chrysoxanthone. As supposed in the Scheme 1, β-methyl-γ-lactone moiety was derived from the tetrahydroxanthone segment, while the absolute configurations of its chiral carbons (marked with the red stars) were not in accordance with that of the obtained tetrahydroxanthones. Hence, more tetrahydroxanthones with other absolute configurations may be discovered in the future.

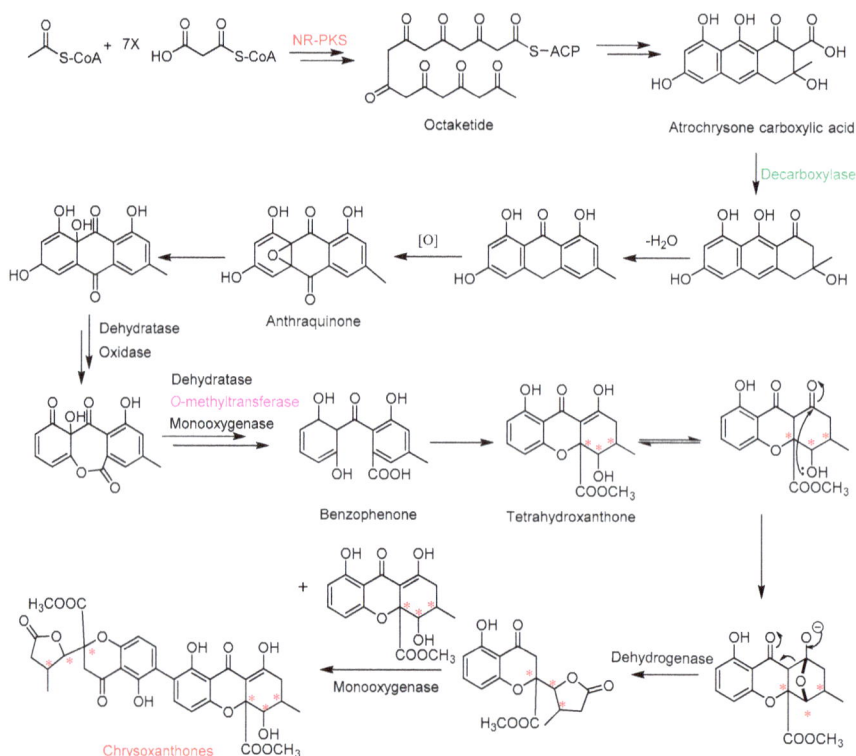

Scheme 1. Plausible biosynthetic pathway of chrysoxanthones. Note: chiral carbons were marked with the red stars.

3. Materials and Methods

3.1. General

Optical rotations were obtained on a JASCO P-2000 digital polarimeter (JASCO, Tokyo, Japan). UV spectra were recorded on a V-650 spectrometer (JASCO, Tokyo, Japan). Circular dichroism (CD) spectra were measured on a JASCO J-815 CD spectrometer (JASCO, Tokyo, Japan). IR spectra were collected on a Nicolet 5700 FT-IR microscope instrument (FT-IR microscope transmission) (Thermo Electron Corporation, Madison, WI, USA). ^{1}H and ^{13}C NMR and 2D NMR spectra were acquired at 500 and 125 MHz on a Bruker AVANCE 500-III instrument (Bruker Biospin Group, Karlsruhe, Germany) in chloroform-*d* with TMS as an internal reference, and chemical shifts were recorded as δ values. HR-ESI-MS data were measured using an Agilent 1100 LC/MSD Trap SL LC/MS/MS spectrometer (Agilent Technologies Ltd., Santa Clara, CA, USA). HPLC analysis of the secondary metabolites of the crude extracts was performed on an Agilent 1200 HPLC system (Agilent Technologies Ltd., Santa Clara, CA, USA), using a Cosmosil C18 column (5 µM, 4.6 × 150 mm, Nacalai Tesque, Inc., Kyoto, Japan). An HPLC system equipped with a Shimadzu LC-6AD pump, a Shimadzu SPD-20A prominence UV-VIS detector (Shimadzu Corporation, Kyoto, Japan), and an Agilent C18 column (5 µM, 250 × 9.6 mm) was used for semipreparative HPLC. Column chromatography was performed with silica gel (60–100 mesh, 200–300 mesh, Qingdao Marine Chemical Ltd., Qingdao, China), RP-18 (40–60 µM, GE Healthcare, Fairfield, CT, USA), and Sephadex LH-20 (18–110 M, GE healthcare, Fairfield, CT, USA).

3.2. Fungal Material and Fermentation

The fungal strain HLS111 was isolated from a *Gelliodes carnosa* sponge collected from Lingshui Bay, Hainan Province, China. It was identified as *P. chrysogenum* on the basis of morphology and internal transcribed spacer (ITS) sequence analysis. The sequence data have been deposited at GenBank as FJ770066 [25]. The fungus HLS111 was first cultivated on YPD agar plates (YPD: yeast extract 10 g, peptone 20 g, glucose 20 g, natural seawater 1 L, pH 7.0–7.2) at 28 °C for 3 days. Then, the mycelia were inoculated into 500 mL Erlenmeyer flasks, each containing 100 mL of liquid YPD medium. The flasks were incubated at 28 °C on a rotary shaker (220 rpm) for 3 days to prepare seed culture. The rice medium was used for solid fermentation, with each 500 mL Erlenmeyer flask containing 100 g rice and 100 mL natural seawater sterilized at 121 °C for 15 minutes. The test-group fermentation was carried out in 25 Erlenmeyer flasks with the HDAC inhibitor VPA at a final concentration of 10 µM, and the control-group fermentation was performed in 3 Erlenmeyer flasks. The flasks were inoculated with 20 mL of seed cultures and incubated at 28 °C for 30 days.

3.3. Extraction, HPLC Analysis, and Isolation

The fermented material was extracted with EtOAc (*v*/*v* 1:3) 3 times by ultrasound, and the organic solvent was evaporated to dryness under vacuum to afford the crude extract (the test group was 45 g and the control group was 4 g). The samples (10 mg/mL) of the test and control groups were analyzed with an Agilent 1200 HPLC system using a Cosmosil C18 column (4.6 × 150 mm, 5 µm) with a solvent gradient from 10% to 100% solvent B (solvent A: H_2O, solvent B: CH_3OH) over the course of 0 to 30 min and 100% solvent B over 30 to 40 min and UV detection at 230 nm at a flow rate of 1 mL/min.

The crude extract of the test group was first fractionated through vacuum liquid chromatography over a silica-gel column (8 × 100 cm, 1100 g), using a gradient elution with petroleum ether-EtOAc-CH_3OH, (petroleum ether; petroleum ether-EtOAc = 100:1, 50:1, 20:1, 10:1, 5:1, 2:1, 1:1; EtOAc; CH_3OH; *v*/*v*, each 2.5 L) to give 10 fractions (Fractions A–J). Fraction E (petroleum ether-EtOAc 10:1, 5.6 g) was further submitted to ODS CC (4 × 60 cm, 200 g) using a stepped gradient elution of CH_3OH-H_2O (20%, 50%, 80%, and 100%, *v*/*v*, each 1.5 L) to afford 4 subfractions E1–E4. Purification of the E4 subfraction (100% CH_3OH, 800 mg) by semipreparative HPLC on an Agilent C18 column (9.6 × 250 mm, 5 µm) with 47% CH_3CN in H_2O as the mobile phase (flow rate 4 mL/min) afforded compounds **1–3** (compound **1**, 3.0 mg, t_R = 34.5 min; compound **2**, 8.3 mg, t_R = 26.7 min; compound **3**, 4.1 mg, t_R = 44.2 min). The isolation of the known compounds was shown in supporting information.

3.4. Spectroscopic Data

Chrysoxanthone A (1). Pale yellow powder; $[\alpha]_D^{20}$ +30.8 (c 0.05, MeOH); UV (MeOH) λ_{max} (log ε) 248.2 (3.6), 335.5 (3.6); CD (c 7.84 × 10^{-4} mol/L, MeOH), λ_{max} ($\Delta\varepsilon$) 219 (−7.1), 264 (−2.7), 316.5 (1.6), 366.5 (4.3) nm; IR (KBr) ν_{max}: 3519, 2959, 1792, 1741, 1613, 1566, 1433, 1360, 1213, 1060, 1022, 819 cm^{-1}; ^{1}H NMR (500 MHz, CDCl$_3$) and ^{13}C NMR (125 MHz, CDCl$_3$) see Table 1; HR-ESI-MS (positive): m/z 639.1716 [M + H]$^+$ (Calcd. for C$_{32}$H$_{31}$O$_{14}$, 639.1714).

Chrysoxanthone B (2). Pale yellow powder; $[\alpha]_D^{20}$ +58.4 (c 0.15, MeOH); UV (MeOH) λ_{max} (log ε) 248.2 (3.6), 335.5 (3.6); CD (c 7.84 × 10^{-4} mol/L, MeOH), λ_{max} ($\Delta\varepsilon$) 225 (−20.9), 329 (6.4) nm; IR (KBr) ν_{max}: 3511, 2958, 1792, 1741, 1613, 1566, 1434, 1359, 1211, 1060, 1023, 819 cm^{-1}; ^{1}H NMR (500 MHz, CDCl$_3$) and ^{13}C NMR (125 MHz, CDCl$_3$) see Table 1; HR-ESI-MS (positive): m/z 639.1712 [M + H]$^+$ (Calcd. for C$_{32}$H$_{31}$O$_{14}$, 639.1714).

Chrysoxanthone C (3). Pale yellow powder; $[\alpha]_D^{20}$ +33.2 (c 0.1, MeOH); UV (MeOH) λ_{max} (log ε) 248.2 (3.6), 335.5 (3.6); CD (c 7.84 × 10^{-4} mol/L, MeOH), λ_{max} ($\Delta\varepsilon$) 242.5 (−6.1), 266 (−6.2), 369 (5.9) nm; IR (KBr) ν_{max}: 3511, 2958, 1792, 1741, 1613, 1566, 1434, 1359, 1211, 1060, 1023, 819 cm^{-1}; ^{1}H NMR (500 MHz, CDCl$_3$) and ^{13}C NMR (125 MHz, CDCl$_3$) see Table 1; HR-ESI-MS (positive): m/z 639.1713 [M + H]$^+$ (Calcd. for C$_{32}$H$_{31}$O$_{14}$, 639.1714).

3.5. ECD Calculation

TD-DFT calculations were executed with Gaussian 09 [26] (Gaussian, Inc., Pittsburgh, PA, USA). Conformational analysis was initially performed using Confab [27] at MMFF94 force field for all relative configurations of each compound. The conformers with lower energies (Boltzmann population of over 1%) were chosen for ECD calculations (Supplementary Materials, Table S1). The conformers were first optimized at PM6 using a semiempirical theory method and further at B3LYP/6-311G** in MeOH using the CPCM polarizable conductor calculation model (Supplementary Materials, Table S2). ECD calculations were conducted in MeOH using TD-DFT, and a total of 50 excited states were calculated. The BP86/6-311G** method was used for compounds **1** and **2**, while the TD-DFT theory level for compound **3**. The ECD and UV-Vis spectra were simulated in SpecDis [28] by overlapping Gaussian functions for each transition, and summed up by Boltzmann weights.

3.6. Biological Assay

Antibacterial and in vitro antitumor assays were performed for the isolated compounds (>90% purity).

3.6.1. Antibacterial Activity

The antibacterial activities (represented by MIC values) of compounds **1–3** against 5 bacteria strains, *B. subitlis* ATCC 63501, *S. aureus* ATCC 29213, *S. epidermidis* ATCC 12228, *E. faecalis* ATCC 29212, and *E. coli* ATCC 25922, were evaluated by a 96-well plate-based assay [29]. The bacterial strains cultured in LB medium were collected at OD$_{600}$ of 0.3~0.5, then diluted to OD$_{600}$ of 5 × 10^{-4}. Aliquots of this suspension (100 µL) were transferred into a 96-well plate. The tested compounds were added into the bacteria suspensions to give the desired concentration, and each concentration had triplicate values. The wells containing the same number of cells but no compounds were set as control groups. Ampicillin was used as the positive control. The 96-well-plates were then incubated at 37 °C for 18 h. The plates were then read using a microplate reader at 600 nm. MIC value was determined by the OD$_{600}$ with the control without bacteria.

3.6.2. Antitumor Activity

The antitumor activities (in vitro, represented by IC_{50} values) of compounds **1–3** against 5 tumor cell lines, human multiform glioblastoma (U87 MG), human nonsmall cell lung cancer (NCI-H1650), human colonic carcinoma (HT29), human renal carcinoma (A498), and human leukemia (HL60), were evaluated with the MTT method, and secalonic acid D was used as the positive control. The detailed methodology for biological testing has already been described in a previous report [30].

4. Conclusions

Three new compounds, chrysoxanthones A–C (**1–3**), along with 17 known compounds (**4–20**) were isolated from the strain *P. chrysogenum* HLS111 cultivated with the histone-deacetylase inhibitor VPA. Our present findings revealed that histone-deacetylase inhibitors could activate diversified compound production or increase the yields of fungal secondary metabolites. Considering the instability and high cost of inhibitors, further deletion of genes related to epigenetic modifications still needs to be conducted.

Supplementary Materials: The following are available online at http://www.mdpi.com/1660-3397/16/10/357/s1: Figure S1: Chemical structures of the known compounds **4–20**; Figure S2: (+)-HRESIMS spectrum of chrysoxanthone A (**1**); Figure S3: IR spectrum of chrysoxanthone A (**1**); Figure S4: ^{1}H NMR spectrum of chrysoxanthone A (**1**) in CDCl$_3$; Figure S5: ^{13}C NMR spectrum of chrysoxanthone A (**1**) in CDCl$_3$; Figure S6: HMBC spectrum of chrysoxanthone A (**1**) in CDCl$_3$; Figure S7: ROESY spectrum of chrysoxanthone A (**1**) in CDCl$_3$; Figure S8: (+)-HRESIMS spectrum of chrysoxanthone B (**2**); Figure S9: IR spectrum of chrysoxanthone B (**2**); Figure S10: ^{1}H NMR spectrum of chrysoxanthone B (**2**) in CDCl$_3$; Figure S11: ^{13}C NMR spectrum of chrysoxanthone B (**2**) in CDCl$_3$; Figure S12: HMBC spectrum of chrysoxanthone B (**2**) in CDCl$_3$; Figure S13: ROESY spectrum of chrysoxanthone B (**2**) in CDCl$_3$; Figure S14: (+)-HRESIMS spectrum of chrysoxanthone C (**3**); Figure S15: IR spectrum of chrysoxanthone C (**3**); Figure S16: ^{1}H NMR spectrum of chrysoxanthone C (**3**) in CDCl$_3$; Figure S17: ^{13}C NMR spectrum of chrysoxanthone C (**3**) in CDCl$_3$; Figure S18: HMBC spectrum of chrysoxanthone C (**3**) in CDCl$_3$; Figure S19: ROESY spectrum of chrysoxanthone C (**3**) in CDCl$_3$; Table S1: Energies of dominative conformers of compounds **1–3** at MMFF94 force field; Table S2: Energies of the conformers of compounds **1–3** at B3LYP/6-311G** in methanol; Table S3: ^{1}H (500 MHz) and ^{13}C NMR (125 MHz) spectroscopic data (δ in ppm, J in Hz) for compound **4** in CDCl$_3$; Figure S20: ^{1}H NMR spectrum of andrastin A (**4**, in keto form) in CDCl$_3$; Figure S21: ^{13}C NMR spectrum of andrastin A (**4**, in keto form) in CDCl$_3$; Figure S22: HSQC spectrum of andrastin A (**4**, in keto form) in CDCl$_3$; Figure S23: HMBC spectrum of andrastin A (**4**, in keto form) in CDCl$_3$; Figure S24: Proposed biosynthetic gene cluster (*pc16g10640-pc16g10860*) of compounds **1–3**; Table S4: Putative ORFs and predicted functions of gene cluster *pc16g10640-pc16g10860*; Table S5: Antitumor activities (in vitro IC_{50}, μM) of compounds **1–3**.

Author Contributions: X.Z. and T.G. designed and performed the experiments, and drafted the manuscript; Y.-H.W. and D.-J.Y. helped to analyze the data; J.-J.C. contributed the antibacterial bioassays; P.Z. conceived, designed, and supervised the study, and revised the manuscript. All authors approved the final manuscript.

Funding: This research was funded by the CAMS Innovation Fund for Medical Sciences (CIFMS) grant number (2017-I2M-4-004).

Acknowledgments: We are grateful to the Department of Instrumental Analysis of our Institute for measurement of the NMR and MS. We acknowledge the research Team of Yan Li, Department of Pharmacology of our Institute for the in vitro antitumor bioassays.

Conflicts of Interest: The authors declare no conflict of interest.

References

1. Van den Berg, M.A.; Albang, R.; Albermann, K.; Badger, J.H.; Daran, J.M.; Driessen, A.J.; Garcia-Estrada, C.; Fedorova, N.D.; Harris, D.M.; Heijne, W.H.; et al. Genome sequencing and analysis of the filamentous fungus *Penicillium chrysogenum*. *Nat. Biotechnol.* **2008**, *26*, 1161–1168. [CrossRef] [PubMed]
2. Asai, T.; Chung, Y.-M.; Sakurai, H.; Ozeki, T.; Chang, F.-R.; Yamashita, K.; Oshima, Y. Tenuipyrone, a novel skeletal polyketide from the entomopathogenic fungus, *Isaria tenuipes*, cultivated in the presence of epigenetic modifiers. *Org. Lett.* **2011**, *14*, 513–515. [CrossRef] [PubMed]

3.   Asai, T.; Morita, S.; Shirata, N.; Taniguchi, T.; Monde, K.; Sakurai, H.; Ozeki, T.; Oshima, Y. Structural diversity of new C13-polyketides produced by *Chaetomium mollipilium* cultivated in the presence of a NAD+-dependent histone deacetylase Inhibitor. *Org. Lett.* **2012**, *14*, 5456–5459. [CrossRef] [PubMed]
4.   Bai, J.; Mu, R.; Dou, M.; Yan, D.; Liu, B.; Wei, Q.; Wan, J.; Tang, Y.; Hu, Y. Epigenetic modification in histone deacetylase deletion strain of *Calcarisporium arbuscula* leads to diverse diterpenoids. *Acta Pharm. Sin. B* **2018**, *8*, 687–697. [CrossRef] [PubMed]
5.   Mao, X.M.; Xu, W.; Li, D.; Yin, W.B.; Chooi, Y.H.; Li, Y.Q.; Tang, Y.; Hu, Y.-C. Epigenetic genome mining of an endophytic fungus leads to the pleiotropic biosynthesis of natural products. *Angew. Chem. Int. Ed.* **2015**, *54*, 7592–7596. [CrossRef] [PubMed]
6.   El-Elimat, T.; Figueroa, M.; Raja, H.A.; Graf, T.N.; Swanson, S.M.; Falkinham, J.O., 3rd; Wani, M.C.; Pearce, C.J.; Oberlies, N.H. Biosynthetically distinct cytotoxic polyketides from *Setophoma terrestris*. *Eur. J. Org. Chem.* **2015**, *2015*, 109–121. [CrossRef] [PubMed]
7.   Arunpanichlert, J.; Rukachaisirikul, V.; Tadpetch, K.; Phongpaichit, S.; Hutadilok-Towatana, N.; Supaphon, O.; Sakayaroj, J. A dimeric chromanone and a phthalide: Metabolites from the seagrass-derived fungus *Bipolaris* sp. PSU-ES64. *Phytochem. Lett.* **2012**, *5*, 604–608. [CrossRef]
8.   Wu, G.; Yu, G.; Kurtan, T.; Mandi, A.; Peng, J.; Mo, X.; Liu, M.; Li, H.; Sun, X.; Li, J.; et al. Versixanthones A–F, cytotoxic xanthone-chromanone dimers from the marine-derived fungus *Aspergillus versicolor* HDN1009. *J. Nat. Prod.* **2015**, *78*, 2691–2698. [CrossRef] [PubMed]
9.   Uchida, R.; Shiomi, K.; Inokoshi, J.; Sunazuka, T.; Tanaka, H.; Iwai, Y.; Takayanagi, H.; Omura, S. Andrastins A–C, new protein farnesyltransferase inhibitors produced by *Penidllium* sp. FO-3929. *J. Antibiot.* **1996**, *49*, 418–424. [CrossRef] [PubMed]
10.   Omura, S.; Inokoshi, J.; Uchida, R.; Shiomi, K.; Masuma, R.; Kawakubo, T.; Tanaka, H.; Iwai, Y.; Kosemura, S.; Yamamura, S. Andrastins A–C, new protein farnesyltransferase inhibitors produced by *Penicillium* sp. FO-3929. I. Producing strain, fermentation, isolation, and biological activities. *J. Antibiot.* **1996**, *49*, 414–417. [CrossRef] [PubMed]
11.   Steyn, P. The isolation, structure and absolute configuration of secalonic acid D, the toxic metabolite of *Penicillium oxalicum*. *Tetrahedron* **1970**, *26*, 51–57. [CrossRef]
12.   Andersen, R.; Buechi, G.; Kobbe, B.; Demain, A.L. Secalonic acids D and F are toxic metabolites of *Aspergillus aculeatus*. *J. Org. Chem.* **1977**, *42*, 352–353. [CrossRef] [PubMed]
13.   Kurobane, I.; Vining, L.C.; MCINNES, A.G. Biosynthetic relationships among the secalonic acids. *J. Antibiot.* **1979**, *32*, 1256–1266. [CrossRef] [PubMed]
14.   Elsässer, B.; Krohn, K.; Flörke, U.; Root, N.; Aust, H.J.; Draeger, S.; Schulz, B.; Antus, S.; Kurtán, T. X-ray structure determination, absolute configuration and biological activity of phomoxanthone A. *Eur. J. Org. Chem.* **2005**, *2005*, 4563–4570. [CrossRef]
15.   Gong, T.; Sun, C.-Y.; Zhen, X.; Qiu, J.-Z.; Yang, J.-L.; Zhu, P. Bioactive metabolites isolated from the fungus *Penicillium chrysogenum* HLS111 associated with the marine sponge *Gelliodes carnosa*. *Chin. Med. Biotechnol.* **2014**, *9*, 174–179.
16.   Mutanyatta, J.; Matapa, B.G.; Shushu, D.D.; Abegaz, B.M. Homoisoflavonoids and xanthones from the tubers of wild and *in vitro* regenerated Ledebouria graminifolia and cytotoxic activities of some of the homoisoflavonoids. *Phytochemistry* **2003**, *62*, 797–804. [CrossRef]
17.   Wang, Q.X.; Bao, L.; Yang, X.L.; Guo, H.; Yang, R.N.; Ren, B.; Zhang, L.X.; Dai, H.Q.; Guo, L.D.; Liu, H.W. Polyketides with antimicrobial activity from the solid culture of an endolichenic fungus *Ulocladium* sp. *Fitoterapia* **2012**, *83*, 209–214. [CrossRef] [PubMed]
18.   Xu, R.; Li, X.-M.; Wang, B.-G. Penicisimpins A–C, three new dihydroisocoumarins from *Penicillium simplicissimum* MA-332, a marine fungus derived from the rhizosphere of the mangrove plant *Bruguiera sexangula* var. *rhynchopetala*. *Phytochem. Lett.* **2016**, *17*, 114–118. [CrossRef]
19.   Nozawa, K.; Nakajima, S. Isolation of radicicol from Penicillium luteo-aurantium, and meleagrin, a new metabolite from Penicillium meleagrinum. *J. Nat. Prod.* **1979**, *42*, 374–377. [CrossRef]
20.   Nagel, D.W.; Pachler, K.G.; Steyn, P.S.; Vleggaar, R.; Wessels, P.L. The chemistry and ^{13}C NMR assignments of oxaline, a novel alkaloid from *Penicillium oxalicum*. *Tetrahedron* **1976**, *32*, 2625–2631. [CrossRef]
21.   Ohmomo, S.; Utagawa, T.; Abe, M. Identification of roquefortine C produced by *Penicillium roqueforti*. *Agric. Biol. Chem.* **1977**, *41*, 2097–2098. [CrossRef]

22. Song, F.; He, H.; Ma, R.; Xiao, X.; Wei, Q.; Wang, Q.; Ji, Z.; Dai, H.; Zhang, L.; Capon, R.J. Structure revision of the penicillium alkaloids haenamindole and citreoindole. *Tetrahedron Lett.* **2016**, *57*, 3851–3852. [CrossRef]
23. Du, L.; Li, D.; Zhu, T.; Cai, S.; Wang, F.; Xiao, X.; Gu, Q. New alkaloids and diterpenes from a deep ocean sediment derived fungus *Penicillium* sp. *Tetrahedron* **2009**, *65*, 1033–1039. [CrossRef]
24. Neubauer, L.; Dopstadt, J.; Humpf, H.-U.; Tudzynski, P. Identification and characterization of the ergochrome gene cluster in the plant pathogenic fungus *Claviceps purpurea*. *Fungal Biol. Biotechnol.* **2016**, *3*, 2. [CrossRef] [PubMed]
25. Liu, W.C.; Li, C.Q.; Zhu, P.; Yang, J.L.; Cheng, K.D. Phylogenetic diversity of culturable fungi associated with two marine sponges: Haliclona simulans and Gelliodes carnosa, collected from the Hainan Island coastal waters of the South China Sea. *Fungal Divers.* **2010**, *42*, 1–15. [CrossRef]
26. Frisch, M.J.; Trucks, G.W.; Schlegel, H.B.; Scuseria, G.E.; Robb, M.A.; Cheeseman, J.R.; Scalmani, G.; Barone, V.; Mennucci, B.; Petersson, G.A.; et al. *Gaussian09, R.A. 1*; Gaussian Inc.: Wallingford, CT, USA, 2009.
27. O'Boyle, N.M.; Vandermeersch, T.; Hutchison, G.R. Confab—Generation of diverse low energy conformers. *J Cheminform.* **2011**, *3*, P32. [CrossRef]
28. Bruhn, T.; Schaumloeffel, A.; Hemberger, Y.; Bringmann, G. SpecDis: Quantifying the comparison of calculated and experimental electronic circular dichroism spectra. *Chirality* **2013**, *25*, 243–249. [CrossRef] [PubMed]
29. Zhang, Q.; Wang, Q.; Xu, S.; Zuo, L.; You, X.; Hu, H.-Y. Aminoglycoside-based novel probes for bacterial diagnostic and therapeutic applications. *Chem. Commun.* **2017**, *53*, 1366–1369. [CrossRef] [PubMed]
30. Zhen, X.; Gong, T.; Liu, F.; Zhang, P.-C.; Zhou, W.-Q.; Li, Y.; Zhu, P. A new analogue ofechinomycin and a new cyclic dipeptide from a marine-derived *Streptomyces* sp. LS298. *Mar. Drugs* **2015**, *13*, 6947–6961. [CrossRef] [PubMed]

marine drugs

MDPI

Article

Bacillibactin and Bacillomycin Analogues with Cytotoxicities against Human Cancer Cell Lines from Marine *Bacillus* sp. PKU-MA00093 and PKU-MA00092

Mengjie Zhou, Fawang Liu, Xiaoyan Yang, Jing Jin, Xin Dong, Ke-Wu Zeng, Dong Liu, Yingtao Zhang, Ming Ma * and Donghui Yang *

State Key Laboratory of Natural and Biomimetic Drugs, Department of Natural Medicines,
School of Pharmaceutical Sciences, Peking University, 38 Xueyuan Road, Haidian District, Beijing 100191,
China; zmj216@bjmu.edu.cn (M.Z.); fawang90@126.com (F.L.); yaner1888@163.com (X.Y.);
jinjing@bjmu.edu.cn (J.J.); lingyu213@sohu.com (X.D.); ZKW@bjmu.edu.cn (K.-W.Z.);
liudong_1982@126.com (D.L.); zytao1988@163.com (Y.Z.)
* Correspondence: mma@bjmu.edu.cn (M.M.); ydhui@bjmu.edu.cn (D.Y.);
  Tel.: +86-10-82805794 (M.M.); +86-10-82801559 (D.Y.)

Received: 27 November 2017; Accepted: 5 January 2018; Published: 10 January 2018

Abstract: Nonribosomal peptides from marine *Bacillus* strains have received considerable attention for their complex structures and potent bioactivities. In this study, we carried out PCR-based genome mining for potential nonribosomal peptides producers from our marine bacterial library. Twenty-one "positive" strains were screened out from 180 marine bacterial strains, and subsequent small-scale fermentation, HPLC and phylogenetic analysis afforded *Bacillus* sp. PKU-MA00092 and PKU-MA00093 as two candidates for large-scale fermentation and isolation. Ten nonribosomal peptides, including four bacillibactin analogues (**1–4**) and six bacillomycin D analogues (**5–10**) were discovered from *Bacillus* sp. PKU-MA00093 and PKU-MA00092, respectively. Compounds **1** and **2** are two new compounds and the ^{1}H NMR and ^{13}C NMR data of compounds **7** and **9** is first provided. All compounds **1–10** were assayed for their cytotoxicities against human cancer cell lines HepG2 and MCF7, and the bacillomycin D analogues **7–10** showed moderate cytotoxicities with IC$_{50}$ values from 2.9 ± 0.1 to 8.2 ± 0.2 μM. The discovery of **5–10** with different fatty acid moieties gave us the opportunity to reveal the structure-activity relationships of bacillomycin analogues against these human cancer cell lines. These results enrich the structural diversity and bioactivity properties of nonribosomal peptides from marine *Bacillus* strains.

Keywords: bacillibactin; bacillomycin; genome mining; marine *Bacillus*; nonribosomal peptides

1. Introduction

Nonribosomal peptides produced by nonribosomal peptide synthetases (NRPS) from *Bacillus* strains are important sources of clinically used agents or bioactive molecules. Bacitracin, one mixture of nonribosomal peptides discovered from *Bacillus subtilis*, is used as an antibacterial drug, with the mode of action of interfering with cell wall and peptidoglycan synthesis [1]. Tyrocidines and gramicidins, nonribosomal peptides discovered from *Bacillus brevis* strains, show antibacterial activities [2,3], and their early biosynthetic research initiate the characterization of the functions of domains in NRPS assembly lines [4–8]. Bacillibactin, one nonribosomal peptide discovered from *Bacillus subtilis*, serves as a catecholic siderophore in iron acquisition that is essential to the host's life [9,10]. Other important nonribosomal peptides produced by *Bacillus* strains include surfactin, fengycin, and iturin, which represent three families of lipopeptides showing antibacterial, antifungal, and hemolytic activities owing to their cell membrane-interacting properties [11,12]. In last two decades, marine *Bacillus* strains as nonribosomal peptide producers have received considerable attention [13]. The marine environments characterized by high salinity, high pressure, and low temperature hold the promise of producing new natural products structurally or bioactively distinct with those from terrestrial environments. Various *Bacillus* species have been isolated from marine sources with different isolation methods [14–16], and new nonribosomal peptides with varied bioactivities have been discovered. Gageopeptides that were discovered from *Bacillus subtilis* collected from Gageocho reef, Republic of Korea were linear lipopeptides showing potent antifungal and antibacterial activities [17]. Bogorol A discovered from *Bacillus* sp. collected from Papua New Guinea was first linear "cationic peptide antibiotics", showing potent antibacterial activity [18]. Bacillistatins discovered from *Bacillus silvestris* collected from southern Chile ocean were cyclodepsipeptides showing potent antitumor activity [19]. More nonribosomal peptides with different bioactivities were summarized in Figure S1, depicting the potential of nonribosomal peptides from marine *Bacillus* strains as natural drug leads.

Genome mining targeting the NRPS genes is a useful tool for the discovery of new nonribosomal peptides. The investigation of sequenced genomic DNA based on bioinformatics analysis has been extensively applied to identify a specific nonribosomal peptide gene cluster and its corresponding natural product [20–22]. The PCR screening using degenerate primers with unsequenced genomic DNAs as the templates has been applied to discover 13 NRPS-containing *Bacillus* strains from 109 marine bacteria collection [14] and grisechelin-type nonribosomal peptides from an actinomycetes collection [23]. We here report the discovery of 10 nonribosomal peptides (**1–10**), including two new compounds (**1 and 2**) from our marine bacteria library (Figure 1), based on the PCR screening with degenerate primers designed from the conserved sequences of adenylation domains (A domain) in six nonribosomal peptides biosynthesis [24]. All the compounds **1–10** were assayed for their cytotoxicities against human cancer cell lines HepG2 and MCF-7, and compound **8** showed best cytotoxicities with IC$_{50}$ values of 2.9 ± 0.1 µM. These results highlight marine *Bacillus* strains as the rich sources of nonribosomal peptides and the power of genome mining in natural product discovery.

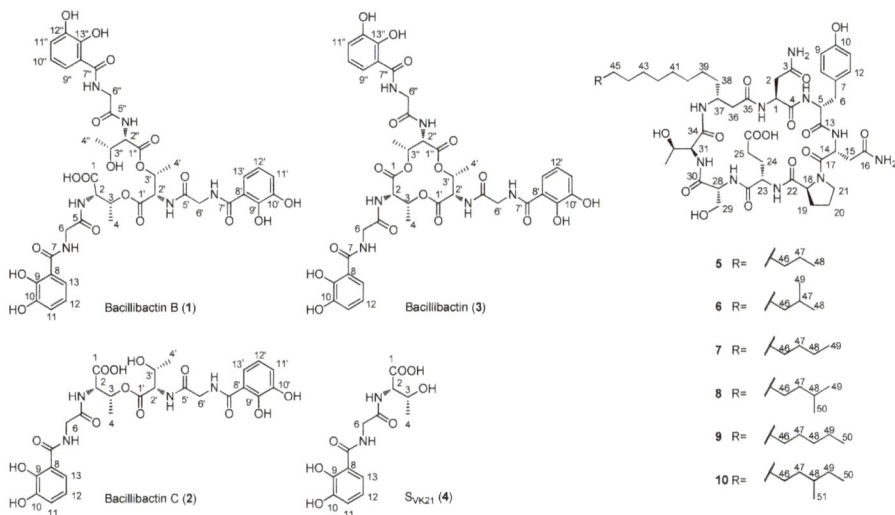

Figure 1. The structures of bacillibactin analogues (**1–4**) and bacillomycin D analogues (**5–10**) discovered from *Bacillus* sp. PKU-MA00093 and PKU-MA00092, respectively.

2. Results and Discussion

2.1. Genome Mining for Nonribosomal Peptides Discovery from Marine Bacteria Library Based on a PCR Screening Method

We have constructed a marine bacteria library containing strains isolated from sponges, corals and deposits collected from South China Sea and south coasts of China. To discover nonribosomal peptides from this library, we carried out a PCR screening targeting adenylation domains (A domain) in nonribosomal peptides biosynthesis with genomic DNAs of strains in the library as the templates, using the degenerate primers (A3F and A7R) that were designed from the conserved sequences of A domains in cephamycin, vancomycin, balhimycin, actinomycin, pristinamycin, and chloroeremomycin biosynthesis [24]. Totally 180 strains in the library were screened and 21 strains were identified as the "positive hits", which gave clear PCR products with expected size of ~700 bp based on the agarose gel electrophoresis analysis (Figure S2). All 21 PCR products were sequenced and confirmed to encode A domains with high sequence identities to known NRPS homologues (Table S1). The high sequence identities (82–99%, Table S1) to known NRPS homologues confirmed our PCR screening's accuracy, but excluded the possibility of new A domain sequences discovered. The same degenerate primers (A3F and A7R) have been used in the screening of NRPS genes from marine Actinobacteria strains [25–27]. These screening gave higher "positive" rates of 92% (22 out of 24) [25], 76% (19 out of 25) [26], and 63% (38 out of 60) [27], when compared to the lower "positive" rate of 12% (21 out of 180) in this screening. The degenerate primers (A3F and A7R) are specific for the screening of NRPS genes from Actinobacteria because they are designed from the conserved sequences of A domains in the biosynthesis of six nonribosomal peptides produced by Actinobacteria strains. We have amplified all the 16S rRNAs sequences of the 21 "positive" strains. The sequencing and BLAST results showed that the 21 "positive" strains consisted of 14 Firmicutes (14 from *Bacillus*) strains and seven Actinobacteria (four from *Rhodococcus*, one from *Streptomyces*, one from *Nocardiopsis*, and one from *Mycobacterium*) stains (Table S1). These results suggest that the majority of the 180 marine bacteria in this screening are not Actinobacteria, which may be the reason for the lower "positive" rate. Other PCR screening for NRPS genes from marine bacteria using different degenerate primers gave different phyla of "positive" strains with different "positive" rates (30% and 14%) [14,28], exemplifying the degenerate primers as another important factor affecting the screening result.

We then fermented the 21 strains in small-scale (50 mL) using four different media (see Materials and Methods), and analyzed the crude extracts by HPLC. The HPLC analysis showed that strains PKU-MA00092 and PKU-MA00093 provided abundant natural products in medium M2, under the UV detection of 210 nm. Thus, strains PKU-MA00092 and PKU-MA00093 were chosen for the large-scale fermentation in medium M2 and natural products isolation. Four bacillibactin analogues (**1–4**), including two new compounds (**1** and **2**) were isolated from a 4.5 L fermentation of PKU-MA00093, and six bacillomycin D analogues (**5–10**) were isolated from a 10.0 L fermentation of PKU-MA00092 (Figure 1). The structures of compounds **1–10** were elucidated based on comprehensive spectroscopic analyses. Bacillibactin and bacillomycin D have been previously isolated from different *Bacillus* strains [9,10,29]. To identify the genuses of strains PKU-MA00092 and PKU-MA00093, the housekeeping 16S rRNA genes of the two strains were amplified by PCR and sequenced. BLAST on the NCBI website showed that the two 16S rRNAs were highly homologous to 16S rRNAs from *Bacillus* strains. The phylogenetic analysis was carried out based on the sequence alignments of the two 16S rRNAs with different *Bacillus* homologues, identifying PKU-MA00092 and PKU-MA00093 as two *Bacillus* strains (Figure S3).

2.2. Structural Elucidation of Compounds **1–10**

Compound **1** was obtained as a light green gum. HRESIMS analysis afforded an $[M - H]^-$ ion at m/z 899.2584, giving the molecular formula of **1** as $C_{39}H_{44}N_6O_{19}$. The ^1H NMR spectrum of **1** resembled that of bacillibactin (**3**) [30,31], except that the resonances at δ_H 5.31 (d, $J = 6.8$ Hz,

H$_3$-3/3′/3″) in **3** split to δ_H 5.33 (br s, H-3), δ_H 5.38 (br s, H-3′) and δ_H 4.19 (br s, H-3″) in **1**, and the resonances at δ_H 4.59 (br s, H$_3$-2/2′/2″) in **3** split to δ_H 4.56 (br d, *J* = 8.0 Hz, H-2), δ_H 4.79 (br d, *J* = 8.0 Hz, H-2′), and δ_H 4.38 (br d, *J* = 8.0 Hz, H-2″) in **1** (Figure S4, Table 1). The ^{13}C NMR spectrum of **1** resembled that of bacillibactin (**3**) [30,31] except that the resonances at δ_C 70.8 (C-3/3′/3″) in **3** split to δ_C 70.7 (C-3/3′) and δ_C 66.3 (C-3″) in **1**, the resonances at δ_C 56.6 (C-2/2′/2″) in **3** split to δ_C 55.1 (C-2), δ_C 54.8 (C-2′) and δ_C 57.4 (C-2″) in **1**, the resonances at δ_C 16.5 (C-4/4′/4″) in **3** split to δ_C 16.4 (C-4), δ_C 16.3 (C-4′) and δ_C 20.2 (C-4″) in **1**, and the resonance attributed to C-1 at δ_C 168.4 in **3** shifted downfield to δ_C 170.9 in **1** (Figure S6, Table 1). These differences in combination with the molecular formula of **1** suggested that **1** was the hydrolyzed product of **3**, which was confirmed by the MS/MS analysis based on the observation of ions [a]$^-$, [b]$^-$, [c]$^-$, and [d]$^-$ at *m/z* 763.2365, 706.2209, 605.1732, and 311.0882, respectively (Figure S11). The linkage relations of 2,3-dihydroxybenzoic acid (DHBA), glycine and threonine in three units were confirmed by COSY, HSQC, and HMBC experiments (Figures 2, S5, S7 and S8). To further establish the absolute configuration of **1**, the Marfey's analysis [32] was carried out. HPLC analysis of the FDAA derivatives of the amino acids in **1** unambiguously established that **1** only contained L-threonine (Figure 3). Thus, compound **1** was identified as the hydrolyzed product of bacillibactin (**3**), and named as bacillibactin B.

Compound **2** was obtained as a light green gum. HRESIMS analysis afforded an [M + Na]$^+$ ion at *m/z* 629.1702, giving the molecular formula of **2** as C$_{26}$H$_{30}$N$_4$O$_{13}$. The ^{1}H NMR spectrum of **2** resembled that of S$_{VK21}$ (**4**) [33] except that the resonances at δ_H 4.20 (d, *J* = 6.0 Hz, H-3) in **4** split to δ_H 5.33 (m, H-3) and δ_H 4.21 (m, H-3′) in **2**, the resonances at δ_H 4.10 (br s, H-2) in **4** split to δ_H 4.61 (dd, *J* = 8.8, 2.5 Hz, H-2) and δ_H 4.39 (dd, *J* = 8.5, 2.5 Hz, H-2′) in **2**, the resonances at δ_H 1.04 (d, *J* = 6.3 Hz, H-4) in **4** split to δ_H 1.15 (d, *J* = 6.4 Hz, H-4) and δ_H 1.05 (d, *J* = 6.2 Hz, H-4′) in **2**, and the resonances at δ_H 7.86 (d, *J* = 8.3 Hz, 2-NH) in **4** split to δ_H 8.36 (d, *J* = 8.8 Hz, 2-NH) and δ_H 8.06 (d, *J* = 8.5 Hz, 2′-NH) in **2** (Figure S12, Table 2). The ^{13}C NMR spectrum of **2** resembled that of S$_{VK21}$ (**4**) [33] except that the resonances at δ_C 66.2 (C-3) in **4** split to δ_C 70.8 (C-3) and δ_C 66.4 (C-3′) in **2**, the resonances at δ_C 57.5 (C-2) in **4** split to δ_C 55.0 (C-2) and δ_C 57.4 (C-2′) in **2**, and the resonances at δ_C 20.1 (C-4) in **4** split to δ_C 16.5 (C-4) and δ_C 20.1 (C-4′) in **2** (Figure S14, Table 2). These differences in combination with the molecular formula of **2** suggested that **2** was the esterification product of two S$_{VK21}$ (**4**) units with the linkage between the two threonine residues. This was confirmed by the MS/MS analysis based on the observation of ions [a + H + Na]$^+$, [b + H + Na]$^+$, and [c + H + Na]$^+$ at *m/z* 493.1541, 436.1345, and 335.0855, respectively (Figure S19). The linkage relations of 2,3-dihydroxybenzoic acid (DHBA), glycine, and threonine in the two units were confirmed by COSY, HSQC, and HMBC experiments (Figures 2, S13, S15 and S16). The two L-threonine residues in **2** were established based on their similar chemical shifts, coupling constants and same biosynthetic pathway with that of **4**. Thus, compound **2** was identified as the esterification product of two S$_{VK21}$ (**4**) units and named as bacillibactin C.

Table 1. The ^{1}H (400 MHz) and ^{13}C NMR (100 MHz) data of compound **1** in DMSO-*d*$_6$.

Position	δ_H (*J* in Hz)	δ_C, Type	Position	δ_H (*J* in Hz)	δ_C, Type	Position	δ_H (*J* in Hz)	δ_C, Type
1		170.9, C	1′		168.0, C	1″		169.3 [b], C
2	4.56, br d (8.0)	55.1, CH	2′	4.79, br d (8.0)	54.8, CH	2″	4.38, br d (7.9)	57.4, CH
3	5.33, br s	70.7, CH	3′	5.38, br s	70.7, CH	3″	4.19, br s	66.3, CH
4	1.09, d (6.0)	16.4 [a], CH$_3$	4′	1.16, d (6.1)	16.3 [a], CH$_3$	4″	1.06, d (5.9)	20.2, CH$_3$
5		168.9 [b], C	5′		169.1 [b], C	5″		169.2 [b], C
6	4.09, m	41.8 [c], CH$_2$	6′	4.09, m	41.9 [c], CH$_2$	6″	4.03, m	42.2 [c], CH$_2$
7		169.49 [b], C	7′		169.52 [b], C	7″		169.7 [b], C
8		115.33 [d], C	8′		115.2 [d], C	8″		115.26 [d], C
9		149.05 [e], C	9′		149.14 [e], C	9″		149.2 [e], C
10		146.1, C	10′		146.1, C	10″		146.1, C
11	6.93, m	118.8, CH	11′	6.93, m	118.8, CH	11″	6.93, m	118.8, CH
12	6.69, m	118.1, CH	12′	6.69, m	118.1, CH	12″	6.69, m	118.1, CH
13	7.31, m	117.7 [f], CH	13′	7.31, m	117.6 [f], CH	13″	7.31, m	117.6 [f], CH
2-NH	8.33, br s		2′-NH	8.47, br s		2″-NH	8.14, br s	

Table 1. *Cont.*

Position	δ_H (*J* in Hz)	δ_C, Type	Position	δ_H (*J* in Hz)	δ_C, Type	Position	δ_H (*J* in Hz)	δ_C, Type
6-NH	9.04, d (5.1)		6′-NH	9.04, d (5.1)		6″-NH	9.13, m	
9-OH	12.31, br s		9′-OH	12.31, br s		9″-OH	12.31, br s	
10-OH	9.23, br s		10′-OH	9.23, br s		10″-OH	9.23, br s	

[a-f] Assignments may be interchanged.

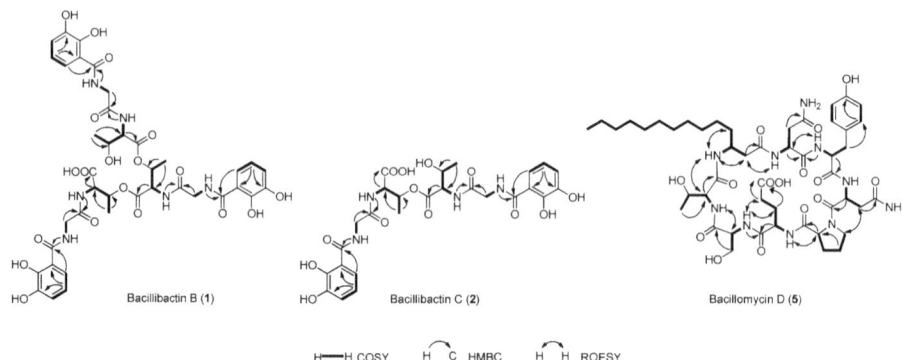

Bacillibactin B (1) Bacillibactin C (2) Bacillomycin D (5)

H——H COSY   H C HMBC   H H ROESY

Figure 2. The key COSY, HMBC, and ROESY correlations of compounds **1**, **2**, and **5**.

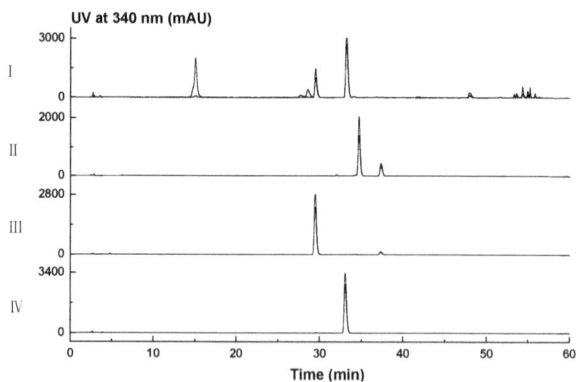

Figure 3. The Marfey's analysis for the determination of absolute configuration of threonine moiety in compound **1**. (I) The HPLC analysis for the FDAA derivative of hydrolysates of compound **1**, showing the peaks for FDAA derivatives of threonine and glycine between 28 and 35 min. (II) The HPLC analysis of FDAA derivative of D-threonine. (III) The HPLC analysis of FDAA derivative of L-threonine. (IV) The HPLC analysis of FDAA derivative of glycine.

Table 2. The ^{1}H (400 MHz) and ^{13}C NMR (100 MHz) data of compound **2** in DMSO-d_6.

Position	δ_H (*J* in Hz)	δ_C, Type	Position	δ_H (*J* in Hz)	δ_C, Type
1		170.7, C	1′		169.4 [a], C
2	4.61, dd (2.5, 8.8)	55.0, CH	2′	4.39, dd (2.5, 8.5)	57.4, CH
3	5.33, m	70.8, CH	3′	4.21, m	66.4, CH
4	1.15, d (6.4)	16.5, CH$_3$	4′	1.05, d (6.2)	20.1, CH$_3$
5		168.9 [a], C	5′		169.0 [a], C
6	4.06, m	41.9 [b], CH$_2$	6′	4.06, m	42.1 [b], CH$_2$

Table 2. *Cont.*

Position	δ_H (*J* in Hz)	δ_C, Type	Position	δ_H (*J* in Hz)	δ_C, Type
7		169.5 [a], C	7′		169.6 [a], C
8		115.3, C	8′		115.3, C
9		149.2, C	9′		149.2, C
10		146.1, C	10′		146.1, C
11	6.93, d (7.6)	118.8, CH	11′	6.93, d (7.6)	118.8, CH
12	6.70, t (7.6)	118.1, CH	12′	6.70, t (7.6)	118.1, CH
13	7.32, d (7.6)	117.7, CH	13′	7.32, d (7.6)	117.7, CH
2-NH	8.36, d (8.8)		2′-NH	8.06, d (8.5)	
6-NH	9.06, br s		6′-NH	9.06, br s	
9-OH	12.36, br s		9′-OH	12.36, br s	
10-OH	9.27, br s		10′-OH	9.27, br s	

[a,b] Assignments may be interchanged.

Compounds **3** and **4** was identified as bacillibactin and S$_{VK21}$, respectively, by comparison of their spectroscopic data with that of the authentic compounds [30,31,33]. The biological role of bacillibactin as the catecholic siderophore in iron acquisition has been characterized in details [10,34]. The biosynthetic research of bacillibactin has revealed how nature utilized NRPS machinery to synthesized this molecule. The standalone adenylation protein in the starter module, DhbE, activates 2,3-dihydroxybenzic acid (DHBA), and loaded this starter unit onto the first peptidyl carrier protein (PCP), initiating the downstream peptide elongation; the crystal structure of DhbE revealed the ten amino acids in the active sites determining the substrate selectivity [35]; the three-molular NRPS catalyzes the condensation between DHBA, glycine, and L-threonine to generate PCP-tethered linear tripeptides intermediate (thioester of compound **4**), then the thioesterase domain (TE) in the last module catalyzes the polymerization of S$_{VK21}$ units to generate the TE-tethered ester of compound **2**, followed by the formation of TE-tethered ester of compound **1** and the intramolecular cyclization to generate bacillibactin (**3**) [35,36]. In this study, we isolated compounds **1**, **2**, and **4** whose structures are corresponding to the three biosynthetic intermediates in bacillibactin biosynthesis. The HPLC analysis of the crude extract of PKU-MA00093 using an increasing gradient of MeOH in H$_2$O (containing no acids or bases) as the moble phase showed the co-occurrence of the peaks for compounds **1**–**4** (data not shown). Therefore, compounds **1**, **2**, and **4** may be isolated as the shunt products released from the NRPS machinery in the biosynthesis of bacillibactin (**3**).

Compound **5** was obtained as a white amorphous powder. HRESIMS analysis afforded an [M − H]$^-$ ion at *m/z* 1029.5260, giving the molecular formula of **5** as C$_{48}$H$_{74}$N$_{10}$O$_{15}$. The ^{1}H NMR and ^{13}C NMR spectra (Figures S27 and S29) resembled that of bacillomycin D [29,37]. The COSY, HSQC, and HMBC experiments (Figures 2, S28, S30 and S31) confirmed the presence of two asparagines, one tyrosine, one proline, one glutamate, one serine, one threonine, and one C$_{14}$-fatty acid moieties in **5**. The correlations of H-36 with 1-NH, H-1 with 5-NH, H-15 with H-21, H-18 with 23-NH, H-24 with 28-NH, H-25 with 28-NH, H-28 with 31-NH, H-31 with 37-NH, 37-NH with H-36, 37-NH with H-38 in the ROESY spectrum (Figures 2 and S32) established the cyclo(Asn-Tyr-Asn-Pro-Glu-Ser-Thr-C$_{14}$-fatty acid) structure of **5**. To confirm that the amino acids in **5** have the same absolute configurations as that in bacillomycin D, the Marfey's analysis was carried out. HPLC analysis of the FDAA derivatives of the amino acids established that **5** contained same L-asparagine, D-asparagine, D-tyrosine, L-proline, L-glutamate, D-serine, and L-threonine residues as bacillomycin D (Figure S49). Thus, compound **5** was identified as bacillomycin D.

The analyses of the ^{1}H NMR and ^{13}C NMR spectra of compounds **6**–**10** showed that they contained the same cyclopeptide skeleton as compound **5**. However, their different molecular weights (see Section 3.6 part in Materials and Methods) determined by ESIMS analyses showed that they contained different fatty acid moieties. Compound **6** contained one C$_{15}$-fatty acid moiety with a branch terminal, which was established by the HRESIMS analysis (Figure S36) and the resonances attributed to two methyl groups at δ_H 0.84 (d, *J* = 6.3 Hz, H$_3$-48, and H$_3$-49) in the ^{1}H NMR spectrum

(Figure S34, Table S4); Compound **7** contained one C_{15}-fatty acid moiety with a linear terminal, which was established by the HRESIMS analysis (Figure S39) and the resonance attributed to the methyl group at δ_H 0.88 (t, J = 7.4 Hz, H$_3$-49) in the ^{1}H NMR spectrum (Figure S37, Table S5); Compound **8** contained one C_{16}-fatty acid moiety with a branch terminal, which was established by the ESIMS analysis (Figure S42) and the resonances attributed to two methyl groups at δ_H 0.84 (d, J = 6.5 Hz, H$_3$-49 and H$_3$-50) in the ^{1}H NMR spectrum (Figure S40, Table S6); Compound **9** contained one C_{16}-fatty acid moiety with a linear terminal, which was established by the ESIMS analysis (Figure S45) and the resonance attributed to the methyl group at δ_H 0.87 (t, J = 6.4 Hz, H$_3$-50) in the ^{1}H NMR spectrum (Figure S43, Table S7); Compound **10** contained one C_{17}-fatty acid moiety with a branch, which was established by the ESIMS analysis (Figure S48) and the resonances attributed to two methyl groups at δ_H 0.84 (t, J = 7.3 Hz, H$_3$-50) and δ_H 0.83 (d, J = 6.3 Hz, H$_3$-51) in the ^{1}H NMR spectrum (Figure S46, Table S8), and the correlations of H$_3$-51 with H-48, H-48 with H$_2$-49, H$_2$-49 with H$_3$-50 in the COSY spectrum (Figure S47). Thus, compounds **6**, **8**, and **10** were identified as iso-C_{15} bacillomycin D [38], iso-C_{16} bacillomycin D [38], and anteiso-C_{17} bacillomycin D [38], respectively. Compounds **7** and **9** were identified as C_{15} bacillomycin D and C_{16} bacillomycin D, respectively. Compounds **7** and **9** were reported previously without NMR data provided [39,40], and we here report the structures of **7** and **9** with fully assigned ^{1}H NMR and ^{13}C NMR data for the first time.

The separation of compounds **5**–**10** is challenging because their structures are highly similar with only the fatty acid moieties different. Due to the difficulties in the separation, compounds **5** and **6** were purified as a mixture in the cytotoxicity assay against Ehrlich carcinoma tumor cell line and hemolytic activity assay [37]. The structural elucidation of compounds **5**–**10** is also challenging, because the large molecular weights and the complex structures of **5**–**10** require large amount of samples for 1D and 2D NMR analyses. Due to the little amount obtained, two bacillomycin D analogues with C_{14} and C_{15} fatty acid moieties [41] and five bacillomycin D analogues with C_{14} to C_{18} fatty acid moieties [42] were proposed only based on LC-MS analyses. The biosynthetic gene cluster of bacillomycin D has been cloned and sequenced [12,43]. The acyl-CoA ligase domain in the first module is responsible for the ligation of the fatty acid moiety to the polyketide synthase (PKS)-NRPS hybrid assembly line. The isolation of compounds **5**–**10** in this study suggested that the acyl-CoA ligase domain possessed substrate promiscuity, which allowed for the ligation of different length and branches of fatty acid moieties to the polyketide-nonribosomal peptide hybrid macrocycle. The substrate promiscuity of the acyl-CoA ligase domain can be employed in combinatorial biosynthesis for the generation of more bacillomycin D analogues.

2.3. Cytotoxicity Assays against Human Cancer Cell Lines

Compounds **1**–**10** were tested for their antitumor activities. Two human cancer cell lines, liver cancer cell line HepG2, and breast cancer cell line MCF7, were used in the assays. Compounds **7**–**10** showed moderate cytotoxicities with IC$_{50}$ values from 2.9 ± 0.1 to 8.2 ± 0.2 µM and compounds **1**–**6** show no cytotoxicities (inhibition rates were below 20% at the concentration of 10 µM) (Table 3). The isolation of **5**–**10** allowed us to analyze the effections of different fatty acid moieties on the cytotoxicities. Compound **5** with a C_{14}-fatty acid moiety showed no cytotoxicities (inhibition rates were below 20% at the concentration of 10 µM); both compounds **6** and **7** contain a C_{15}-fatty acid moiety, however, compound **7** with a linear C_{15}-fatty acid moiety showed moderate cytotoxicities and **6** with a branched C_{15}-fatty acid moiety lost the cytotoxicities, suggesting that a linear C_{15}-fatty acid moiety is important to the cytotoxicities; compounds **8** and **9** containing a C_{16}-fatty acid moiety showed most potent cytotoxicities; compound **10** containing one more methyl group attached at C-49 when compared to **8** or attached at C-48 when compared to **9**, showed decreased cytotoxicity against MCF7 and lost the cytotoxicity against HepG2, suggesting a C_{17}-fatty acid moiety with terminal branch is disadvantageous to the cytotoxicities. In conclusion, bacillomycin D analogues with a C_{16}-fatty acid moiety may possess best cytotoxicities against HepG2 and MCF7 cancer cell lines. The bacillomycin D analogues have been mostly assayed for their antifungal activities [44], especially against fungi of plant pathogens [38,40,45].

For their antitumor activities, a mixture of compounds **5** and **6**'s moderate cytotoxicities have been reported against mice Ehrlich's carcinoma cells [37]. In another report, one lipopeptide was proposed as bacillomycin D (**5**) only based on LC-MS analysis and showed inhibitory activities against three human cancer cell lines A549, A498 and HCT-15 [46]. The cytotoxicities against human cancer cell lines HepG2 and MCF7 and the structure-activity relationships that we first revealed here enriched our understanding of the diversity of bacillomycin D analogues' bioactivities.

Table 3. The cytotoxicity assays of compounds **7–10**.

	Cytotoxicity (IC$_{50}$ Values in µM)				
Cell Lines	7	8	9	10	Taxol
HepG2	8.2 ± 0.2	5.1 ± 0.2	4.9 ± 0.2	>10	0.075 ± 0.004
MCF7	4.2 ± 0.1	2.9 ± 0.1	3.3 ± 0.1	7.2 ± 0.2	0.043 ± 0.003

3. Materials and Methods

3.1. General Experimental Procedures

Optical rotations were measured on an Autopol III automatic polarimeter (Rudolph Research Analytical, Hackettstown, NJ, USA). UV spectra were collected on a Cary 300 spectrometer (VARIAN, Palo Alto, CA, USA). IR spectra were collected on a Nicolet Nexus 470 FT-IR spectrometer (Thermo Scientific, Waltham, MA, USA). ^{1}H and ^{13}C NMR spectra were collected on a Bruker Avance-400FT NMR spectrometer (Bruker Corporation, Billerica, MA, USA). HRESIMS and MS/MS spectra were collected on a Waters Xevo G2 Q-TOF spectrometer (Waters, Milford, MA, USA). ESIMS spectra were collected on a Waters SQD Single Quadruple Mass spectrometer (Waters, Milford, MA, USA). HPLC analysis was performed on an Agilent 1260 series (Agilent Technologies, Santa Clara, CA, USA) with a diode array detector (Agilent Technologies, Santa Clara, CA, USA) and a C$_{18}$ RP-column (Eclipse XDBC$_{18}$, 150 × 4.6 mm, 5 µm, Agilent Technologies, Santa Clara, CA, USA). Semi-preparative HPLC was performed on a SSI 23201 system (Scientific Systems Inc., State College, PA, USA) with a YMC-Pack ODS-1 column (250 × 10 mm, 5 µm, YMC Co., Ltd., Shimogyo-ku, Kyoto, Japan). MPLC was performed on a LC3000 series (Beijing Tong Heng Innovation Technology, Beijing, China) with a Claricep^{TM} Flash i-series C$_{18}$ cartridge (20–35 µm, 40 g, Bonna-Agela, Wilmington, DE, USA). Size exclusion chromatography was carried out using a Sephadex LH-20 (Pharmacia Fine Chemicals, Piscataway, NJ, USA) column.

3.2. Marine Bacterial Strains Isolation

Total 15 Sponge samples were collected from Yongxin Island in the South China Sea and were deposited at the State Key Laboratory of Natural and Biomimetic Drugs, Peking University, China. The isolation of the marine bacterial strains was carried out according to the published procedures [47]. The sponge samples were thoroughly washed with sterilized water containing 3% sea salt for three times, and cut into pieces of about 1 cm^3, then homogenized in a sterilized mortar with 9 mL sterilized water containing 3% sea salt. The homogenates were incubated in a water bath at 55 °C for 6 min, and put into biosafety cabinet for 30 min, and the supernatants were serially diluted by sterilized water containing 3% sea salt and plated in triplicate on agar plates for each sponge sample. Four media, M1 (yeast extract 1 g, peptone 5 g, beef extract 1 g, FePO$_4$ 0.01 g, agar 18 g, and sea salt 33 g in 1.0 L distilled water, pH 7.4), M2 (glycerol 6 mL, arginine 1 g, K$_2$HPO$_4$ 1 g, MgSO$_4$ 0.5 g, agar 18 g, sea salt 33 g in 1.0 L distilled water, pH 7.2), M3 (soluble starch 20 g, KNO$_3$ 1 g, K$_2$HPO$_4$ 0.5 g, MgSO$_4$·7H$_2$O 0.5 g, NaCl 0.5 g, FeSO$_4$·7H$_2$O 0.01 g, agar 18 g, sea salt 33 g in 1.0 L distilled water, pH 7.2), and M4 (sodium acetate trihydrate 5 g, peptone 0.5 g, yeast extract 0.5 g, glucose 0.5 g, sucrose 0.5 g, sodium citrate 0.05 g, malic acid 0.05 g, NH$_4$NO$_3$ 1 g, NH$_4$Cl 0.2 g, KH$_2$PO$_4$ 0.5 g, agar 18 g, sea salt 33 g in 1.0 L distilled water, pH 7.6), were used for bacterial strains isolation. Each plate was supplemented with cycloheximide (50 µg/mL) and nalidixic acid (50 µg/mL) to suppress the growth of fungi and Gram-negative bacteria, and incubated at 28 °C for four weeks. Different colonies of strains were carefully selected and restreaked on agar plates with medium M1 for further proliferation and dereplication. Totally 180 bacterial strains were isolated and stored at −80 °C in 20% glycerol.

3.3. PCR Screening for Potential Producers of Nonribosomal Peptides

The genomic DNAs of all 180 strains were extracted following standard protocols [48]. The forward primer (5′-GCSTACSYSATSTACACSTCSGG-3′) and the reverse primer (5′-SASGTCVCCSGTSCGGTAS-3′), designed from the conserved sequences of adenylation domains (A domains) in the biosynthesis of cephamycin, vancomycin, balhimycin, actinomycin, pristinamycin, and chloroeremomycin, were used in the screening [24]. A 20 µL PCR system consisting of 10 µL Easy Taq Polymerase (Beijing TransGen Biotech, Beijing, China), 2 µL of forward and reverse primer mixture (each for 10 µM), 1 µL genomic DNA, and 7 µL sterilized water, were used. The PCR program was performed with an initial denaturation at 95 °C for 5 min, followed by 30 cycles of denaturation at 95 °C for 30 s, annealing at 57 °C for 1 min and extension at 72 °C for 1 min, followed by incubation at 72 °C for 10 min. The PCR products were analyzed by agarose gel electrophoresis (Figure S2), recovered with a gel purification kit (Beijing TransGen Biotech, Beijing, China) and sequenced to afford 21 "positive" strains.

3.4. Small-Scale Fermentation, HPLC

For each of the 21 "positive" strains, 50 µL of spore suspension was inoculated into 50 mL of seed medium (medium M1 broth), and incubated with a HYG-C shaker (Suzhou Peiying Laboratory Equipment, Suzhou, China) at 28 °C, 200 rpm for three days. Three milliliter of the resultant seed culture was inoculated into 50 mL of production media, and the fermentation continued at 28 °C, 200 rpm for seven days. Four different production media, media M1, M2, and M3 broth, and M5 (soluble starch 10 g, casein 0.3 g, K_2HPO_4 2 g, KNO_3 2 g, $MgSO_4·7H_2O$ 0.05 g, NaCl 2 g, $FeSO_4·7H_2O$ 0.01 g, $CaCO_3$ 0.02 g, sea salt 33 g in 1.0 L distilled water, pH 7.2), were used in the small-scale fermentation. The Diaion HP20 (2 g/100 mL, Mitsubishi Chemical Corporation, Tokyo, Japan) and Amberlite XAD-16 (2 g/100 mL, Sigma-Aldrich, St. Louis, MO, USA) resins were added 10 h before the fermentation finished. The resins and cell mass were harvested by centrifugation and extracted with MeOH. The MeOH extracts were concentrated and analyzed by HPLC. The HPLC analysis was carried out with a flow rate of 1 mL/min with UV detection at 210 nm, using a gradient elution program from 5% MeOH in H_2O to 100% MeOH over 45 min.

3.5. Phylogenetic Analysis

The phylogenetic analysis of strains PKU-MA00092 and PKU-MA00093 was carried out by sequence alignments of their 16S rRNAs with different *Bacillus* homologues. The primers 27F (5′-AGAGTTTGATCMTGGCTCAG-3′) and 1492R (5′-TACGGYTACCTTGTTACGACTT-3′) [49] were used to amplify the 16S rDNA genes (GeneBank accession number MF774821 for PKU-MA00092 and KY780192 for PKU-MA00093). Homologous genes were searched using BLAST on the NCBI website and the phylogenetic tree was generated with Mega 5.1 using the Neighbor-Joining algorithm.

3.6. Large-Scale Fermentation and Isolation

The large-scale fermentation of *Bacillus* sp. PKU-MA00093 and *Bacillus* sp. PKU-MA00092 in medium M2 broth were carried out with similar procedures used in the small-scale fermentation. For *Bacillus* sp. PKU-MA00092 or PKU-MA00093, 50 µL of spore suspension was inoculated into 50 mL of seed medium (medium M1 broth, pH 7.4) and incubated with HYG-C shakers at 28 °C, 200 rpm for three days. Twelve milliliter of the resultant seed culture was inoculated into 200 mL of medium M2 broth (pH 7.2) in 1 L Erlenmeyer flasks, and the fermentation continued at 28 °C, 200 rpm for seven days. The Diaion HP20 (2 g/100 mL), and Amberlite XAD-16 (2 g/100 mL) resins were added 10 h before the fermentation finished. The resins and cell mass were harvested by centrifugation and extracted with MeOH.

A 4.5 L fermentation of *Bacillus* sp. PKU-MA00093 gave a 4.3 g of MeOH extract. The MeOH extract was concentrated in a vacuum and suspended in deionized water, extracted with ethyl acetate for three times. The ethyl acetate extracts were combined (1.4 g) and loaded onto MPLC using step-gradient elution of MeOH in H_2O (10%, 20%, 35%, 50%, 70%) to yield five fractions (F1–F5). Fraction F4 (147 mg) was purified by semi-preparative HPLC with MeOH/H_2O (44/56, v/v) as the mobile phase to yield **1** (4.6 mg); fraction F2 (169 mg) was purified by semi-preparative HPLC with MeOH/H_2O (37/63, v/v) as the mobile phase to yield compound **2** (7.8 mg) and **4** (1.4 mg); fraction F3 (231 mg) was purified by semi-preparative HPLC with MeOH/H_2O (40/60, v/v) as the mobile phase to yield **3** (17.4 mg). All of the semi-preparative HPLCs were carried out at a flowrate of 2 mL/min and under the detection of 248 nm.

A 10.0 L fermentation of *Bacillus* sp. PKU-MA00092 gave a 1.5 g of MeOH extract. The MeOH extract was concentrated and loaded onto MPLC using step-gradient elution of MeOH in H_2O (20%, 40%, 60%, 80%) to yield four fractions (F1–F4). Fraction F4 (368 mg) was purified repeatedly by semi-preparative HPLC with CH_3CN/H_2O (45/55 v/v) containing 0.01% TFA as the mobile phase, to afford compounds **5** (36.5 mg), **6** (15.7 mg), **7** (11.4 mg), **8** (13.3 mg), **9** (14.5 mg) and **10** (3.1 mg), at a flowrate of 2 mL/min and under the detection of 210 nm.

Bacillibactin B (**1**): light green gum; $[\alpha]_D^{25}$ +37 (*c* 0.23, MeOH); UV (MeOH) λ_{max} (log ε): 207 (1.47), 248 (0.40), 312 (0.16) nm; IR (KBr) ν_{max} 3317, 1740, 1642, 1541, 1266, 1023, 745 cm^{-1} (Figure S10); ^{1}H NMR and ^{13}C NMR data, Table 1; HRESIMS *m/z* 899.2584 [M − H]$^-$ (calcd. for $C_{39}H_{43}N_6O_{19}$, 899.2580).

Bacillibactin C (**2**): light green gum; $[\alpha]_D^{25}$ +40 (*c* 0.27, MeOH); UV (MeOH) λ_{max} (log ε): 206 (2.22), 249 (1.12), 312 (0.54) nm; IR (KBr) ν_{max} 3320, 1738, 1644, 1540, 1264, 1022, 746 cm^{-1} (Figure S18); ^{1}H NMR and ^{13}C NMR data, Table 2; HRESIMS *m/z* 605.1740 [M − H]$^-$ (calcd. for $C_{26}H_{29}N_4O_{13}$, 605.1731); *m/z* 629.1702 [M + Na]$^+$ (calcd. for $C_{26}H_{30}N_4O_{13}Na$, 629.1705).

Bacillibactin (**3**): light green gum; ^{1}H NMR and ^{13}C NMR data, Supplementary Table S2; HRESIMS *m/z* 883.2621 [M + H]$^+$ (calcd. for $C_{39}H_{43}N_6O_{18}$, 883.2631).

S_{VK21} (**4**): light green gum; ^{1}H NMR and ^{13}C NMR data, Supplementary Table S2; HRESIMS *m/z* 311.0878 [M − H]$^-$ (calcd. for $C_{13}H_{15}N_2O_7$, 311.0878).

Bacillomycin D (**5**): white amorphous powder; ^{1}H NMR and ^{13}C NMR data, Supplementary Table S3; HRESIMS *m/z* 1029.5260 [M − H]$^-$ (calcd. for $C_{48}H_{73}N_{10}O_{15}$, 1029.5254).

iso-C_{15} Bacillomycin (**6**): white amorphous powder; ^{1}H NMR and ^{13}C NMR data, Supplementary Table S4; HRESIMS *m/z* 1043.5421 [M − H]$^-$ (calcd. for $C_{49}H_{75}N_{10}O_{15}$, 1043.5411).

C_{15} Bacillomycin (**7**): white amorphous powder; ^{1}H NMR and ^{13}C NMR data, Supplementary Table S5; HRESIMS *m/z* 1043.5411 [M − H]$^-$ (calcd. for $C_{49}H_{75}N_{10}O_{15}$, 1043.5411).

iso-C_{16} Bacillomycin (**8**): white amorphous powder; ^{1}H NMR and ^{13}C NMR data, Supplementary Table S6; ESIMS *m/z* 1057.34 [M − H]$^-$.

C_{16} Bacillomycin (**9**): white amorphous powder; ^{1}H NMR and ^{13}C NMR data, Supplementary Table S7; ESIMS *m/z* 1057.43 [M − H]$^-$.

anteiso-C_{17} Bacillomycin (**10**): white amorphous powder; ^{1}H NMR data, Supplementary Table S8; ESIMS *m/z* 1071.45 [M − H]$^-$.

3.7. Marfey's Analysis for Compounds **2** *and* **5**

The Marfey's analysis of compounds **2** and **5** was carried out with same procedure [32,50]. The peptide (1.0 mg) was hydrolyzed with 500 μL of 6 N HCl in a sealed glass vial (1.5 mL) at 110 °C for 18 h, and dried with N_2 gas. The hydrolysate was dissolved in 50 μL of H_2O, followed by the addition of 100 μL FDAA (1-fluoro-2,4-dinitrophenyl-5-L-alanine amide, 1% in acetone, 3.7 μmol) and 20 μL of 1 M NaHCO$_3$. The mixture was heated at 45 °C for 1 h, followed by the addition of 10 μL of 2 M HCl to terminate the reaction. The FDAA-derivatives were dried with N_2 gas and dissolved in 1 mL of 50% CH_3CN in H_2O for HPLC analysis. The HPLC was carried out with a C_{18} column (50 × 4.6 mm, 5 μm, Grace Corporate, Columbia, MD, USA) and with a program of linear gradient of 10% CH_3CN in H_2O containing 0.1% TFA to 40% CH_3CN in H_2O containing 0.1% TFA over 40 min, at a flow rate of 1.0 mL/min, and under the detection of 340 nm. The FDAA-derivatives of standard L or D-amino acids were used as comparison.

*3.8. Cytotoxicity Assays for Compounds **1–10***

HepG2 and MCF7 cell lines were purchased from Peking Union Medical College, Cell Bank (Beijing, China). In this study, the cells were grown in Dulbecco's modified Eagle's medium (Hyclone, Waltham, MA, USA), supplemented with 10% fetal bovine serum (FBS, PAN Biotech, Aidenbach, Germany), penicillin (100 U/mL), and streptomycin (100 μg/mL) in a humidified incubator containing 95% air and 5% CO_2 at 37 °C. Cell viability was detected by MTT colorimetric assay. Briefly, the cells were treated with different concentrations of compounds (in DMSO stock) for 48 h, with taxol as the positive control. Then, culture supernatants were replaced with medium containing 0.5 mg/mL MTT for 4 h. The supernatant was removed and 100 μL of dimethyl sulfoxide was added. Absorbance was detected at 550 nm and cell viability was expressed as the mean percentage of the absorbance in treated vs. control cells. Different concentrations of compounds ranging from 0.1 to 40 μM were tested and each of the concentration was tested in parallel triplicate. Six concentration points in the slope region were used in the calculation of IC_{50} values. The IC_{50} values were calculated from the concentration–response curves that were generated by nonlinear regression with Graph Pad Prism 5 software (Graph Pad Software, Inc., San Diego, CA, USA).

Supplementary Materials: The following are available online at www.mdpi.com/1660-3397/16/1/22/s1, Figure S1: nonribosomal peptides from marine-derived *Bacillus* species, Figure S2: the agarose gel electrophoresis analysis of the 21 "positive" PCR products, Figure S3: the phylogenetic analysis of strains PKU-MA00092 and PKU-MA00093, Figures S4–S11: the NMR, MS and IR spectra of compound **1**, Figures S12–S19: the NMR, MS and IR spectra of compound **2**, Figures S20–S23: the NMR and MS spectra of compound **3**, Figures S24–S26: the NMR and MS spectra of compound **4**, Figures S27–S33: the NMR and MS spectra of compound **5**, Figures S34–S36: the NMR and MS spectra of compound **6**, Figures S37–S39: the NMR and MS spectra of compound **7**, Figures S40–S42: the NMR and MS spectra of compound **8**, Figures S43–S45: the NMR and MS spectra of compound **9**, Figures S46–S48: the NMR and MS spectra of compound **10**, Figure S49: the Marfey's analysis of compound **5**. Table S1: the homologues of the 21 "positive" strains and their PCR products, Tables S2–S8: the ^{1}H NMR and ^{13}C NMR data of compounds **3–10**.

Acknowledgments: This research was financially supported by the National Natural Science Foundation of China (grant number 81573326 and grant number 81673332).

Author Contributions: M.Z. and F.L. contributed equally; M.M. and D.Y. conceived and designed the experiments; M.Z., F.L., X.Y., J.J. and X.D. performed the experiments; K.-W.Z., D.L., Y.Z., M.M. and D.Y. analyzed the data; M.M. wrote the manuscript.

Conflicts of Interest: The authors declare no conflict of interest.

References

1. Johnson, B.A.; Anker, H.; Meleney, F.L. Bacitracin: A new antibiotic produced by a member of the *B. subtilis* group. *Science* **1945**, *102*, 376–377. [CrossRef] [PubMed]
2. Danders, W.; Marahiel, M.A.; Krause, M.; Kosui, N.; Kato, T.; Izumiya, N.; Kleinkauf, H. Antibacterial action of gramicidin S and tyrocidines in relation to active transport, in vitro transcription, and spore outgrowth. *Antimicrob. Agents Chemother.* **1982**, *22*, 785–790. [CrossRef] [PubMed]
3. Liou, J.W.; Hung, Y.J.; Yang, C.H.; Chen, Y.C. The antimicrobial activity of gramicidin A is associated with hydroxyl radical formation. *PLoS ONE* **2015**, *10*, e0117065. [CrossRef] [PubMed]
4. Felnagle, E.A.; Jackson, E.E.; Chan, Y.A.; Podevels, A.M.; Berti, A.D.; McMahon, M.D.; Thomas, M.G. Nonribosomal peptide synthetases involved in the production of medically relevant natural products. *Mol. Pharm.* **2008**, *5*, 191–211. [CrossRef] [PubMed]

5.   Mootz, H.D.; Marahiel, M.A. The tyrocidine biosynthesis operon of *Bacillus brevis*: Complete nucleotide sequence and biochemical characterization of functional internal adenylation domains. *J. Bacteriol.* **1997**, *179*, 6843–6850. [CrossRef] [PubMed]

6.   Linne, U.; Stein, D.B.; Mootz, H.D.; Marahiel, M.A. Systematic and quantitative analysis of protein-protein recognition between nonribosomal peptide synthetases investigated in the tyrocidine biosynthetic template. *Biochemistry* **2003**, *42*, 5114–5124. [CrossRef] [PubMed]

7.   Krätzschmar, J.; Krause, M.; Marahiel, M.A. Gramicidin S biosynthesis operon containing the structural genes *grsA* and *grsB* has an open reading frame encoding a protein homologous to fatty acid thioesterases. *J. Bacteriol.* **1989**, *171*, 5422–5429. [CrossRef] [PubMed]

8.   Conti, E.; Stachelhaus, T.; Marahiel, M.A.; Brick, P. Structural basis for the activation of phenylalanine in the non-ribosomal biosynthesis of gramicidin S. *EMBO J.* **1997**, *16*, 4174–4183. [CrossRef] [PubMed]

9.   May, J.J.; Wendrich, T.M.; Marahiel, M.A. The dhb operon of *Bacillus subtilis* encodes the biosynthetic template for the catecholic siderophore 2,3-dihydroxybenzoate-glycine-threonine trimeric ester bacillibactin. *J. Biol. Chem.* **2001**, *276*, 7209–7217. [CrossRef] [PubMed]

10. Segond, D.; Abi Khalil, E.; Buisson, C.; Daou, N.; Kallassy, M.; Lereclus, D.; Arosio, P.; Bou-Abdallah, F.; Nielsen Le Roux, C. Iron acquisition in *Bacillus cereus*: The roles of IlsA and bacillibactin in exogenous ferritin iron mobilization. *PLoS Pathog.* **2014**, *10*, e1003935. [CrossRef] [PubMed]

11. Inès, M.; Dhouha, G. Lipopeptide surfactants: Production, recovery and pore forming capacity. *Peptides* **2015**, *71*, 100–112. [CrossRef] [PubMed]

12. Roongsawang, N.; Washio, K.; Morikawa, M. Diversity of nonribosomal peptide synthetases ivolved in the biosynthesis of lipopeptide biosurfactants. *Int. J. Mol. Sci.* **2011**, *12*, 141–172. [CrossRef] [PubMed]

13. Mondol, M.A.M.; Shin, H.J.; Islam, M.T. Diversity of secondary metabolites from marine *Bacillus* species: Chemistry and biological activity. *Mar. Drugs* **2013**, *11*, 2846–2872. [CrossRef] [PubMed]

14. Zhang, W.; Li, Z.; Miao, X.; Zhang, F. The screening of antimicrobial bacteria with diverse novel nonribosomal peptide synthetase (NRPS) genes from South China Sea sponges. *Mar. Biotechnol.* **2009**, *11*, 346–355. [CrossRef] [PubMed]

15. Toledo, G.; Green, W.; Gonzalez, R.A.; Christoffersen, L.; Podar, M.; Chang, H.W.; Hemscheidt, T.; Trapido-Rosenthal, H.G.; Short, J.M.; Bidigare, R.R.; et al. High throughput cultivation for isolation of novel marine microorganisms. *Oceanography* **2006**, *19*, 120–125. [CrossRef]

16. Margassery, L.M.; Kennedy, J.; O'Gara, F.; Dobson, A.D.; Morrissey, J.P. Diversity and antibacterial activity of bacteria isolated from the coastal marine sponges *Amphilectus fucorum* and *Eurypon major*. *Lett. Appl. Microbiol.* **2012**, *55*, 2–8. [CrossRef] [PubMed]

17. Tareq, F.S.; Lee, M.A.; Lee, H.S.; Lee, Y.J.; Lee, J.S.; Hasan, C.M.; Islam, M.T.; Shin, H.J. Non-cytotoxic antifungal agents: Isolation and structures of gageopeptides A-D from a *Bacillus* strain 109GGC020. *J. Agric. Food Chem.* **2014**, *62*, 5565–5572. [CrossRef] [PubMed]

18. Barsby, T.; Kelly, M.T.; Gagné, S.M.; Andersen, R.J. Bogorol A produced in culture by a marine *Bacillus* sp. reveals a novel template for cationic peptide antibiotics. *Org. Lett.* **2001**, *3*, 437–440. [CrossRef] [PubMed]

19. Pettit, G.R.; Knight, J.C.; Herald, D.L.; Pettit, R.K.; Hogan, F.; Mukku, V.J.R.V.; Hamblin, J.S.; Dodson, M.J., II; Chapuis, J.C. Isolation and structure elucidation of bacillistatins 1 and 2 from a marine *Bacillus silvestris*. *J. Nat. Prod.* **2009**, *72*, 366–371. [CrossRef] [PubMed]

20. Lautru, S.; Deeth, R.J.; Bailey, L.M.; Challis, G.L. Discovery of a new peptide natural product by *Streptomyces coelicolor* genome mining. *Nat. Chem. Biol.* **2005**, *1*, 265–269. [CrossRef] [PubMed]

21. Gross, H.; Stockwell, V.O.; Henkels, M.D.; Nowak-Thompson, B.; Loper, J.E.; Gerwick, W.H. The genomisotopic approach: A systematic method to isolate products of orphan biosynthetic gene clusters. *Chem. Biol.* **2007**, *14*, 53–63. [CrossRef] [PubMed]

22. Wyatt, M.A.; Wang, W.; Roux, C.M.; Beasley, F.C.; Heinrichs, D.E.; Dunman, P.M.; Magarvey, N.A. *Staphylococcus aureus* nonribosomal peptide secondary metabolites regulate virulence. *Science* **2010**, *329*, 294–296. [CrossRef] [PubMed]

23. Xie, P.; Ma, M.; Rateb, M.E.; Shaaban, K.A.; Yu, Z.; Huang, S.X.; Zhao, L.X.; Zhu, X.; Yan, Y.; Peterson, R.M.; et al. Biosynthetic potential-based strain prioritization for natural product discovery: A showcase for diterpenoid-producing actinomycetes. *J. Nat. Prod.* **2014**, *77*, 377–387. [CrossRef] [PubMed]

24. Ayuso-Sacido, A.; Genilloud, O. New PCR primers for the screening of NRPS and PKS-I systems in actinomycetes: Detection and distribution of these biosynthetic gene sequences in major taxonomic groups. *Microb. Ecol.* **2005**, *49*, 10–24. [CrossRef] [PubMed]

25. Jiang, S.; Sun, W.; Chen, M.; Dai, S.; Zhang, L.; Liu, Y.; Lee, K.J.; Li, X. Diversity of culturable actinobacteria isolated from marine sponge *Haliclona* sp. *Antonie Van Leeuwenhoek* **2007**, *92*, 405–416. [CrossRef] [PubMed]

26. Abdelmohsen, U.R.; Yang, C.; Horn, H.; Hajjar, D.; Ravasi, T.; Hentschel, U. Actinomycetes from Red Sea sponges: Sources for chemical and phylogenetic diversity. *Mar. Drugs* **2014**, *12*, 2771–2789. [CrossRef] [PubMed]

27. Gontang, E.A.; Gaudêncio, S.P.; Fenical, W.; Jensen, P.R. Sequence-based analysis of secondary-metabolite biosynthesis in marine actinobacteria. *Appl. Environ. Microbiol.* **2010**, *76*, 2487–2499. [CrossRef] [PubMed]

28. Tambadou, F.; Lanneluc, I.; Sablé, S.; Klein, G.L.; Doghri, I.; Sopéna, V.; Didelot, S.; Barthélémy, C.; Thiéry, V.; Chevrot, R. Novel nonribosomal peptide synthetase (NRPS) genes sequenced from intertidal mudflat bacteria. *FEMS Microbiol. Lett.* **2014**, *357*, 123–130. [CrossRef] [PubMed]

29. Oleinikova, G.K.; Kuznetsova, T.A.; Huth, F.; Laatsch, H.; Isakov, V.V.; Shevchenko, L.S.; Elyakov, G.B. Cyclic lipopeptides with fungicidal activity from the sea isolate of the bacterium *Bacillus subtilis*. *Russ. Chem. Bull.* **2001**, *50*, 2231–2235. [CrossRef]

30. Budzikiewicz, H.; Bössenkamp, A.; Taraz, K.; Pandey, A.; Meyer, J.M. Corynebactin, a cyclic catecholate siderophore from *Corynebacterium glutamicum* ATCC 14067 (*Brevibacterium* sp. DSM 20411). *Z. Naturforschung* **1997**, *52*, 551–554. [CrossRef]

31. Li, J.; Liu, S.; Jiang, Z.; Sun, C. Catechol amide iron chelators produced by a mangrove-derived *Bacillus subtilis*. *Tetrahedron* **2017**, *73*, 5245–5252. [CrossRef]

32. Marfey, P. Determination of D-amino acids. II. Use of a bifunctional reagent, 1,5-difluoro-2,4-dinitroenzene. *Carlsberg Res. Commun.* **1984**, *49*, 591–596. [CrossRef]

33. Temirov, Y.V.; Esikova, T.Z.; Kashparov, I.A.; Balashova, T.A.; Vinokurov, L.M.; Alakhov, Y.B. A catecholic siderophore produced by the thermoresistant *Bacillus licheniformis* VK21 Strain. *Russ. J. Bioorg. Chem.* **2003**, *29*, 542–549. [CrossRef]

34. Dertz, E.A.; Xu, J.; Stintzi, A.; Raymond, K.N. Bacillibactin-mediated iron transport in *Bacillus subtilis*. *J. Am. Chem. Soc.* **2006**, *128*, 22–23. [CrossRef] [PubMed]

35. May, J.J.; Kessler, N.; Marahiel, M.A.; Stubbs, M.T. Crystal structure of DhbE, an archetype for aryl acid activating domains of modular nonribosomal peptide synthetases. *Proc. Natl. Acad. Sci. USA* **2002**, *99*, 12120–12125. [CrossRef] [PubMed]

36. Raza, W.; Wu, H.; Shah, M.A.A.; Shen, Q. A catechol type siderophore, bacillibactin: Biosynthesis, regulation and transport in *Bacillus subtilis*. *J. Basic Microbiol.* **2008**, *48*, 1–12. [CrossRef] [PubMed]

37. Oleinikova, G.K.; Dmitrenok, A.S.; Voinov, V.G.; Chaikina, E.L.; Shevchenko, L.S.; Kuznetsova, T.A. Bacillomycin D from the Marine Isolate of *Bacillus subtilis* KMM 1922. *Chem. Nat. Compd.* **2005**, *41*, 461–464. [CrossRef]

38. Tanaka, K.; Ishihara, A.; Nakajima, H. Isolation of *anteiso*-C17, *iso*-C17, *iso*-C16, and *iso*-C15 bacillomycin D from *Bacillus amyloliquefaciens* SD-32 and their antifungal activities against plant pathogens. *J. Agric. Food Chem.* **2014**, *62*, 1469–1476. [CrossRef] [PubMed]

39. Birgit, K.; Andrea, O.; Birgit, H.; Peter, B.; Jewgenij, M.; Winfried, E. Peptides, Their Production and Use and Producing Microorganism. Patent PCT/ DE1964/1213 A1, 16 April 1998.

40. Kim, J.D.; Jeon, B.J.; Han, J.W.; Park, M.Y.; Kang, S.A.; Kim, B.S. Evaluation of the endophytic nature of *Bacillus amyloliquefaciens* strain GYL4 and its efficacy in the control of anthracnose. *Pest Manag. Sci.* **2016**, *72*, 1529–1536. [CrossRef] [PubMed]

41. Yuan, J.; Li, B.; Zhang, N.; Waseem, R.; Shen, Q.; Huang, Q. Production of bacillomycin- and macrolactin-type antibiotics by *Bacillus amyloliquefaciens* NJN-6 for suppressing soilborne plant pathogens. *J. Agric. Food Chem.* **2012**, *60*, 2976–2981. [CrossRef] [PubMed]

42. Jemil, N.; Manresa, A.; Rabanal, F.; Ben Ayed, H.; Hmidet, N.; Nasri, M. Structural characterization and identification of cyclic lipopeptides produced by *Bacillus methylotrophicus* DCS1 strain. *J. Chromatogr. B* **2017**, *1060*, 374–386. [CrossRef] [PubMed]

43. Zhao, P.; Quan, C.; Jin, L.; Wang, L.; Guo, X.; Fan, S. Sequence characterization and computational analysis of the non-ribosomal peptide synthetases controlling biosynthesis of lipopeptides, fengycins and bacillomycin D, from *Bacillus* amyloliquefaciens Q-426. *Biotechnol. Lett.* **2013**, *35*, 2155–2163. [CrossRef] [PubMed]

44. Tabbene, O.; Azaiez, S.; Grazia, A.D.; Karkouch, I.; Slimene, I.B.; Elkahoui, S.; Alfeddy, M.N.; Casciaro, B.; Luca, V.; Limam, F.; et al. Bacillomycin D and its combination with amphotericin B: Promising antifungal compounds with powerful antibiofilm activity and wound-healing potency. *J. Appl. Microbiol.* **2016**, *120*, 289–300. [CrossRef] [PubMed]
45. Moyne, A.L.; Shelby, R.; Cleveland, T.E.; Tuzun, S. Bacillomycin D: An iturin with antifungal activity against *Aspergillus flavus*. *J. Appl. Microbiol.* **2001**, *90*, 622–629. [CrossRef] [PubMed]
46. Hajare, S.N.; Subramanian, M.; Gautam, S.; Sharma, A. Induction of apoptosis in human cancer cells by a *Bacillus* lipopeptide bacillomycin D. *Biochimie* **2013**, *95*, 1722–1731. [CrossRef] [PubMed]
47. Xi, L.; Ruan, J.; Huang, Y. Diversity and biosynthetic potential of culturable actinomycetes associated with marine sponges in the China Seas. *Int. J. Mol. Sci.* **2012**, *13*, 5917–5932. [CrossRef] [PubMed]
48. Kieser, T.; Bibb, M.J.; Buttner, M.J.; Chater, K.F.; Hopwood, D.A. *Practical Streptomyces Genetics*; The John Innes Foundation: Norwich, UK, 2000.
49. Heuer, H.; Krsek, M.; Baker, P.; Smalla, K.; Wellington, E.M.H. Analysis of actinomycete communities by specific amplification of genes encoding 16S rRNA and gel-electrophoretic separation in denaturing gradients. *Appl. Environ. Microbiol.* **1997**, *63*, 3233–3241. [PubMed]
50. Daletos, G.; Kalscheuer, R.; Koliwer-Brandl, H.; Hartmann, R.; de Voogd, N.J.; Wray, V.; Lin, W.; Proksch, P. Callyaerins from the marine sponge *Callyspongia aerizusa*: Cyclic peptides with antitubercular activity. *J. Nat. Prod.* **2015**, *78*, 1910–1925. [CrossRef] [PubMed]

marine drugs

MDPI

Brief Report

Discovery of an Unusual Fatty Acid Amide from the $ndgR_{yo}$ Gene Mutant of Marine-Derived *Streptomyces youssoufiensis*

Jing Hou [1], Jing Liu [1], Lu Yang [1], Zengzhi Liu [1], Huayue Li [1,2,*], Qian Che [1,2], Tianjiao Zhu [1,2], Dehai Li [1,2] and Wenli Li [1,2,*]

[1] Key Laboratory of Marine Drugs, Ministry of Education, School of Medicine and Pharmacy, Ocean University of China, Qingdao 266003, China; 17864275159@163.com (J.H.); liujing900908@163.com (J.L.); xihongshi94@163.com (L.Y.); liuzz1990@outlook.com (Z.L.); cheqian1396@sina.com (Q.C.); zhutj@ouc.edu.cn (T.Z.); dehaili@ouc.edu.cn (D.L.)
[2] Laboratory for Marine Drugs and Bioproducts of Qingdao National Laboratory for Marine Science and Technology, Qingdao 266237, China
* Correspondence: lihuayue@ouc.edu.cn (H.L.); liwenli@ouc.edu.cn (W.L.); Tel./Fax: +86-532-8203-1813 (H.L. & W.L.)

Received: 19 December 2018; Accepted: 24 December 2018; Published: 28 December 2018

Abstract: $NdgR_{yo}$, an IclR-like regulator, was selected as the target gene to activate new secondary metabolites in the marine-derived *Streptomyces youssoufiensis* OUC6819. Inactivation of the $ndgR_{yo}$ gene in *S. youssoufiensis* OUC6819 led to the accumulation of a new fatty acid amide (**1**), with an unusual 3-amino-butyl acid as the amine component. Moreover, its parent fatty acid (**2**) was also discovered both in the wild-type and $\Delta ndgR_{yo}$ mutant strains, which was for the first time isolated from a natural source. The structures of compounds **1** and **2** were elucidated by combination of LC-MS and NMR spectroscopic analyses. This study demonstrated that the $ndgR_{yo}$ homologs might serve as a target for new compound activation in *Streptomyces* strains.

Keywords: $NdgR_{yo}$; IclR family regulator; *Streptomyces*; fatty acid amide; genome mining

1. Introduction

Marine *Streptomyces* have evolved unique abilities to adapt to the marine environment, which ensures their survival in extreme habitats (e.g., low temperature, high pressure, and poor nutrients) and provides a variety of novel secondary metabolites [1]. With the increase in the number of sequences deposited in microbial genome databases, an increasing number of secondary metabolite biosynthetic gene clusters have been disclosed; however, the majority of them are silent or barely expressed under ordinary laboratory conditions [2]. Thus, activation of silent gene clusters has become an effective strategy for natural product discovery, attracting more and more scientists to this research field.

Secondary metabolism of *Streptomyces* is controlled by a complicated and elaborate regulatory network [3,4]. The precursors for the biosynthesis of secondary metabolites are usually derived from primary metabolism. Manipulation of the regulators in central metabolism has a far-reaching impact on the production of secondary metabolites [5]. The IclR-like global regulator, $ndgR$, is a representative of this metabolism that is involved in amino acid metabolism and conserved among *Streptomyces* species as well as other actinomycetes [6]. Disruption of $ndgR$ in *Streptomyces coelicolor* led to defective differentiation and enhanced actinorhodin production in minimal media containing certain amino acids [7]. In *Streptomyces clavuligerus*, deletion of *areB*, a homolog of $ndgR$, resulted in increased production of clavulanic acid and cephamycin C [8].

In our effort to discover novel natural products from the marine-derived *Streptomyces youssoufiensis* OUC6819 by genome mining [9,10], the $ndgR_{yo}$ gene was selected as a target for compound activation. The disruption of the $ndgR_{yo}$ gene caused accumulation of a new fatty acid amide (**1**) that shares similar UV spectrum with its parent fatty acid (**2**) present in the wild-type strain (Figure 1). Herein, we describe the isolation, structure elucidation as well as biological evaluation of compounds **1** and **2** from the $\Delta ndgR_{yo}$ mutant of *S. youssoufiensis* OUC6819.

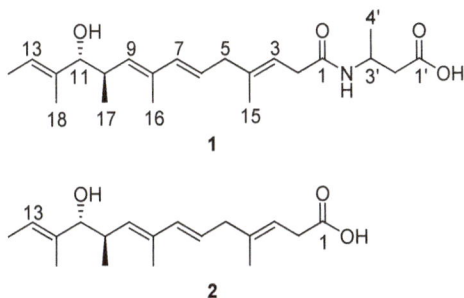

Figure 1. Structures of compounds **1** and **2**.

2. Results and Discussion

The $ndgR_{yo}$ gene was identified from the *S. youssoufiensis* OUC6819 genome using the local BlastP program. NdgR$_{yo}$ harbors a Helix-Turn-Helix motif at the N-terminus, and displays 57% identity to the NdgR from *S. coelicolor* (NP_629686.1). A positive cosmid, pWLI551, was obtained through genome library screening (Table S2). The $\Delta ndgR_{yo}$ mutant was obtained using a PCR-targeting strategy, as described in the Section 3.4. The fermentation broths of the wild-type and the $\Delta ndgR_{yo}$ mutant strains were extracted with ethyl acetate, and were subsequently subjected to HPLC analysis (Figure 2). The newly accumulated compound **1** in the $\Delta ndgR_{yo}$ mutant showed similar UV spectra with that of **2** (Figure 2), indicating they might belong to the same compound class. Large scale fermentation of the $\Delta ndgR_{yo}$ mutant resulted in the isolation of compounds **1** and **2**, followed by identification by NMR spectroscopy.

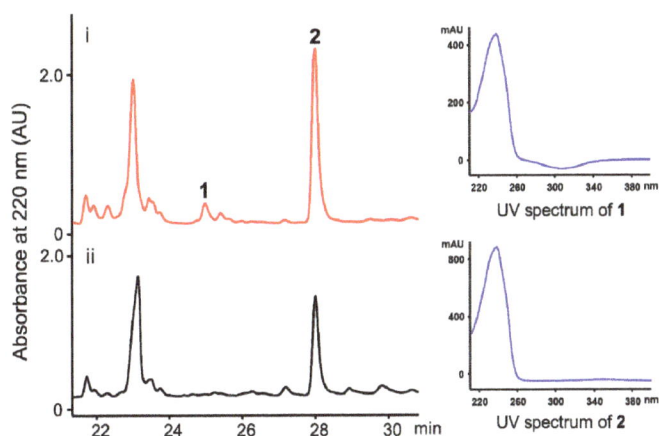

Figure 2. HPLC traces of the fermentation products from the *S. youssoufiensis* OUC6819 strains. (i) The $\Delta ndgR_{yo}$ mutant; (ii) the wild-type strain. Compound **1** was newly accumulated in the $\Delta ndgR_{yo}$ mutant strain. Compound **2** was produced in both the wild-type and $\Delta ndgR_{yo}$ mutant strains, and shares similar UV spectrum with **1**.

Compound **1** was isolated as a yellow oil. The molecular formula of **1** was established as $C_{22}H_{35}NO_4$ (five degrees of unsaturation), as determined by HR-ESIMS data (m/z 378.2654 $[M + H]^+$, calcd 378.2644) (Figure S1). The structure of **1** was determined from the 1D and 2D NMR (COSY, HSQC, HMBC, and NOESY) data (Figures S2–S7). The ^1H and HSQC spectra of **1** disclosed six methyl groups (δ_H 1.64, 1.66, 1.78, 0.83, 1.63, and 1.21), three methylenes (δ_H 2.95, 2.82, 2.54), three methines (δ_H 2.71, 3.72, and 4.26), and five olefinic protons (δ_H 5.36, 5.57, 6.12, 5.33, and 5.46). The COSY spectrum established five spin systems of H-2 (δ_H 2.95)/H-3 (δ_H 5.36), H-5 (δ_H 2.82)/H-6 (δ_H 5.57)/H-7 (δ_H 6.12), H-9 (δ_H 5.33)/H-10 (δ_H 2.71)/H-11 (δ_H 3.72)/H-17 (δ_H 0.83), H-13 (δ_H 5.46)/H-14 (δ_H 1.64), and H-2' (δ_H 2.54)/H-3' (δ_H 4.26)/H-4' (δ_H 1.21) (Figure 3). The HMBC correlations from H-13 and H-9 to the hydroxylated carbon C-11 (δ_C 82.3), from H-11 to C-12 (δ_C 136.7), from H-7 to C-9 (δ_C 134.4), from H-6 to C-8 (δ_C 133.7), from H-2 to C-4 (δ_C 138.4) and a carbonyl carbon C-1 (δ_C 172.4), from H-5 to C-3 (δ_C 117.3), from H-3' to C-1, and from H-2' to C-1' (δ_C 173.2) established the main carbon chain of **1** (Figure 3). The HMBC correlations from the methyl protons H-15 (δ_H 1.66) to C-5 (δ_C 42.5), H-16 (δ_H 1.78) to C-9, H-17 to C-11, and H-18 (δ_H 1.63) to C-13 (δ_C 121.4), together with the COSY correlations of H-13/H-14 and H-3'/H-4', confirmed the location of six methyl groups (Figure 3). The ^{13}C chemical shifts of C-3' (δ_C 42.2) and C-1, together with the HR-ESIMS data of **1**, revealed the presence of an amide group. Moreover, the configurations of the four double bonds were confirmed to be *E* by NOESY correlations between H-3/H-5, H-5/H-7, H-7/H-9, H-11/H-13, H-2/H-15, H-6/H-16, and H-10/H-16 (Figure 4). The relative configuration between H-10 and H-11 was proposed to be *trans* by the large coupling constant value of 8.4 Hz. Then, the absolute configurations of C-10 and C-11 were determined by comparison of the experimental ECD spectra of **1** and **2** (Figure S15) with calculated ECD spectra of the (2*E*,4*E*,8*E*)-7-methoxy-4,6,8-trimethyldeca-2,4,8-triene moiety reported in the literature [11], which showed high agreement with the 10*R*, 11*R* calculated model. Thus, compound **1** was identified as 3-((3*E*,6*E*,8*E*,10*R*,11*R*,12*E*)-11-hydroxy-4,8,10,12-tetramethyltetradeca-3,6,8,12-tetraenamido) butanoic acid, a new branched-chain fatty acid amide with 3-amino-butyl acid as the amine component. The ^1H and ^{13}C chemical shifts of compound **1** were shown in Table 1.

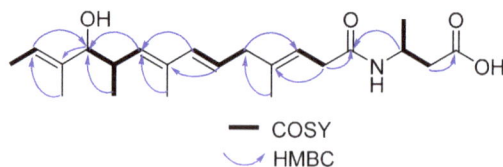

Figure 3. ^1H-^1H COSY and key HMBC correlations of **1** in CD$_3$OD.

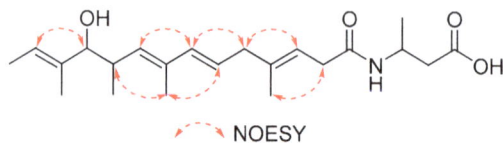

Figure 4. Key NOESY correlations of **1** in CD$_3$OD.

Compound **2** was isolated as a yellow oil. The molecular formula of **2** was established as $C_{18}H_{28}O_3$ (five degrees of unsaturation), as determined by HR-ESIMS data (m/z 310.2397 $[M + NH_4]^+$, calcd 310.2382) (Figure S8). The structure of **2** was determined from the 1D and 2D NMR (COSY, HSQC, HMBC, and NOESY) data (Figures S9–S14). According to the ^1H and ^{13}C NMR data (Table 1), compound **2** lacked the 3-amino-butyl acid moiety compared to **1**. Moreover, the NOESY correlations (H-3/H-5, H-5/H-7, H-7/H-9, H-11/H-13, H-2/H-15, H-6/H-16, and H-10/H-16) as well as the experimental ECD spectrum revealed that compound **2** displays the same absolute configurations with **1**. Thus, compound **2** was identified to be the parent fatty acid of **1**, named (3*E*,6*E*,8*E*,10*R*,11*R*,12*E*)-11-hydroxy-4,8,10,12-tetramethyltetradeca-3,6,8,12-tetraenoic acid, which is available from Aurora Fine Chemicals in the United States. Noticeably, this is the first time that compound **2** has been isolated from a natural source.

Table 1. ^{1}H (600 MHz) and ^{13}C (150 MHz) NMR chemical shifts of **1** and **2** in CD$_3$OD.

Position	1		2	
	δ_H (*J* in HZ)	δ_C	δ_H (*J* in HZ)	δ_C
1		172.4		174.9
2	2.95 (2H, d, 7.2)	35.0	3.06 (2H, d, 7.2)	33.0
3	5.36 (1H, t, 6.6)	117.3	5.39 (1H, t, 6.6)	116.7
4		138.4		137.7
5	2.82 (2H, d, 7.2)	42.5	2.82 (2H, d, 7.2)	42.5
6	5.57 (1H, dt, 15.6, 7.2)	124.4	5.56 (1H, dt, 15.6, 7.2)	124.4
7	6.12 (1H, d, 15.6)	136.7	6.13 (1H, d, 15.6)	136.7
8		133.7		133.7
9	5.33 (1H, d, 9.0)	134.4	5.33 (1H, d, 9.6)	134.4
10	2.71 (1H, m)	36.1	2.71 (1H, m)	36.1
11	3.72 (1H, d, 8.4)	82.3	3.72 (1H, d, 7.8)	82.3
12		136.7		136.7
13	5.46 (1H, q, 6.6)	121.4	5.47(1H, q, 6.0)	121.4
14	1.64 (3H, d, 6.6)	11.6	1.64 (3H, d, 6.0)	11.7
15	1.66 (3H, s)	15.0	1.66 (3H, s)	15.0
16	1.78 (3H, m)	11.6	1.78 (3H, m)	11.7
17	0.83 (3H, d, 6.6)	16.6	0.83 (3H, d, 7.2)	16.7
18	1.63 (3H, s)	9.7	1.63 (3H, s)	9.7
1'		173.2		
2'	2.54 (1H, dd, 15.6, 6.0) 2.42 (1H, dd, 15.6, 6.0)	40.0		
3'	4.26 (1H, m)	42.2		
4'	1.21 (3H, d, 6.6)	18.8		

In the antibacterial activity evaluation of compounds **1** and **2**, neither of them showed obvious inhibitory effects, in the range of concentrations tested, against the multi-drug resistant (MDR) strains, including *Enterococcus faecalis* CCARM 5172, *Enterococcus faecium* CCARM 5203, *Staphylococcus aureus* CCARM 3090, *Escherichia coli* CCARM 1009, and *Salmonella typhimurium* CCARM 8250. Some branched-chain oleic acid derivatives exhibited growth inhibition against MCF-7 and HT-29 cells [12]. Therefore, we also tested the cytotoxicity of compounds **1** and **2** against these two cell lines, but they showed null activity up to the concentration of 50 μM (data not shown).

Fatty acid amides are a class of compounds formed from a fatty acid and an amine that play an important role in intracellular signaling, many of which in nature have ethanolamine as the amine component [13–15]. Inactivation of the *ndgR$_{yo}$* gene in *S. youssoufiensis* OUC6819 led to the isolation of a new fatty acid amide and (**1**) and its parent branched-chain fatty acid (**2**). The 3-amino-butyl acid moiety in **1** is likely to come from L-glutamate [16]. As the IclR-like global regulator, NdgR, is generally involved in amino acid metabolism [6], we proposed that *ndgR$_{yo}$* in *S. youssoufiensis* OUC6819 might contribute to the generation of the unusual branched-chain fatty acid amide (**1**) with an amino acid as the amine component.

3. Materials and Methods

3.1. General Experimental Procedures

1D (^{1}H and ^{13}C) and 2D (COSY, HSQC, HMBC, and NOESY) NMR spectra were recorded on Bruker Avance III 600 spectrometers at 298 K. The mixing time used for the NOESY spectrum was 142 ms. Chemical shifts were reported with reference to the respective solvent peaks and residual solvent peaks (δ_H 3.31 and δ_C 49.0 for CD$_3$OD). HR-ESIMS data were obtained on a Q-TOF Ultima

Global GAA076 MS spectrometer. HPLC was performed on an Agillent 1260 Infinity apparatus equipped with a diode array detector (DAD).

3.2. Bacterial Strains and Culture Conditions

Escherichia coli DH5α served as the host for general subcloning [17]. *Escherichia coli* ET12567/pUZ8002 was used as the cosmid donor host for *E. coli-Streptomyces* intergenic conjugation [18]. *Escherichia coli* BW25113/pIJ790 was used for λRED-mediated PCR-targeting [19]. The *S. youssoufiensis* OUC6819 was isolated from reed rhizosphere soil collected from the mangrove conservation area of Guangdong province, China [9]. *E. coli* strains were routinely cultured in Luria–Bertani (LB) liquid medium at 37 °C, 200 rpm, or LB agar plate at 37 °C. *Streptomyces* strains were grown at 30 °C on R2YE medium for sporulation and ISP4 for conjugation, and were cultured in tryptic soy broth (TSB) medium for genomic DNA preparation. Fermentation medium consists of 1% soluble starch, 2% glucose, 4% corn syrup, 1% yeast extract, 0.3% beef extract, 0.05% $MgSO_4 \cdot 7H_2O$, 0.05% KH_2PO_4, 0.2% $CaCO_3$, and 3% sea salt, pH = 7.0.

3.3. DNA Isolation and Manipulation

Plasmid extractions and DNA purifications were carried out using standardized commercial kits (OMEGA, Bio-Tek, Guangzhou, China). PCR reactions were carried out using Pfu DNA polymerase (TIANGEN, Beijing, China). Oligonucleotide synthesis and DNA sequencing were performed by TSINGKE company (Qingdao, China).

3.4. Gene Inactivation

Positive cosmids harboring the $ndgR_{yo}$ gene were screened against the genomic library of *S. youssoufiensis* OUC6819 by using PCR with primers listed in Table S1. One cosmid, pWLI551 (Table S2), was obtained and then confirmed by DNA sequencing in TSINGKE company (Qingdao, China). The amplified *aac(3) IV-oriT* resistance cassette from pIJ773 was transformed into *E. coli* BW25113/pIJ790 containing pWLI551 to replace an internal region of the target gene, the PCR primers are listed in Table S3. The mutant cosmid was constructed and introduced into *S. youssoufiensis* OUC6819 by conjugation from *E. coli* ET12567/pUZ8002 according to the reported procedure, using *S. youssoufiensis* OUC6819 ultrasonic fragmented mycelia as acceptors [20]. The desired mutants were selected by the apramycin-resistant and kanamycin sensitive phenotype, and were confirmed by PCR (Figure S16), using the primers listed in Table S4.

3.5. Isolation and Purification of the Compounds

The fermentation broth (50 mL) of the *S. youssoufiensis* OUC6819 strains was extracted twice with an equal volume of ethyl acetate, and subsequently subjected to the HPLC analysis. Analytical HPLC was performed on a YMC-Pack ODS-A column (5 μm, 4.6 × 150 mm) developed with a linear gradient from 20% to 100% B/A in 45 min (phase A: H_2O; phase B: 100% acetonitrile) at the wavelength of 220 nm. The culture broth (15 L) of a scaled-up culture of the Δ*ndgR_{yo}* mutant was extracted with ethyl acetate and evaporated at room temperature, which was partitioned between 90% methanol and *n*-hexane to remove nonpolar components. Compounds **1** (3 mg) and **2** (10 mg) were obtained by separation of the methanol layer with a linear gradient from 70% to 90% B at a flow rate of 2.0 mL/min using a YMC-Pack ODS-A column (5 μm, 120 Å, 250×10 mm; wavelength 220 nm).

Compound 1: Yellow oil; $[\alpha]_D$ −8.1 (c 0.1, MeOH); CD (MeOH) λ_{max} ($\Delta\varepsilon$) 202.5 (+5.23), 234.5 (−3.99) nm; ^{1}H and ^{13}C NMR data, see Table 1; HR-ESIMS m/z 378.2654 [M + H]$^+$ (calcd for $C_{22}H_{36}NO_4$, 378.2644).

Compound 2: Yellow oil; $[\alpha]_D$ −2.8 (c 0.1, MeOH); CD (MeOH) λ_{max} ($\Delta\varepsilon$) 200.5 (+4.66), 231.5 (−3.51) nm; ^{1}H and ^{13}C NMR data, see Table 1; HR-ESIMS m/z 310.2397 [M + NH$_4$]$^+$ (calcd for $C_{18}H_{32}NO_3$, 310.2382).

3.6. Nucleotide Sequence Accession Number

The nucleotide sequence of $ndgR_{yo}$ in this paper has been deposited in the GenBank database, and the accession number is MH252211.

4. Conclusions

A new fatty acid amide (**1**) with an unusual 3-amino-butyl acid as the amine component, together with its parent fatty acid (**2**) were isolated from the $\Delta ndgR_{yo}$ mutant strain of *S. youssoufiensis* OUC6819. Compounds **1** and **2** displayed neither inhibitory effects against the five MDR bacterial strains, nor cytotoxicity against MCF-7 and HT-29 cancer cell lines. This study demonstrated that the $ndgR_{yo}$ homologs might serve as a target for activation of structurally novel secondary metabolites in the *Streptomyces* strains.

Supplementary Materials: The following are available online at http://www.mdpi.com/1660-3397/17/1/12/s1, Supporting Figures S1–S16 and Tables S1–S4, including HR-ESIMS, NMR and ECD spectra of compounds **1** and **2**, plasmids and primer lists, and biological assay methods.

Author Contributions: J.H., J.L. and Z.L. performed the experiments. J.H. wrote the draft manuscript. L.Y. was involved in NMR analysis. Q.C., T.Z. and D.L. isolated the *Streptomyces* strain. W.L. and H.Y. supervised the whole work and wrote the manuscript. All authors read and approved the final manuscript.

Acknowledgments: This work was supported by the National Natural Science Foundation of China under Grants 31570032, 31711530219, 41506157 and 21502180; and the NSFC-Shandong Joint Foundation under Grants U1706206 and U1406403.

Conflicts of Interest: The authors declare no conflict of interest.

References

1. Manivasagan, P.; Venkatesan, J.; Sivakumar, K.; Kim, S.K. Pharmaceutically active secondary metabolites of marine actinobacteria. *Microbiol. Res.* **2014**, *169*, 262–278. [CrossRef] [PubMed]
2. Rutledge, P.J.; Challis, G.L. Discovery of microbial natural products by activation of silent biosynthetic gene clusters. *Nat. Rev. Microbiol.* **2015**, *13*, 509–523. [CrossRef] [PubMed]
3. Lu, C.; Liao, G.; Zhang, J.; Tan, H. Identification of novel tylosin analogues generated by a *wblA* disruption mutant of *Streptomyces ansochromogenes*. *Microb. Cell Fact.* **2015**, *14*, 173. [CrossRef] [PubMed]
4. Li, Y.; Tan, H. Biosynthesis and molecular regulation of secondary metabolites in microorganisms. *Sci. China Life Sci.* **2017**, *60*, 935–938. [CrossRef] [PubMed]
5. Liu, G.; Chater, K.F.; Chandra, G.; Niu, G.; Tan, H. Molecular regulation of antibiotic biosynthesis in *Streptomyces*. *Microbiol. Mol. Biol. Rev.* **2013**, *77*, 112–143. [CrossRef] [PubMed]
6. Kim, J.N.; Jeong, Y.; Yoo, J.S.; Roe, J.H.; Cho, B.K.; Kim, B.G. Genome-scale analysis reveals a role for NdgR in the thiol oxidative stress response in *Streptomyces coelicolor*. *BMC Genom.* **2015**, *16*, 116. [CrossRef] [PubMed]
7. Yang, Y.H.; Song, E.; Kim, E.J.; Lee, K.; Kim, W.S.; Park, S.S.; Hahn, J.S.; Kim, B.J. NdgR, an IclR-like regulator involved in amino-acid-dependent growth, quorum sensing, and antibiotic production in *Streptomyces coelicolor*. *Appl. Microbiol. Biotechnol.* **2009**, *82*, 501–511. [CrossRef] [PubMed]
8. Santamarta, I.; Lópezgarcía, M.T.; Pérezredondo, R.; Koekman, B.; Martín, J.F.; Liras, P. Connecting primary and secondary metabolism: AreB, an IclR-like protein, binds the ARE$_{ccaR}$ sequence of *S. clavuligerus* and modulates leucine biosynthesis and cephamycin C and clavulanic acid production. *Mol. Microbiol.* **2007**, *66*, 511–524. [CrossRef] [PubMed]
9. Che, Q.; Li, T.; Liu, X.; Yao, T.; Li, J.; Gu, Q.; Li, D.; Li, W.; Zhu, T. Genome scanning inspired isolation of reedsmycins A-F, polyene-polyol macrolides from *Streptomyces* sp. CHQ-64. *RSC Adv.* **2015**, *5*, 22777–22782. [CrossRef]
10. Yao, T.; Liu, Z.; Li, T.; Zhang, H.; Liu, J.; Li, H.; Che, Q.; Zhu, T.; Li, D.; Li, W. Characterization of the biosynthetic gene cluster of the polyene macrolide antibiotic reedsmycins from a marine-derived *Streptomyces* strain. *Microb. Cell Fact.* **2018**, *17*, 98. [CrossRef] [PubMed]
11. Han, X.; Liu, Z.; Zhang, Z.; Zhang, X.; Zhu, T.; Gu, Q.; Li, W.; Che, Q.; Li, D. Geranylpyrrol A and piericidin F from *Streptomyces* sp. CHQ-64 $\Delta rdmF$. *J. Nat. Prod.* **2017**, *80*, 1684–1687. [CrossRef] [PubMed]

12. Dailey, O.D., Jr.; Wang, X.; Chen, F.; Huang, G. Anticancer activity of branched-chain derivatives of oleic acid. *Anticancer Res.* **2011**, *31*, 3165–3169. [PubMed]
13. Jain, M.K.; Ghomashchi, F.; Yu, B.Z.; Bayburt, T.; Murphy, D.; Houck, D.; Solowiej, J.E. Fatty acid amides: scooting mode-based discovery of tight-binding competitive inhibitors of secreted phospholipases A2. *J. Med. Chem.* **1992**, *35*, 3584–3586. [CrossRef] [PubMed]
14. Blancaflor, E.B.; Kilaru, A.; Keereetaweep, J.; Khan, B.R.; Faure, L.; Chapman, K.D. N-Acylethanolamines: lipid metabolites with functions in plant growth and development. *Plant J.* **2014**, *79*, 568–583. [CrossRef] [PubMed]
15. Tuo, W.; Leleu-Chavain, N.; Spencer, J.; Sansook, S.; Millet, R.; Chavatte, P. Therapeutic potential of fatty acid amide hydrolase, monoacylglycerol lipase, and N-acylethanolamine acid amidase inhibitors. *J. Med. Chem.* **2016**, *60*, 4–46. [CrossRef] [PubMed]
16. Takaishi, M.; Kudo, F.; Eguchi, T. A unique pathway for the 3-aminobutyrate starter unit from L-glutamate through β-Glutamate during biosynthesis of the 24-membered macrolactam antibiotic, incednine. *Org. Lett.* **2012**, *14*, 4591–4593. [CrossRef] [PubMed]
17. Maniatis, T.; Fritsch, E.F.; Sambrook, J. *Molecular Cloning: A Laboratory Manual*; Cold Spring Harbor: New York, NY, USA, 1982.
18. Paget, M.S.; Chamberlin, L.; Atrih, A.; Foster, S.J.; Buttner, M.J. Evidence that the extracytoplasmic function sigma factor σ^E is required for normal cell wall structure in *Streptomyces coelicolor* A3(2). *J. Bacteriol.* **1999**, *181*, 204–211. [PubMed]
19. Gust, B.; Challis, G.L.; Fowler, K.; Kieser, T.; Chater, K.F. PCR-targeted *Streptomyces* gene replacement identifies a protein domain needed for biosynthesis of the sesquiterpene soil odor geosmin. *Proc. Natl. Acad. Sci. USA* **2003**, *100*, 1541–1546. [CrossRef] [PubMed]
20. Liu, Z.; Li, T.; Liu, J.; Yao, T.; Zhang, H.; Xia, J.; Li, H.; Che, Q.; Li, W. Development of the genetic system of mangrove derived *Streptomyces* sp. OUC6819. *Chin. J. Mar. Drugs* **2016**, *35*, 53–59.

MDPI

St. Alban-Anlage 66

4052 Basel

Switzerland

Tel. +41 61 683 77 34

Fax +41 61 302 89 18

www.mdpi.com

Marine Drugs Editorial Office

E-mail: marinedrugs@mdpi.com

www.mdpi.com/journal/marinedrugs

www.ingramcontent.com/pod-product-compliance
Lightning Source LLC
Chambersburg PA
CBHW051850210326
41597CB00033B/5845